T0292092

FRATERNITY AMONG THE FRENCH PEASANTRY

Sociability and Voluntary Associations in the Loire Valley, 1815–1914

The individualism of the French peasantry during the nineteenth century has frequently been asserted as one of its most striking characteristics. In this new study, based upon meticulous archival research, Alan Baker demonstrates that such a portrayal distorts the extent to which peasants both continued with traditional, and developed new, forms of collective action. He examines representations of the peasantry and discusses the discourse of fraternity in nineteenth-century France in general before considering specifically the historical development, geographical diffusion and changing functions of fraternal voluntary associations in Loir-et-Cher, straddling the middle Loire valley, between 1815 and 1914. Alan Baker focuses principally upon associations aimed at reducing risk and uncertainty (insurance associations, mutual aid societies, and voluntary fire-fighting corps), and upon associations intended to provide agricultural protection (syndicates and co-operatives). A wide range of new voluntary associations were established in Loir-et-Cher – and indeed throughout rural France – during the nineteenth century. Their historical geography throws new light upon the sociability, upon the changing *mentalités*, of French peasants, and upon the role of fraternal associations in their struggle for survival.

ALAN R. H. BAKER is a lecturer in the Department of Geography, University of Cambridge and Fellow of Emmanuel College, Cambridge. He has written widely in the field of historical geography and his publications include, as co-editor, *Explorations in Historical Geography* (1984) and *Ideology and Landscape in Historical Perspective* (1992). In 1997 he was honoured by the French government as a *Chevalier dans l'Ordre des Palmes Académiques* for his research on the historical geography of France.

Cambridge Studies in Historical Geography 28

Cambridge Studies in Historical Geography encourages exploration of the philosophies, methodologies and techniques of historical geography and publishes the results of new research within all branches of the subject. It endeavours to secure the marriage of traditional scholarship with innovative approaches to problems and to sources, aiming in this way to provide a focus for the discipline and to contribute towards its development. The series is an international forum for publication in historical geography which also promotes contact with workers in cognate disciplines.

For a full list of titles in the series, please see end of book.

FRATERNITY AMONG THE FRENCH PEASANTRY

Sociability and Voluntary Associations
in the Loire Valley, 1815–1914

ALAN R. H. BAKER

CAMBRIDGE
UNIVERSITY PRESS

PUBLISHED BY THE PRESS SYNDICATE OF THE UNIVERSITY OF CAMBRIDGE
The Pitt Building, Trumpington Street, Cambridge, United Kingdom

CAMBRIDGE UNIVERSITY PRESS
The Edinburgh Building, Cambridge CB2 2RU, UK
40 West 20th Street, New York NY 10011–4211, USA
477 Williamstown Road, Port Melbourne, VIC 3207, Australia
Ruiz de Alarcón 13, 28014 Madrid, Spain
Dock House, The Waterfront, Cape Town 8001, South Africa

http://www.cambridge.org

© Alan R. H. Baker 1999

First published 1999
First paperback edition 2004

Typeset in Times NR 10/12 pt in QuarkXPress™ [SE]

A catalogue record for this book is available from the British Library

Library of Congress cataloguing in publication data
Baker, Alan R. H.
Fraternity among the French peasantry: sociability and voluntary associations in the
Loire valley, 1815–1914 / Alan R. H. Baker.
 p. cm. (Cambridge studies in historical geography, 28)
Includes bibliographical references and index.
ISBN 0 521 64213 2 (hardback)
1. Friendly societies – France – Loir-et-Cher – History – nineteenth century.
2. Peasantry – France – Loir-et-Cher – History – nineteenth century.
I. Title II. Series.
HS1508.F8B35 1999
334′.7′094453 – dc21 98-35994 CIP

ISBN 0 521 64213 2 hardback
ISBN 0 521 60271 8 paperback

In memory of Edie and Reg Baker, my parents,
and Clifford Darby, my mentor

Contents

Figures

Photographs

Preface

In researching and writing this book, I have had generous help from many colleagues and friends in Britain, in France and elsewhere. They are too numerous to be mentioned here individually but I am grateful to every one of them. In addition, I am indebted to the published and unpublished work of many historians, geographers and other scholars, as citations in the notes and bibliography indicate.

I wish especially to acknowledge the professionalism of the staffs of libraries in Cambridge, Paris and Blois who have provided considerable support for my work. I owe particular thanks to the staff of the Archives Départementales de Loir-et-Cher for their advice and assistance during my many visits to that records' office, where I was able to search not only catalogued but also uncatalogued materials. I am also specifically indebted to Mike Young for drawing all of the maps and diagrams and to James Youlden for photographic reproduction of the illustrations.

Drafts of parts of this book have been commented upon critically by colleagues both at local and regional seminars and at national and international conferences, and they have always benefited from such exposure. In addition, I am grateful to the three anonymous readers commissioned by Cambridge University Press to review a draft of this research monograph: revising it in the light of their comments revealed how carefully and constructively they had undertaken their assignments.

Much of the preparatory work for this book has been achieved in Blois and its environs. It has benefited from contact with some of Loir-et-Cher's local and regional historians, notably Maurice Gobillon, André Prudhomme, Jean Vassort and especially Régis Bouis. It has also gained from personal contacts and friendships developed over the years with many residents of the Beauceron commune of Mulsans, notably Jeanne Piedallu and the Guillard family. Living for extended periods each year among some of the descendants of the people whose historical geography is the concern of this book provided insights into French rural life not obtainable from the documentary records upon which it is principally based.

My research visits to France have been supported financially by the University of Cambridge; Emmanuel College, Cambridge; the Centre National de la Recherche Scientifique, Paris; the British Academy; and the Leverhulme Trust. To each of these I express my thanks. Part of chapter 6 is based upon an article published in the *Journal of Historical Geography,* part of chapter 7 upon an article in the *Würzburger Geographische Arbeiten,* and part of chapter 8 upon an article in the *Agricultural History Review.* I am grateful to the publishers of those journals for permission to draw upon those articles here.

The final stages of production of the book have benefited enormously from the critical advice of Richard Fisher, Deputy Group Director (Social Sciences) at Cambridge University Press, and from the detailed attention given to the copy and proofs by Hilary Hammond. The index has been compiled by Simon Cross. I am very grateful to each of them.

Finally, this book would never have been completed without the constant encouragement of my wife, Sandra, and without the forebearance of my two sons, Andrew and Jeremy. But it would certainly not have even been begun without the earlier support of my parents, Edie and Reg Baker, and without the initial stimulus of my intellectual mentor, Clifford Darby: it is, therefore, dedicated to their memories.

Mulsans, Loir-et-Cher
20 September 1997

1

Peasants and peasantry in nineteenth-century France

The rural community

There are many myths about the French peasantry. They have been both assiduously cultivated and painstakingly up-rooted.[1] More than forty years ago Harvey Goldberg dissected what he termed 'the myth of the happy [French] peasant', that of the idea of an independent, land-owning, stable, contented peasantry living and working in a balanced national economy and rural democracy.[2] There was also, as Gordon Wright has pointed out, the myth that France's agrarian history really began with the Revolution of 1789 because it completely transformed the country's rural structure; the myth that the peasantry, both before and after the Revolution, constituted a solid and largely undifferentiated bloc with common interests and aspirations; and the myth that the excessive parcellation of the fields of France was a product of the Napoleonic Code which abolished primogeniture in favour of partible inheritance.[3] Then there is the myth of France as a peasants' republic:[4] although the peasantry could have been the masters of France, because in 1848 when universal male suffrage was proclaimed they and their dependants constituted more than half of the country's population and even by 1939 they were still the largest single social grouping, they made little use of their potential political power because they were fighting on other fronts.[5] There is also, of course, the carefully and selectively nurtured myth of peasant traditionalism and backwardness,[6] or at least of the peasantry's espousal of conservative attitudes towards property, religion and the family.[7] A related and equally pervasive idea – that of the emergence of the French peasant during the nineteenth century as a sturdy individualist and of the related decline of the rural community – is the focus of this present study.

If the problem of the rural community lies at the heart of the history of the French peasantry,[8] then the question of peasant individualism lies at the heart of the rural community itself. The cult of individualism – of *liberté* – might have triumphed with the Revolution of 1789, but the potentially conflicting

1

concept of community – of *fraternité* – had roots in the ancien régime and came into flower well after the Revolution had passed into historical memory.

The rural community towards the end of the ancien régime

By the eighteenth century, there was a great diversity among the French peasantry but a sense of community was constructed locally because of shared experiences.[9] All peasants were, for example, subjects of the kings of France and so all were tied into a single political system, with its instabilities, and they were all linked into a fragile rural economy, with its environmental hazards and economic uncertainties.[10] Risk and insecurity were the norm, crises not exceptional. Shared perceptions of these external threats contributed towards the creation of an internal sense of community. A commonly shared experience of opposition to 'outsiders', for example, the lord, the tax-collector or the money-lender, helped to underscore the 'insiders' sense of social cohesion and local community. As Marc Bloch once wrote: 'It was above all by opposing its enemies that the small collectivity of the countryside acquired a firmer consciousness of itself.'[11] Similarly, a shared awareness of the scarcity of resources and of the need for the collective management of some of them provided rural communities with a strong social and economic basis.

The foundations of the rural community were the collective ownership and use of communal goods, in particular collective constraints upon private property (such as prohibiting enclosure and prescribing a crop rotation) for the benefit of the group as a whole, collective rights of usage over forests (for grazing some livestock and for gathering wood) and fields (for pasturing the fallow and for gleaning the stubble), and collective regulation of farming (such as determining which lands were to lie temporarily uncultivated, fixing the dates of harvests, and managing the pasturing of common lands).[12] There were, of course, spatial variations and temporal evolutions in the precise character of such collective regulations and practices, but fundamentally the rural community was characterised by the duties it imposed on all or most of its members and by the constraints it imposed on individual property.[13] Co-operation, mutual aid and a sense of fraternity bred of practical necessity were embedded in the rural community of the ancien régime. This is not to deny that, as Philip Hoffman has recently emphasised, such communities were also infused with internal social and economic divisions.[14] A community's consciousness of its interdependence, of its collective unity, no doubt owed something to the all-embracing role of the Church and to the assembly of villagers (*communauté d'habitants*) which determined issues of concern to the community (such as managing public property, like roads and bridges and the church, and appointing public officials, like a shepherd, hayward, schoolmaster, or collectors of tithes). But the rural community at the end of the eighteenth century owed its unity to its economic system more than to its religious or administrative institutions.[15] The Revolution, with its principled emphasis upon both individualism and fraternity, as well as upon equality, challenged the existing order of the rural community.

The rural community and the Revolution of 1789

The Revolution, while destroying the seigneurial regime and abolishing feudal rights, elevated the principles of liberalism and individualism, asserting the total right to property in both the 1789 and 1793 Declarations of Rights: it established the freedom of an individual to enclose and to farm his property, and it restricted collective rights of use over such property. The private ownership of property came to be regarded as the foundation of post-revolutionary French society, as the basis of social and economic progress, as the key to modernisation. For example, the right to vote in the early nineteenth century was closely linked to property ownership, being restricted not only to men of a certain age but also of a given liability to direct tax on wealth, itself derived principally from property. Hence the myth emerged of a property owning democracy, based upon an individualism which has been seen as giving free play to the development of the capitalist mode of production in the countryside and hastening the disintegration of the rural community.[16]

With hindsight, contemporary perceptions of the significance of the land settlement of the French Revolution may be viewed as part of a false consciousness. The current assessment of the situation has been summarised recently by P. M. Jones in his wide-ranging analysis of the peasantry and the French Revolution:

Contemporaries imagined that the sale of church property and the confiscated lands of *émigré* noblemen, plus the clearing of wastes and the division of common land, had brought into being a class of freehold peasant proprietors. That belief became a key component of the nineteenth-century republican myth . . . we now know that peasant land ownership was well-entrenched before 1789. All the Revolution did was to accelerate existing trends, but the transformation was scarcely dramatic because the quantity of property changing hands represented only a small percentage of the total land surface of the country.[17]

Moreover, while individual property rights were enshrined in the Rural Code of 1791, so too were collective rights over private property.

The ideology of individualism, of *liberté*, suffused rural France at the turn of the eighteenth and nineteenth centuries but it was embraced much more by those who were already owners of viably sized farms than it was by the unpropertied and by the owners of small farms who were traditionally dependent upon usage of some lands in common in order to extend their resource base. The concept of collectivism, of *fraternité*, appears to have featured less in discourse although it continued to be of significance in practice. Sections of the peasantry successfully resisted the doctrinaire enthusiasms of bourgeois revolutionaries on the question of collective rights. Unhindered grazing, gleaning and scavenging affected the immediate livelihoods of many more of the rural population than did the attribution of tithes, or the sale of church and *émigré* estates, so that the Rural Code was more a pragmatic compromise than an ideologically grounded, revolutionary statement. As Jones has argued,

> it offered landowners a (largely unenforceable) right to enclose, while allowing peasants an (eminently enforceable) right to graze their stock as tradition dictated. In effect, therefore, agrarian *liberté* was postponed indefinitely. All attempts to revise the Rural Code in a direction favourable to landowners stumbled against the political argument: the great majority of peasants relied on collective rights and would not be parted from them without resistance.[18]

The theoretical tension between individualism and collectivism found practical expression within rural communities. Although the former might have been dominant ideologically and was intended to mark a discontinuity in rural society, the latter's practical role persisted but it can hardly be claimed that it provided an underlying continuity in the collective consciousness of rural communities. The tension between individual and collective rights was in practice a major source of conflict within rural societies, undermining their cohesion.

There were other ways in which a rural community might either have retained or had reinforced a consciousness of itself. Without doubt, the most important institutionally was the Revolution's administrative reorganisation, creating a hierarchical system which linked the thousands of localities into a single state. The unifying social role traditionally played by the Church was, at least potentially, assumed by the State. Elected municipal councils now ordered and controlled much of the daily life of a community. Because of the range of their powers, the councils contributed to the creation of a sense of local identity, of locality and community, even though the right to stand for election and the right to vote were restricted to adult males with certain qualifications and the mayors of the councils were not elected but appointed by a higher authority. A municipal council was simultaneously both an expression of a self-governing local society and a local agent of central, state authority. Although by no means actively embracing all of the population of a commune, such a council nonetheless served to produce and reproduce a sense of place and of community.[19]

Other mechanisms and manifestations of community can also be recognised during the revolutionary period. To a degree, the peasantry were politicised by the Revolution. In each *communauté d'habitants* or *paroisse*, all men aged twenty-five and more who paid some tax, however little, had been given the opportunity to have their grievances recorded in *cahiers de doléances*, in the process learning a remarkable lesson in nascent democracy.[20] Also of potential significance were the local Jacobin clubs which mushroomed in France during the early revolutionary era, attracting patriots determined to defend the gains of the Revolution against a possible aristocratic reaction. They spread from towns and into the countryside, diffusing down the settlement hierarchy so that ultimately, according to Jones, about one in six or seven communes possessed a Jacobin club. Their impact was very uneven regionally as well as socially, and they were fundamentally a bourgeois conception; nonetheless, they became an integral part of the politicisation of many rural communities, but revealing conflicts within those communities rather than necessarily promoting their social cohesion.[21] Exceptionally, a basic and somewhat vague form of rural communism was advocated by François-Noël Babeuf, a socialist revolutionary, to no great effect at the time but sowing the seeds of a political movement which was to come to fruition in some regions of France during the nineteenth century.[22]

The Revolution did, then, go some way towards encouraging a sense of community within rural societies. In particular, it provided the experience of some form of local, municipal self-government. But the countervailing tendencies were probably more powerful: the peasant of 1815 was more likely to have been aware of the growing power of the State and of its use of the local council to exercise its own central control; and he was likely to have been increasingly conscious of the growing internal conflict with the community, of the developing power struggle for the ownership and use of scarce resources, in effect of the evolving class struggle within the countryside, of growing social differentiation. The rural community had by no means completely disintegrated, but it had been seriously undermined and a new sense of individualism – of what Alain Corbin, in a wider social context, has called 'individuation' – emerged to contest it.[23]

The rural community during the nineteenth century

The nature of the rural community during the nineteenth century remains problematic. Some thirty years ago, Wright warned that 'a social historian or social scientist would do well to avoid the quicksands of the [French] peasant problem', given that 'rural France is almost infinitely diverse, and almost any generalisation about the peasantry becomes partially false as soon at it is formulated'.[24] Generalisation about the French rural community in the nineteenth century is certainly fraught with difficulties: to differences from place to place and changes from period to period have to be added complexities arising from the many economic, social and political conflicts both among and within those rural communities. Peter McPhee's cameo of rural France in the 1840s admirably celebrates such differences while highlighting the shared dimensions of all rural communities.[25] One such dimension, the tension between individualism and collectivism, will be the main concern of this book, which focuses upon practical expressions of the principle of *fraternité*. But before addressing that issue directly and empirically through a series of studies of fraternal associations in one region of France, a better general understanding will initially be sought in this chapter of the *mentalités* of the nineteenth-century French peasantry in general as portrayed by some of its contemporaries and as perceived by some of its modern observers. The following chapter will then examine the theoretical discourse within which the concept of fraternity was situated and had to find practical expression.

Perhaps at this point it is necessary to state that the term 'peasantry' is being used here – following many precedents – in a broad rather than a narrow sense, as a general term referring to agriculturally dependent populations within rural French communities rather than to specific, narrowly defined, economic and social groups within them. Of course, such rural communities contained within them 'peasants' whose circumstances differed considerably, for example in terms of the sizes, tenures and family based nature of their farms as well as in their degrees of self-sufficiency or market orientation. Such distinctions will, of course, be recognised when they illuminate the basic problem being addressed. But the term 'peasantry' has such a general usage in the literature that its continued employment here requires little justification.

There exist a number of surveys of the *mentalités* of French rural society during the nineteenth century.[26] There is consequently no need to undertake a comprehensive enquiry here. Instead, the focus will be on the contemporary and current portrayals and perceptions of the relative roles of individualism and collectivism as components of peasant *mentalités*.

Contemporary portrayals of the peasantry

Word pictures

The peasantry have left us with very few self-portraits: they existed predominantly within an oral rather than a written culture. But, for varied reasons, they attracted the attention of better-educated non-peasants who produced many descriptions of the peasantry in words, pictures and numbers. While all of those representations do not tell a single and unchanging story, collectively they constructed a generally unfavourable image of the French peasantry during the nineteenth century. It is, it needs to be remembered, an image of an 'Other' created almost entirely by a non-peasant – and sometimes non-French – 'Self'.[27] During the opening decades of the nineteenth century, the French peasantry had a 'bad press', being criticised by disciples of the Enlightenment, by agronomists fascinated by the English model of agricultural improvement, and by government officials seeking good harvests as a way of promoting public order.[28]

Two foreign influential image makers have been Arthur Young and Karl Marx. During his extensive, three-years', tour of France on the eve of the political revolution of 1789, Young observed its countryside as an agronomist fully acquainted with the pace and character of England's agricultural revolution. He was critical of many aspects of French agriculture, including its indifferent and often absentee landlords, the continuance of feudal ties, the persistence of fallow in field rotations and of uncultivated wasteland generally. He decried the absence of an enclosure movement, the predominance of small farms and the partible inheritance practices which produced them, and he was dismayed by the depths of rural poverty and by the extent to which rural populations were controlled by superstition rather than driven by the spirit of improvement. For example, of Brittany he wrote: 'The country has a savage aspect; husbandry not much further advanced, at least in skill, than among the Hurons, which appears incredible amidst enclosures; the people almost as wild as their country . . .' Young's account of his *Travels in France* was first published at Bury St Edmunds in 1792 and then, in a French translation, at Paris in the following year.[29] It has become embedded into the historiography of French agricultural history: in 1976 Young was five times cited as an authority in Maurice Agulhon's magisterially edited essays on French rural history between 1789 and 1914, whereas Karl Marx received a single mention.[30]

'Barbarism within civilisation' was Marx's dismissive description of French peasants.[31] Duggett has argued that although such an epigram should not be taken too seriously, it is clear that Marx despised French peasants,[32] essentially because their individual self-sufficiency was a severe brake upon the development of any sense of community or class. In his 1852 analysis of the class struggle in France during the preceding four years, Marx wrote:

The peasants who farm their own small holdings form the majority of the French population. Throughout the country, they live in almost identical conditions, but enter into very little relationships one with another. Their mode of production isolates them, instead of bringing them into mutual contact. The isolation is intensified by the inadequacy of the means of communication in France, and by the poverty of the peasants. Their farms are so small that there is practically no scope for a division of labour, no opportunity for scientific agriculture. Among the peasantry, therefore, there can be no multiplicity of development, no differentiation of talents, no wealth of social relationships. Each family is almost self-sufficient, producing on its own plot of land the greater part of its requirements, and thus providing itself with the necessaries of life through an interchange with nature rather than by means of intercourse with society. Here is a small plot of land, with a peasant farmer and his family; there is another plot of land, another peasant with wife and children. A score or two of these atoms make up a village, and a few score of villages make up a department. In this way, the great mass of the French nation is formed by the simple addition of like entities, much as a sack of potatoes consists of a lot of potatoes huddled into a sack.[33]

For Marx, the French peasantry could be seen as a class to the extent that 'millions of families live in economic circumstances which distinguish their mode of life, their interests, and their culture, from those of other classes' but 'insofar as the tie between the peasants is merely one of propinquity, and insofar as the identity of interests has failed to find expression in a community, in a national association, or in political organisation, these peasant families do not form a class'.[34] Although Duggett has shown that Marx was ambivalent towards peasants in general rather than, as Mitrany has argued, 'against' them,[35] it is clear that for Marx peasants in mid-nineteenth-century France were individuals lacking an awareness of their social and political potential as a community or as a class.

Images of the peasantry as 'barbarians' are also evident in French fiction. During the nineteenth century there are identifiable in novels about the rural world two contrasting representations, one 'romantic' and the other 'realistic', although the difference between them is not as wide as such a binary opposition might at first suggest.[36] All of the novels about the rural world were set in particular geographical localities and were concerned with historical actualities; virtually none of their authors had first-hand knowledge or experience of agricultural practices and problems; and almost all of the novels provide a view of a rural world seen from an urban perspective, contributing to a wider discourse about the distinction between 'countryside' and 'town', between peasants and other social classes, in France in the nineteenth century. Novelists were certainly not 'indifferent' to the peasant world.[37] Rémy Ponton has shown that pastoral novels like those of George Sand provided an idealised view of the rural world, a picture of a peasant utopia, of a society characterised by wisdom, balance and a purity of sentiments.[38] Such simplicity and naiveté in an apparently timeless, apolitical, world constructed a countryside which was, as Sand admitted, 'a perfumed Eden where souls tormented and tossed by the tumult of the world can seek refuge'.[39] A fictional tranquil countryside was provided by these novelists as a counterpoint to real, turbulent towns: conservative peasants seemed to be more acceptable as subjects for novels than did radical workers. Such a pastoral vision of the rural world came to be incorporated into primary school reading books during the Third Republic, Ponton argues, because it served to inculcate the values of sobriety, thrift, diligence and fraternity which were vital to the maintenance of republican order.[40] These values seemed to make the 'imagined' peasantry the foundation of a stable French society.

But there were also other imaginations at work and even George Sand in her memoirs referred to the peasants as 'ces êtres vulgaires' (these unrefined beings) whom she had idealised 'en sens inverse dans leur laideur ou leur bêtise' (contrariwise for their ugliness and their stupidity).[41] The image of a barbarian peasantry was widely held and promoted by urbane authors, many of whom adopted a 'realist' stance towards their subjects. In effect, this meant painting a 'black' picture of the peasants' way of life, of their continual conflict not only with the forces of nature and with outsiders such as representatives of the State and of the Church, but also among themselves. For example, Stendhal's peasants in *Le rouge et le noir* (1830) were greedy and brutal; Balzac's in *Les paysans* (1844) were materialist, selfish, immoral savages, self-confessed stupid animals; in Flaubert's *Madame Bovary* (1857) they are gullible, subservient to bourgeois officialdom; Maupassant's Normandy peasants in his many short stories were unscrupulous, constantly thirsting for alcohol and hungry for sex; and Zola's peasants in *La terre* (1877) were described specifically by the village schoolmaster as brutes and generally characterised as fighting among themselves for the possession of land, of women and of money.[42] This bleak picture of the peasantry portrays them as being essentially selfish, avaricious, suspicious, land-hungry individualists with little sense of community, of solidarity or even of social responsibility. Elements of that nineteenth-century picture were still discernible in rural novels of the 1950s and 1960s: they emphasised both the intimate, sexually-charged relationship between farmers and the lands they cultivated, and the independent, even autarchic, nature of a peasant existence characterised by liberty of the individual peasant and the absence of social constraints.[43]

At the turn of the nineteenth century portrayals of the rural world had been nuanced by novels – such as René Bazin's *La terre qui meurt* (1899) and *La blé qui lève* (1907), Eugène Le Roy's *Jacquou le croquant* (1899), and Emile Guillamin's autobiographical *La vie d'un simple* (1904) – which emphasised its internal class struggles, highlighting the different perspectives and interests of large landowners, of share-croppers, of tenant farmers, of small proprietors and of landless labourers, as well as the threats of rural depopulation and of urban modernisation upon very different French farming communities. These accounts tended to be more sympathetic to the rural world and one of these authors was the first French peasant novelist. Emile Guillamin was a share-cropper and autodidact who provided in *La vie d'un simple* an exceptional, insider's view of a peasant world which acts in some respects as a useful corrective to some of the pictures drawn by non-peasant observers.[44] Guillamin's country folk were (as Eugen Weber has noted) rough but not savage, truculent but not callous, indeed strangely pacific, perhaps in reaction to Zola's savage brutes in *La terre*.[45] Guillamin's peasants were, however, individualists: he himself came to believe in the value of collective action as a non-revolutionary rural reform and in his *Le syndicat de Baugignoux* (1912) he provided an account of the protest movement which he led with the share-croppers of Bourbonnais, a movement which was unsuccessful because – in Guillamin's view – peasants were considered by others to be and also considered themselves to be socially inferior.[46] Collective action, it could be argued, foundered on the rock of individualism. Pierre-Jakez Hélias, in his vivid autobiographical account of life in a Breton village in the early years of the twentieth century, confirms the dominance of individuals and of families, as well of course as the school and the church, on the social stage.[47]

The overall picture portrayed by novels located in different places in France and situated in different periods of the nineteenth century is of a rural society founded upon competition rather than upon co-operation among individuals, of a peasantry grounded both in an environmental and social conflict and in a self-preservationist conservatism. That picture finds affirmation – in both positive and negative terms – in accounts by contemporary commentators and historians. For example, Hannah Lynch, in her account of French life in town and countryside at the end of the nineteenth century, noted the peasants' 'sturdy passion for independence. It is this passion that enables them to scrape, and serve, and suffer privation with dignity and patience. However meagre their resources may be, they are content with their lot, provided the roof they sleep beneath is their own, the land they till their own, the goat, the pig, the poultry theirs to do what they will with . . . the distinctive characteristic of the peasant is an indomitable spirit of independence'.[48] A similar sentiment was expressed by Mary Duclaux (née Robinson) at about the same time. Having commented sympathetically upon the plight of many small farmers in France, she concluded:

Unfortunately, the peasant is, as a rule, intellectually idle, incapable of combination, suspicious, and impatient of new-fangled ideas; he finds it simpler to sell his goods to the buyer from Paris as his father did before him, than to combine with his neighbours in an agricultural syndicate. The principle of solidarity has scarcely penetrated into rustic parts, but the need of resisting the low prices imposed by the large farms using machine labour will certainly, in time, teach the peasant many things. Let his mind once grasp the idea of a common prosperity – where Tom's good luck is not ensured by the misfortunes of Dick and Harry, but all are implicated in the well being of each – let him forget to suspect and learn to combine; from that day forth his social future and well-being are assured.[49]

The independence of the French peasant – as well as his industry, self-denial and frugality – was similarly stressed by the distinguished English agricultural historian, Rowland Prothero, in his survey of French farming in the early nineteenth century. Prothero also saw the morcellation of farms and the parcellation of fields as militating against peasant co-operation.[50] Similarly, although H. W. Wolff commented in his paper to the Royal Agricultural Society of England in 1900 that agricultural syndicates had brought a remarkable change to the face of French agriculture during the previous fifteen years or so, in applying the principle of combination to the furtherance of common interests in agriculture, he emphasised that it was 'one thing to make admission [to agricultural syndicates] easy and quite another to induce a sufficient number of the French peasantry to join, many of whom are backward beyond anything that we can conceive, and all of whom are wanting in personal initiative and expect to be pushed to whatever they are to do by some superior person'.[51]

From these multiple and sometimes contradictory literary characterisations of French peasants it is now time to turn to related pictorial representations of them.

Paintings

Many paintings of peasants were produced during the nineteenth century as pictorial records and impressions of French rural cultures and they became a part of the wider political discourse on the role of the peasantry within French society.[52]

Images of the peasantry in paintings were – like those in novels – numerous and sometimes inconsistent but they do provide a resplendent gallery of striking portrayals of the individuals and communities who are the focus of this study. There have been a number of attempts made recently to distil the central themes of French peasant paintings from a vast literature analysing the works of individual artists.[53] These will be drawn upon here as providing one set of especially valuable observations of peasant *genres de vie* in general and of the relative roles of individualism and collectivism in particular.

Before selecting a few significant themes in French peasant painting during the nineteenth century, it must be emphasised that a single, monolithic and integrated image of the peasant in such work is neither to be expected nor to be found. Peasants were painted both as specific persons and as general types.[54] There are, unsurprisingly, some contradictory, opposing constructions to be addressed. To take first a specific example, in the work of Jean-François Millet (1814–75) have been detected different sets of concerns and subjects. Millet's paintings, as considered by Jean-Claude Chamboredon, include both pastoral and realist representations of peasant and rural landscapes, both idyllic scenes in a familiar tradition and realist pictures of peasants in a new form as socialist workers; Millet's paintings also, Chamboredon argues, construct an opposition between an independent peasantry of small owner-occupiers from regions of woodlands and enclosed fields and an agricultural proletariat from open-field regions of large farms. Even so, these dichotomies are collapsed by Chamboredon into a single view of a rural world in which the morality of labour and a fatalistic resignation to one's position in life are the dominant social values: thus he identifies in Millet's paintings of the late 1840s to the early 1870s the portrayal of a new peasant utopia.[55] More generally, Brettell and Brettell identified a different dichotomy, that between the image of peasants as little more than animals, as people close to nature and without culture, and that which saw peasant culture as authentic, even as the foundation of national civilisation, free from the artifice of urban life. They argued, however, that the former construction of the peasant was more prevalent in literature than in the pictorial arts, while the latter construction found equal expression in both.[56] That point can, of course, be questioned. Geneviève

Lacambre has pointed out that although the emerging novelty of paintings in the nineteenth century was their portrayal of peasants at work, depicting them with a new grandeur and realism, ultimately the paintings became more romanticised and even nostalgic depictions of regionally specific peasant types.[57]

Paintings of peasants recorded rural cultures historically and geographically. In so doing, they contributed both to the search for a national past and to an acknowledgement of the regional diversity of France. Paintings and lithographs of rural France – of its toilers and tradesmen, of its costumes and its customs, of its landscapes and its light – came to be given titles which identified precisely the locality, region or department represented. As such, paintings were part of a wider study and representation of the folklore and popular traditions of France, part of a quest for a French national identity which recognised cultural regionality, a study which also harnessed to it the new art of photography.[58]

Of that there can be little question. But to portray peasants as possessing their own cultures need not – indeed, did not – mean representing them as free from nature. On the contrary, peasants were frequently pictured as heroic individuals closely linked to nature, wresting their livelihoods from a reluctant earth. Perhaps the epitome of this image is that of Millet's *The Sower* (1850), boldly striding alone across a field, broadcasting the seeds whose product will be the bread of life for future generations.[59] Millet's famous sower – like so many of his peasants – was a solitary, individual rural worker. But even pictures of groups of peasants could emphasise their isolation from each other. Gustave Courbet's (1819–77) *Peasants of Flagey Returning from the Fair* (1850) depicted a procession of people all of whom (except two socially superior people on horseback) were, in the words of Brettell and Brettell, 'self absorbed and hence isolated from the collective experience chosen by the painter'.[60] Similarly, Jules Breton's (1827–1906) peasants were depicted as strong individuals, even when they are shown apparently working communally: such was the case, for example, in *The Gleaners* (1854).[61]

The powerful connection between peasants and nature, between individuals and their physical environments, was a fundamental thread running through much of realist painting. Agricultural work in all of its variety constitutes a major theme of these works. Labouring men and women are seen engaging with the earth: the paintings stress the physicality of existence, the often painful, certainly continuous, struggle for survival. For example, Millet's paintings of peasant life, such as *The Winnower* (1847–48), *Man Turning over the Soil* (1847–50) and *Man with the Hoe* (1860–62), 'clearly express a deep sympathy for the harsh fate of the great majority of the working population in rural France. Such images show men and women who toil at thankless and back-breaking tasks and remain desperately poor.'[62] As John Berger has pointed out, Millet identified with his peasants, with their labours, in part because he knew personally what it was like to be one: his paintings were in part expressions of his own experience. Berger considered that Millet, without a trace of sentimentality, painted the truth about peasants as he knew it, whereas the public – or certain sections of it – foisted their own interpretations upon his pictures and even requisitioned them for false preaching, as in the case of *The Angelus* (1855–7), which came to be widely perceived as portraying a passive couple whereas it contains many indications of potential conflict between them, for while the woman is submissive to religion the man is restless and protesting.[63] But Millet's peasants, and indeed the peasants of Vincent van Gogh, were not explicitly the revolutionaries identified often in paintings by Courbet, whose self-confessed commitment to socialism strongly influenced the reception of his work.[64]

Paintings of peasants in part reflected the cultural values of the bourgeoisie, while their iconographies were certainly refracted through bourgeois values. Brettell and Brettell have argued that the fundamental values identifiable in peasant paintings of the nineteenth century are those of work, family, religion and patriotism:

As a kind of Everyman, the peasant served to render absolute these values and further to suggest they were generally human rather than bourgeois . . . In fact, peasants, like their urban counterparts whether bourgeois or proletariat, had no unified system of values, and the fact that such a large percentage of peasant pictures embody one of the four values mentioned above is proof of the ideological importance of those pictures to their audience. The peasant image served to teach 'real' values to the bourgeoisie, while the peasant himself was plagued by many of the same ills – unemployment, malaise, ambition, and alcoholism – as was the bourgeoisie.[65]

In addition to exhibiting those four fundamental values, the peasant is also represented in paintings as inhabiting an apparently timeless, almost unchanging, realm in which the rhythms of his life are determined by the seasonal cycle of agricultural activities. But indications of the peasant's encounter with modernisation are not entirely absent: some pictures represent peasant responses to war and conscription, to class struggle, to the developing market economy, and to mechanisation – but generally doing so within the framework of the normative values of work, family, religion and patriotism.[66]

What, then, of the signs of collectivism or fraternalism in peasant paintings? Contemporary radical critics, when defending Millet and Courbet, claimed that the tradition of peasant naturalism in art was one that stemmed from seventeenth-century Holland and was one that spoke for individualism and for democratic values, rather than for government or religious authority. Painting rural life and landscape was pitted against painting religious and historical subjects.[67] Raymond Grew was astonished by how few paintings there are depicting a group of peasants acting together. He noted that most of Millet's peasant figures stand alone, their relationship to others unstated even when a second or third figure is placed in equal isolation within the same frame. There is, admittedly, some but not much ambiguity: 'Millet's gleaners can be seen as working together or in lonely desperation; they bend over the same field much as the survivors of the Medusa cling to the same raft'.[68] What Millet's gleaners exhibit is the struggle for personal survival, not a spirit of solidarity. His peasant paintings came to be accepted as expressions of the bourgeois morality of labour, of working hard to carve out a place in society by individual labour.[69] Given that Courbet's realism was allied to his socialism, we might expect his paintings of peasants to have signalled not only what he himself described as their unending misery and destitution as individuals.[70] John Berger has claimed that within Courbet's paintings one can indeed discover 'a sense of potential Arcadia' and that 'Courbet's socialism was expressed in his work by its quality of uninhibited Fraternity.'[71] Berger's claim here is neither supported nor convincing. Courbet sought only practical, not utopian, solutions to social problems, and he placed a very high value on the freedom of the individual – including that of his own independence as an artist. Courbet painted peasants who were hard-working, proud, dignified and free individuals.[72]

For Monica Juneja, a distinctive aspect of the iconography of rural life in nineteenth-century France was 'the emergence of the peasant as an individual whose identity is defined by his act of labour'. She argued that many paintings representing the agricultural proletariat were composed of certain basic formal elements: 'a vast, open field, belonging to a large, prosperous farm and populated by numerous individuals often distributed in an almost decorative manner over the expansive format of the canvas'. Juneja argued that the portrayal of agricultural labour on a large scale during the closing decades of the nineteenth century, as in Lhermite's *La paye des moissonneurs* (1882), idealised fraternal ties between employers and workers, reminiscent of a pre-capitalist era – this at a time when a new collective class consciousness was being forged by farm labourers on the anvil of the agricultural crisis.[73] Juneja directly considered how artists addressed changing forms of peasant sociability during the second half of the nineteenth century. She concluded:

Paintings chose to focus on forms of social interaction essentially centred on the family – the extended family, especially in pictures of wedding feasts or the single family reunion – or on social activities which united the entire agricultural community, such as collective feasting after the harvest. Only towards the end of the century do we have manifestations of forms of sociability which expressed a new, nationalist spirit – as in Lhermite's *Un quatorze juillet villageois* – for by this time the anti-clerical republican municipalities had attempted to substitute religious processions by civic festivities. Most of these images are rich in symbols of domesticity and consolation, as well as in folkloric details such as traditional costumes, pottery and musical instruments drawn from the abundant repertoire of illustrated literature on the provinces; they ensconce the peasant within the security of the familial or work space, reinforcing ideals of communal solidarity and the permanence of familial bonds.[74]

In short, insofar as paintings portrayed peasant sociability at all they did so in its traditional, rather than in any of its modern, forms.[75] Brettell and Brettell, in their massive survey of peasant paintings in the nineteenth century, emphasised that many pictures portraying peasant labour showed single figures, or at most two involved in a single task. In this feature, the paintings stem from 'the first and greatest image of peasant labour, Millet's *The Sower* (1850)'. Such images, because of the isolation of the worker, 'possess an emblematic quality'.[76] Brettell and Brettell pointed out that such paintings reflect what many contemporary commentators considered to be the key problem of peasant politics, 'the possessive and individual spirit of the peasantry. For many nineteenth-century intellectuals, the peasant had no ability to organise, was jealous of everyone else, and was opposed to any form of collective action.'[77] Paintings show few signs of the politicisation of the peasantry or of their embrace of new forms of fraternalism.

Three broad generalisations can now be emphasised. First, all of these literary and pictorial portrayals of the peasantry both reflected contemporary attitudes and reflexively contributed to the production and reproduction of those attitudes themselves. Topographies, novels and paintings – perhaps, in a society not wholly literate, especially paintings – were themselves an integral part of the political discourse about the peasantry. Second, peasants were represented principally as individuals, or at best as members of a family, rather than as members of wider social groups. Third, these image makers portrayed the peasant world as if it were virtually timeless. They at best neglected, at worst ignored, the transformation – some would say modernisation – of rural France which was going on around them: they portrayed and represented essentially a traditional, a pre-modern, peasantry. The extent to which the influence of these contemporary portrayals has persisted into the modern historiography of the French peasantry now needs to be considered.

Current perceptions of the peasantry

While there has been a considerable and lively debate about the precise pace and detailed character of the development of capitalist agriculture and the 'modernisation' of rural society in France during the nineteenth century, there has been less critical examination of the alleged decline of the rural community. These issues now deserve attention, first in terms of the general debates about the changing character of agriculture and of rural society in France during the nineteenth century, and then in terms of the specific claims which have been made about individualism and collectivism as changing components of peasant *mentalités*. There is neither the space nor the need to map in detail here the whole gamut of agricultural and social changes in rural France during the nineteenth century. All that is necessary is to provide a sketch of the controversies about them, as the framework within which to consider the specific problem of fraternal associations.

The development of agriculture?

Whether or not France witnessed an 'agricultural revolution' during the eighteenth and nineteenth centuries has been much debated. The pace, location, character, and underpinning processes of agricultural changes have all been hotly contested.[78] The long-running controversy has resulted to some extent from the limitations of the statistical evidence upon which it is based, but it has also reflected the difficulty of coming to agreed conclusions for France as a whole in view of the considerable regional and local variations which have to be accommodated within a national picture. Given that to those geographical variations have also to be added both temporal fluctuations and sectoral differences, it would be very surprising indeed – even disturbing – if the nature of agricultural change in France during the eighteenth and nineteenth centuries had not been a matter of contention. Nonetheless, two broadly differing sets of views – one essentially orthodox, one essentially revisionist – can be identified.

In brief, the long-standing orthodox view has been that the relatively slow development of the French economy as a whole was a consequence in considerable measure of the relatively slow development of the agricultural sector in particular: the Industrial Revolution in France was delayed by the retarded nature of its Agricultural Revolution, specifically by the slow increase in agricultural productivity needed to feed an increasingly industrial and urban population and by the failure of the agricultural sector to release labour and capital to the manufacturing and service sectors of the economy.[79] Much attention came to be focused upon the rate of diffusion of new crops (especially artificial meadows, but also potatoes), of improvements to livestock and heavier stocking of animals (and hence greater production of organic manure), of the use of chemical fertilisers, and of the mechanisation of farming. There has been considerable debate, even within this orthodox account, about when agricultural productivity in France began significantly to increase, with dates being proposed which vary between the early 1700s and the 1840s.[80] The most recent, massive and quantitative analysis of agricultural production, productivity and growth in France between 1810 and 1990 by Jean-Claude Toutain has identified a growing regional specialisation and a developing national and international commercialisation of French agriculture. Nonetheless, Toutain argued that, while in terms of agricultural productivity differences among the regions of France diminished between 1840 and 1929 (but were accentuated thereafter), when looking at the national picture one cannot justifiably speak of an Agricultural Revolution in France before 1960.[81] Despite the scale of Toutain's study, the central problem remains unresolved – in part because Toutain failed to define what for him would constitute such a phenomenon.

There has been more agreement among historians about the failure of French agriculture to submit to the structural reorganisations (principally enclosures and the consolidation of farm holdings) which were such an important part of agricultural improvements in England in the eighteenth and nineteenth centuries.[82] The predominance of small, fragmented, labour-intensive but inefficient, conservatively operated family farms producing mainly for subsistence or at best for only local markets has been seen as the fundamental brake on the development of French agriculture. The very substantial and 'backward' peasant sector has been blamed in its turn for a much wider economically retarded performance.[83]

That pessimistic view of French agriculture in general and of the peasantry in particular has come increasingly to be challenged by revisionist historians and historical geographers.[84] For example, Philip Hoffman has argued for considerable economic growth in the French countryside between 1450 and 1815 and Paul Bairoch has argued that wheat yields in France between 1800 and 1880 were not significantly different from the European average and that the rate of growth of productivity in French agriculture between 1830 and 1880 was almost twice the European average.[85] Pautard has also identified the latter period as one both of significant overall national agricultural growth and of a diminution in the regional differences in agricultural productivity, as the more laggard regions narrowed the gap between themselves and the more advanced regions during this period of overall growth.[86] Until almost mid-century the most agriculturally significant changes were restricted to a few regions, notably to the Paris Basin with its large urban markets, its better-than-average communications and its proximity to centres of agricultural innovation in England and Holland.[87] But by the 1880s agricultural improvements had come to affect most regions of France. The growth of urban and industrial markets and improved access to them not only by better roads but also by the new railways and developing rural tramways encouraged the commercialisation, and with it the specialisation, of agriculture. The rate of annual growth of agricultural productivity in France between 1830 and 1880 was, it seems, well above the European average, although it then fell back to the average.[88] Moreover, since the 1850s, some regions of the country had been releasing labour to the growing urban and industrial economy and had, as a consequence, improved their own agricultural productivity through the wider adoption of labour-saving machinery.[89]

Importantly, this revisionist case also includes a more positive view of the peasantry, of small-scale farming, than that embedded within the orthodox narrative. In its most extreme form, this case involves a Marxist rejection of the commonly held assumption that the development of agrarian capitalism required large-scale farming. Instead, it argues that because in France large landowners were not practising farmers but merely landlords the route to capitalist development lay with small independent producers.[90] While this is to ignore the role of tenanted farms on the large estates, it does nonetheless reinstate the peasant proprietor as an active agent in the development of capitalist agriculture. In those regions favoured by their proximity or improved accessibility to growing markets, or by soils and physical conditions, small farms became increasingly commercialised and specialised. Smallholdings were especially suited to viticulture, to market gardening, to fruit growing and to livestock production. Thus even peasant farmers, originating within a 'natural' economy, could insert themselves – and did so increasingly – within the growing 'capitalist' economy.[91]

The revisionist school also reverses another orthodoxy, by arguing that the slow growth of the French population in general and of its urban-industrial population in particular – in effect, the slow growth of the market for agricultural produce – provided little stimulus to improve farming productivity, either through new methods or through structural change.[92] It also argues that to the extent that French peasants in some regions stayed on the land rather than migrating into towns and their industries, they did so because the slow growth of the manufacturing and service sectors in those regions created few new employment opportunities for them.[93]

There are, therefore, as Annie Moulin's survey has emphasised, now grounds for believing that neither French agriculture as a whole nor its peasant sector in particular were as backward as it has so often been argued.[94] As Ronald Hubscher has recently concluded, in France between 1800 and 1914 both agriculture in general and the peasant world in particular were 'profoundly transformed' under the impact of the developing global economy and society. The changing rural landscapes of France were far removed from the increasingly conventional and unreal portrayal in literature and art of an unchanging, fossilised peasantry, an image grounded in an agrarian mythology and a bucolic vision of life in the fields.[95]

The 'modernisation' of rural society?

A vivid and influential account of the 'modernisation' of rural France during the nineteenth century has been provided by Eugen Weber in his *Peasants into Frenchmen: the Modernisation of Rural France, 1870–1914*, published in 1976.[96] This is an evocative account which draws upon archives, folklore, ethnography and local historical studies to construct a lively picture of the collapse of traditional peasant culture and the rise of modern rural society. It is a fascinating portrayal, full of anecdotes and surprising details. In many ways Weber's *Peasants into Frenchmen* appears to follow the advice offered by David Pinkney forty years ago that foreign historians of France should not try to match native historians in basic archival research but should instead concentrate on synthesis and interpretation, bringing to their work the advantages of their detachment, freshness of perspective and capacity for surprise.[97]

Put simply, Weber's thesis was that well into the nineteenth century most of rural France was comprised of local, at best regional, autarchic peasant economies, diverse and almost unchanging, poor and primitive, with little contact with each other and even less with towns and markets. The village or commune, at most the *pays*, constituted the limits of social intercourse for most peasants, whose knowledge and experience were fundamentally rooted in their immediate locality. This 'traditional' rural society was, Weber claimed, 'modernised' from the 1880s onwards, under the increasingly transformative impact of roads and railways, primary schools, and military service. These processes of change incrementally integrated local, essentially rural, communities into a national, essentially urban, culture. Peasants became Frenchmen. Such a broad-sweeping thesis and the way in which it was presented by Weber command admiration. Many parts of the general thesis resonated with what more cautious, less adventurous scholars knew about particular places or processes within nineteenth-century rural France. For example, his emphasis on the role of population migration in the disintegration of traditional peasant communities has been convincingly confirmed.[98] Weber's thesis has become probably one of the most influential interpretations of French rural history. It has certainly had a major impact upon nineteenth-century French studies, both in the original and in its French translation.[99]

There have, however, been several and severe criticisms of Weber's thesis, notably those by Ted Margadant, Charles Tilly and Peter McPhee.[100] Collectively, these critics argue that Weber provided a misconceived and misleading impression of French rural society in the nineteenth century. Questions have been raised, for example, about Weber's identification of the timing and of the generality of the changes described, as well as about his explicit use of limited evidence and his implicit use of discredited theory.

Many of the changes which Weber identified as being crucial only from the 1880s were in fact operative in many parts of France much earlier in the nineteenth century and possibly even before then. There was, in Tilly's phrase, 'no solid cake of custom to break'.[101] As Margadant put it, a more balanced comparison of rich lowland with poor highland regions would suggest a different periodisation of market development and social change in the French countryside.[102] Indeed, a related major criticism is that Weber ignored both the regional diversity of social, economic and political change and the many regional monographs by geographers and historians which have explicated it. As far as his sources were concerned, Weber relied principally and uncritically upon (often titillating) anecdotal evidence and upon observations of rural society by those bourgeois writers who held pessimistic, derogatory views about the peasantry. He undertook no systematic, rigorous analysis of statistical observations of the history and geography of 'modernisation' (such as literacy, income, population mobility, or prices) for France as a whole. As far as theory is concerned, Tilly criticised Weber for his use of the concept of 'modernisation'. The problem here is twofold. First, Weber uncritically adopts the Third Republic's own view of the civilising mission of the State, through its control of schools and of military service and its very considerable influence upon communications, and assumes rather than demonstrates that the mission succeeded between 1870 and 1914. Second, Weber implicitly and uncritically adopts the liberal, progressive version of 'modernisation theory'. Tilly argued that 'many of the most concrete changes in the social life of nineteenth-century Europe did not follow the paths required by theories of modernisation'. While Tilly saw nothing wrong in the use of the term 'modernisation' for merely descriptive purposes, he did object to 'the elevation of the idea of modernisation into a model of change – especially a model in which expanded contact with the outside world alters people's mentalities, and altered mentalities produce a break with traditional forms of behaviour'. In Tilly's view 'the magic mentalism' was 'not only wrong but unnecessary', because for him an analysis of 'the interaction of two deeper and wider processes' – the growth of national states and the expansion of capitalism – offered 'a far more adequate basis for the understanding of change in nineteenth-century Europe'.[103] Similarly, from a revisionist Marxist perspective McPhee has argued that Weber 'grossly exaggerated the extent of an essentially autarchic peasantry', that a more fruitful way of understanding the peasantry is as a transitional element of an agrarian social formation, and that as independent small and middling producers they comprised a progressive element in the countryside.[104] Again, and similarly from a Marxist perspective, Magraw has challenged Weber's whole thesis and suggested that what some might see as 'progress' others might see as internal colonialism or even as cultural genocide.[105]

More recently, another major challenge to Weber's thesis has been provided by James Lehning in a study of cultural contact in the department of Loire. Lehning has shown how the very concept of 'peasant' in France was itself a cultural construction whose meaning changed during the course of the nineteenth century. He argued that the early nineteenth-century version of 'peasant' – isolated, religious, ignorant, violent and bordering on savage – 'was no longer useful for French culture' in the second half of the century. The economic, social and political conflicts of late nineteenth-century France required, according to Lehning, a different 'peasant', 'one who, with help from the state, could be peaceful, literate, secularised, a patriotic republican, and the repository of French values against the radical working class'. Lehning concludes that it was not so much a matter of converting peasants into Frenchmen but of (local and regional) peasants becoming French peasants.[106]

Embedded within this broad debate about the 'modernisation' of the rural society has been a more specific one of particular relevance to this present study of fraternalism in the countryside, a debate about the politicisation of France's peasantry. As Edward Berenson pointed out in his admirable critical review of this latter controversy, to some extent its origins lie in the ambivalent attitude which Marx held about French peasants, viewing them principally as isolated from currents within society at large and detached from politics but also recognising the existence of a peasant 'revolutionary', one who was able to 'strike out beyond the condition of his existence'. But the debate conducted by historians over the last four decades has had such a long run in part, Berenson suggested, because the term 'politicisation' has itself been ill-defined and so employed with different meanings. For some 'politicisation' has meant the development of political consciousness, for others engagement in political movements, for still others the adoption of political forms of protest, such as demonstrations, petitioning, and voting.[107] In essence, the debate has been between those – like Maurice Agulhon, Ted Margadant, Raymond Huard and Edward Berenson – who have claimed many of the peasantry had become politicised by mid-century (and especially during the period 1848–51) and those – notably Eugen Weber but also others, like P. M. Jones – who have argued that, while the mid-century saw the adjustment of peasant allegiances from one set of notables to another, a genuine political awakening was not to be witnessed among the peasantry until the 1870s or even 1880s.[108] As with almost any aspect of French history during the nineteenth century, some of the differences are more apparent than real, because some of them are regional – the pace and character of the politicisation of the peasantry varied from place to place, reflecting different economic, social and cultural circumstances. While the debate can be reconciled to some extent in this way, it is not fully resolved because those who have argued for peasant conservatism have challenged the politicisation thesis in general rather than in particular places.

But the timing and the spacing of peasant politicisation are less important for this present study than is its character. There is broader agreement among historians about the processes which promoted a heightened political awareness among the peasantry, taking 'political awareness' here to mean initially an appreciation of the wider structures and conjunctures within which local peasant communities acted, an appreciation which could be followed ultimately by an understanding of the ways in which peasants as agents could interact positively with those structures and conjunctures, could endeavour to exert increasing control over them. The tangibly increasing impact upon peasant communities of the growth of the nation-state and of the spread of a market economy made them gradually more aware of the extent to which their livelihoods were dependent upon circumstances emanating from beyond their own immediate localities. Such processes introduced yet more uncertainties into an already insecure peasant existence. That in turn, it could be argued, would have led peasants gradually to adopt new strategies for managing such risks, as they would no doubt have been initially perceived before they came to be seen as opportunities. One of the strategies potentially available for managing such risks was increasing engagement with, and even in, the political process, an engagement which could itself take a variety of forms, including voting, seeking election to local councils, and the creation of fraternal associations aiming to defend and, if possible, promote the interests of particular social and/or economic groups.

Debate here has centred around the extent to which such politicisation emerged and evolved spontaneously within rural communities and the extent to which it was exogenously imposed upon them by urban activists and influences. No doubt once again the initial answer must be 'to differing extents at different periods and in different regions', but it should also be possible to develop a more nuanced response which recognises the role of local, regional and national (political) issues in the life of every rural community in France during the nineteenth century. In this context, Tony Judt's study of socialism in Provence is exemplary.[109] Those of the conservative school of thought have argued that rural politics reflected local, not national, concerns: for example, Michael Burns has claimed that even at the end of the century 'rural politics remained greatly influenced by personalities, family ties, local interests and local intimidation'.[110] But that has also been true of rural politics in France during the second half of the twentieth century[111] and it hardly comes as a surprise in the nineteenth century. The blatant importance of local issues and of local personalities in rural politics during the nineteenth century must not blind us to the local significance also of regional and even national issues and personalities. A reading of the minutes of almost any commune council soon reveals the complex interweaving of such issues and personalities: each council was constantly being required to respond to matters raised by the department's administration and by the national government, and in doing so its locally specific responses inevitably inserted its councillors and commune into a wider political network.[112] There can be no doubt that national and regional issues resonated locally.[113]

This does, however, raise the question of the extent to which peasant politicisation should be seen as merely a pragmatic response to changing circumstances and to an enhanced awareness of the nature of those circumstances and of the means by which they might be managed, and the extent to which it should also or alternatively be seen as a principled commitment to a particular ideology. Of central concern here are the processes by which rural communities and individuals came to adopt particular practices and ideologies. In the literature of the politicisation debate there is a consensus that ideas, especially socialist and republican ideas, were diffused from towns into countrysides. The implication is that urban principles were transformed into rural practices.

The collapse of the peasant community and the rise of peasant individualism?

The history of the French rural community after the Revolution has been classically portrayed as one of bitter struggle between two conflicting forms of economy, the 'natural' and the 'capitalist'. More than forty years ago, Albert Soboul argued that 'from the moment that agricultural production was integrated into the capitalist economy, the community was doomed: after more than a century of resistance it disintegrated completely'.[114] His powerful thesis was that throughout the nineteenth century the small peasants fought vigorously for their own right to live and for the common rights of usage over fields, heaths and woods which safeguarded that living. He saw the agrarian unrest of 1848–51 as 'the last violent episode, the last paroxysm of the traditional peasantry on the verge of disappearance'. In the course of those struggles, peasants acquired a consciousness of their interests as a class. The appearance of class consciousness among the peasantry was, in Soboul's view, accompanied by the disappearance of the rural community as it became increasingly integrated into the capitalist economy.[115] Soboul provided a classic portrayal of 'the decline of the French village community' and the growth of 'unchained individualism'. His view, that 'the rural community was bound to disappear in the course of the great revolution which integrated agricultural production into the capitalist economy', has underpinned much of the discourse on the French peasantry for an entire generation of scholars.[116]

For example, Harriet Rosenberg, in her study of the French Alpine community of Abriès, has argued that local autonomy and agency were eroded during the nineteenth century with the growth of the State and the spread of capitalism.[117] More generally, Theodore Zeldin's panoramic survey of the *mentalités* of French peasants from 1848 onwards concluded that the old community spirit was collapsing, that it was breaking up, that collective controls and traditions of co-operation were gradually abandoned: 'The peasants had long traditions of mutual assistance, and in some regions even of organised co-operation in farming. The nineteenth century was in some ways a gap, a period in which these traditions were suppressed while the peasants struggled for ownership of the land; but towards the end of it there was some revival of co-operation.' Zeldin argued that although some co-operatives were established from the late nineteenth century onwards, 'the peasants rejected any collective discipline'.[118]

In effect, rural France during the nineteenth century came to be seen as having witnessed the triumph of individualism over collectivism. For example, Yves-Marie Bercé has argued that from the sixteenth century to the eighteenth century a solidaristic, self-interested peasant community responded to external threats by means of riots, rebellions and revolutions. But during the nineteenth century, Bercé argued, the unified interest and solidarity of the community declined and undermined the foundations for peasant revolts.[119]

Even Maurice Agulhon, in his magisterial survey of *la civilisation paysanne* in France, argued that during the nineteenth century under the impact of modernising influences rural municipalities became less community-minded: he cites by way of example the role of public spaces for collective activities, including the playing of games, pointing to the break in continuity which existed between the *communal* of the eighteenth century and the *stade* of the twentieth century.[120] The Revolution of 1789 had been based on the premise that modernity and progress had private property ownership and use as a necessary condition, whereas conservatism and routine flowed from collective ownership and use.[121] That premise, according to this school of thought, gradually eroded the rural community during the nineteenth century.

That argument has been extended by those who claim that the peasantry, lacking a sense of community, class or solidarity, were slow to politicise and to organise themselves into associations. For example, Suzanne Berger has argued that in Brittany peasants were 'against politics' even in the late nineteenth and early twentieth century, and that their social and political milieu militated against voluntary associations. 'The traditional features of peasant life were obstacles to active participation in an organisation: the constraints of an undifferentiated workday, loyalty to the Church, and a network of social relations which extended no further than the village. In addition, there were attitudes of apathy, jealous egalitarianism, and defensive individualism, all of which supported a weak participant role that rural organisations could have transformed only by attacking traditional society at its roots.'[122] Gordon Wright argued that the peasantry took little active part in politics until the Great Depression of the 1930s and that few of the agricultural associations or syndicates formed before then were organised by peasants themselves, 'most were organised by local crusaders from the landed aristocracy or the bourgeoisie'.[123]

As has already been emphasised, a very influential account of the French peasantry during the nineteenth century has been that provided some twenty years ago by Eugen Weber – and it is an account which represents peasants fundamentally as individuals rather than as communities or other social groupings.[124] Weber's panoramic perspective included within its compass autarky but not associations, conscription but not co-operatives, furniture but not fraternities, sewing machines and suicides but not societies or syndicates, and undergarments but not unions. In his portrayal of 'traditional' societies Weber did, of course, have something to say about those activities which involved opportunities for socialisation, such as fairs and markets, baptisms, burials and funerals, religious festivals, and *veillées*. But his massive survey did not recognise, or even suggest, that the 'modernisation' of rural France might have involved the emergence of new social groupings, new forms of sociability, new expressions of fraternity, within and among communities. Weber's primary concern, of course, was with the processes by which peasants were transformed into Frenchmen, a stance which left little, if any, room for new social groupings intermediate between the individual and the State. He argued, instead, for the supreme importance of what he termed 'local solidarities' at mid-century and their replacement by the end of the century of a consciousness that 'great local questions no longer found their origin or solution in the village, but had to be resolved outside and far from it'. Weber's view was that the seemingly monolithic traditional rural society in France was based on the community, with no break existing between the group and the individual, and only when and as that society 'began to disintegrate could individuals begin to see themselves as separate or separable from the group'. Only then could new rival solidarities – 'like the new-fangled one of class' – develop.[125] But he had nothing more to say about voluntary associations based upon class or any other perceived collective interest.

A similar emphasis upon the growth of individualism is to be seen in Annie Moulin's more recent survey of peasantry and society in France since 1789. She argued that solidarities and divisions in the period from 1789 to 1815 were based upon the two basic social units of the family and the local community, but that the period from 1815 to 1870 was marked by a 'weakening of the old economic bonds between members of the community, victim, it is clear, of the relentless rise of agrarian individualism'. Communal institutions, she suggested, 'fell into decline' and in some areas, like the Beauce, 'systems of mutual aid between farmers quickly vanished'. But within the period from 1870 to 1914 Moulin also recognises the development – in the face of economic difficulties – of a new collectivist spirit, expressed in part in the creation of 'mutual-insurance societies which offered some security against fire, loss of harvest or animal sickness' and in part in the setting-up of syndicates for the bulk purchase of agricultural supplies. Moulin claimed that most of the initiatives taken to organise such associations came from outside the peasantry itself – in effect, from the rural elite of landowners, doctors and lawyers, supported by local administrations and especially by their professors of agriculture. She also insisted that the roots of the new collectivism were hardly deep among the peasantry, claiming that in 1914 some three-quarters of all peasants played no part in this co-operative and syndicalist movement.[126] Moulin offered, then, a more ambivalent, but potentially more nuanced, recognition of the relative roles of individualism and collectivism.

Indeed, a more complex interpretation of the transformation of rural society in France during the nineteenth century has gradually been emerging. It has come increasingly to be argued that one can identify a growing sense of community within rural France during the nineteenth century and that there was during its second half even a great flowering of its social and cultural life. To some extent this has been attributed to the growing role of the commune and its council as a locally influential institution and to the developing politicisation of the peasantry. Agulhon, for example, has argued that the law of

1831 played a crucial role in that process, because it lowered significantly the qualifying wealth level of the right to vote and so brought a much larger fraction of the rural population into local democracy and, more importantly, because it required mayors and their deputies – although still nominated by an external authority (the prefect) – to be selected from among the elected councillors, which at least theoretically made it possible for a commune to prevent an unpopular noble from gaining access to local power.[127] During the course of the century, the municipal council became increasingly active in the ordering of local life, and in so doing provided a focus for the production and reproduction of a sense of community even while the locality was being increasingly integrated into the French nation.[128] Councils contributed to the general process of economic, social and political transformation which has now been so widely identified in rural France during the 1850s, 1860s and 1870s, a period which some scholars now identify as witnessing 'the peak of rural civilisation', with locally and regionally diverse and vibrant economies and cultures responding to, and being transformed by, modernising influences, by the growing range and expanding volume of flows of people, of ideas, of capital, and of commodities.[129] The growing electoral participation of the peasantry was both a sign and a means of their gradual integration into the French nation-state.[130]

The gamut of social, economic, political and cultural changes which impacted upon the French countryside during the nineteenth century – at a rate and in a combination which varied from period to period as well as from place to place – meant that peasant *mentalités* were likely to change. Old ways of doing and thinking were increasingly challenged, old solidarities and allegiances based largely upon family and community came increasingly under attack. New solidarities – based, for example, upon occupation, class or gender – became possibilities, a new sense of 'collectivity' could emerge, even while old solidarities persisted.[131] Peasant *mentalités* might have changed slowly, but there is accumulating evidence to indicate that they did both resist *and* adjust to the forces of change during the nineteenth century. There did develop, for example, explicit class struggles within the countryside, especially in those regions in which a rural proletariat – such as waged farm labourers, vineyard workers, and wood-cutters – were numerous and could be organised into unions.[132]

In his recent (1996) *tour d'horizon* of France between 1814 and 1914, Robert Tombs has provided a thick interpretation of rural society which is much better balanced than the thin description which has for so long been based upon Eugen Weber's thesis. In general, Tombs recognised that '"the peasant" as a type, whether stigmatised or idealised, was a creation of non-peasants'. In particular, while acknowledging that the nineteenth century saw the development of a sharper individual consciousness, of greater self-awareness, of individuation, and that social identity was derived principally from families, Tombs also emphasised the positive role of the small rural communities within which individuals and families 'lived, moved and had their being'. Both the significance of community and the sense of community were, he argued, stronger in regions of large nucleated villages than in those of isolated dispersed farmsteads. 'Communal social activities – festivals, religious confraternities, and during the nineteenth century drinking clubs, political groups and sports clubs – were much more developed in the urban villages. Sociability was sparser in the dispersed habitat.' In addition, Tombs recognised that from the mid-1880s rural France saw the development and spread of agricultural syndicates, partly in response to the growing economic pressures of the period and partly in relation to trade union legislation in 1884 permitting the creation of associations in defence of professional interests. Tombs argued that these syndicates, for buying farming supplies and for processing and selling farming products, were vitally important for the structure of peasant agriculture in the long run: they 'determined the survival and future of peasant agriculture, as well as constituting a new set of institutions within rural society and giving expression to peasant identity'.[133]

Peasant collectivism and risk

It is not the purpose of this present study to create yet another myth about the French peasantry by making extravagant claims for the role of fraternalism in the French countryside during the nineteenth century. It is, however, based on the premise that the role of fraternalism has been much neglected and merits closer attention. A reconstruction of the historical geography of fraternal associations – of their historical development, geographical distribution and contribution to peasant culture – will be used to examine the fundamental tension between individualism and collectivism in rural France during the nineteenth century. Both co-operation and conflict, both 'socialism' and 'capitalism', both fraternalism and individualism can be expected to have been threads within the rich tapestry of nineteenth-century rural France. The central concern of this study is to trace the hitherto neglected thread of fraternalism running through that fabric, not to remove from it the thread of individualism.

At the same time, this study will consider the extent to which peasant associations may be viewed as one form of dealing with risk. French peasants – like those elsewhere – were accustomed to confronting both man-made and natural hazards, both edicts of governments and acts of God.[134] John Berger has described the peasantry as 'a class of survivors'.[135] Against the many threats they faced, both cultural and natural, individual peasants had traditionally sought group protection within the immediate family and the proximate rural community, but also in the more remote Church. The insecurity of life for many in rural France had for long been countered by a peasant fatalism itself fostered by the Church. The peasant's sense of helplessness in relation to natural disasters, of his subordination to natural forces, was closely bound up with a rural popular religion whose practices were interpreted, in part at least, as providing some assurance and protection against the many adversities which could befall him, his family and his property – and when such protection failed, as inevitably it often did, religion encouraged as an alternative a fatalistic acceptance of those adversities, as acts of God. Gradually during the nineteenth century, and especially from mid-century, the spread of schooling and the growth of a scientific culture undermined the traditional role of popular religion and of its priests, while the insecurity of the peasant lifestyle was itself diminished by economic developments founded in part upon scientific advances. This gradual transformation of rural *men-talités* in nineteenth-century France has come to be widely accepted as an orthodoxy.[136] But scant attention has been paid to the extent to which, as faith in the protective powers of religion declined, rural dwellers embraced secular forms of defence by establishing or joining associations intended to provide a degree of protective insurance for their property and for their own persons and families. Another concern of this study, therefore, is the extent to which religious assurance came to be replaced by secular insurance as a means of coping with risk and uncertainty in rural France during the nineteenth century.

The nation of France was a cultural construction – but behind the idea of a unified France lay the reality of its diverse regions and its mosaic of localities (*pays*), as G. Bruno's classic school textbook, *Tour de France par deux enfants*, made very clear when it was first published in 1877.[137] Not a few attempts to launch grand generalisations about France have been wrecked on the rocks of regional and local differences and specificities. Many scholars for many years have been pleading the case for a better understanding of the regional nuances of French national history. While few would agree with C. K. Warner's extreme, clearly unachievable but perhaps merely incautious, claim that we need to study *every* village and region in France from pre-industrial times to the present because 'only this way can we hope to get an accurate catalogue of [peasant] *mentalités*',[138] there are many who have argued that the problems of economic, social, political and cultural change in France need to be studied from the standpoint of the region rather than – or at least as well as – that of the nation.[139] There is a very strong tradition within French scholarship of the production of historical regional monographs, such as Georges Dupeux's classic study of the social and political history of Loir-et-Cher between 1848 and 1914, and of geographical regional monographs, such as Roger Dion's classic account of the Val de Loire.[140] Furthermore, with the recent burying of the 'Pinkney thesis' by its own creator, foreign historians (and so too historical geographers) of France are advised no longer to concentrate on broad syntheses based upon detachment but instead to engage in analysis based upon detailed archival research.[141]

There are, then, both sound reasons and excellent precedents for essaying a regional historical geography of fraternal associations in one French department during the nineteenth century. That of Loir-et-Cher has been selected in part because it lay astride the often recognised frontier between the 'two Frances' and thus can be said to have had contacts in both of them.[142] Without claiming that Loir-et-Cher was wholly representative of either or of both of the 'two Frances', it will be possible to argue that Loir-et-Cher's experience of fraternal associations was of course singular while at the same time having some commonalities with other parts of France. The findings of this analytical regional study will be considered within their national context and some comparative remarks will be offered in a concluding synthesis. Loir-et-Cher was also chosen for study because it was the location of France's first agricultural syndicate, founded before legislation was passed which was to provide the framework for the development of such associations. But before turning to the regional case of Loir-et-Cher, it is necessary to consider the national theoretical discourse within which its fraternal associations were situated.

2

The theory and practice of fraternal association in nineteenth-century France

As thinking in France as a whole moved during the nineteenth century from religious traditionalism to secular rationalism, so its rural world came increasingly to be impacted by the Revolutionary concepts of liberty, equality and fraternity. There was, unsurprisingly, a gap which was only slowly narrowed between the philosophical debates about the meanings of those concepts by the intellectual elite and their practical application in the everyday lives of the peasantry. But in order better to understand the latter in its detail, it is necessary first to comprehend the former in general outline. The term 'association' was to acquire a precise, legal meaning in France by 1900, for an association would by then be an historically constructed, legal reality, but the idea of association had been a major concern of political theorists for at least the preceding century and the question of association would remain as an essential part of political debate for the succeeding century, through to the present day. The significance of fraternal associations must, therefore, be contextualised within both historical and modern discourses about the idea of association, about the principle and practice of fraternity.

The discourse of fraternity

The Industrial and the French Revolutions combined to engineer, directly or indirectly, the end of traditional, holistic societies throughout much of Europe. In so doing, they demanded a general reappraisal of the relations between the individual and the society of which s/he was a part, and a specific reassessment of the status of secondary groupings (such as associations) which were intermediate between those two primary social structures. The theoretical status of such groupings and their practical role in the 'progression' towards a capitalist economy and a democratic society became subjects of a major discourse in France.

The Revolutionary period

The Revolution of 1789 involved, *inter alia*, the construction of a problematic about the relations between the private and the public spheres of life, about the relations between the individual and the State, which structured the ideological context within which organised groups of individuals – fraternal associations – were developed during the succeeding decades and even during the following two centuries. The republican ideal – of squaring the triangle of liberty, equality and fraternity – created and constituted a discourse outside of which it would be difficult and inappropriate to try to understand the historical geography of voluntary associations in nineteenth-century France.

The Revolution of 1789 was fundamentally a celebration of the rights of the individual, of the freedom of the individual. The tradition of political theory upon which the Revolution drew emphasised the primacy of the individual over the group within society. For example, the seventeenth-century English philosopher, Thomas Hobbes, even argued that groups constituted a threat to the State but it became a generally held liberal view that rights of individuals were more important than those of groups, even though such groups could clearly exist within a free, democratic society. Of course, as G. Sabine pointed out in comparing French and Anglo-Saxon political theory, 'liberty is not simply an attribute of an individual; it is a relationship between a person and the complex of societies to which he belongs. The extent to which freedom of association can be generally and effectively achieved, and the extent to which association can preserve individual spontaneity, are the measures of liberty in any society.'[1] In eighteenth-century France, Jean-Jacques Rousseau was the principal political theorist of the liberal democratic school but he viewed groups and associations with considerable hostility, arguing that such 'partial societies' would impede the full expression of the general will. Rousseau's theory, as set out in *Du contrat social* (1762), provided a powerful argument in support of imposing constraints on the freedom of individuals to associate and, although it was certainly not the only reason for legal restrictions on the rights of associations, it very probably provided an intellectual rationale and a social respectability for such a policy of control and at times of repression. A discourse which prioritised the freedom of the individual, and which provided governments fearful of non-conforming and potentially revolutionary forces with a ready-made justification for the explicit repression of associations, was a serious check on the effective development of such associations for at least half a century after the Revolution. Only gradually was the revolutionary discourse of liberty challenged during the nineteenth century by the discourse of fraternity.

The nineteenth century

Although relegated to the background during the revolutionary period and for many decades thereafter, the concept of fraternity, often expressed as sociability, can be identified as a theoretical thread which ran through the entire fabric of the nineteenth century: the private virtue of politeness and respect for other individuals was generalised into a public virtue of tolerance of other groups within a politically-pluralist society. Fraternity was often portrayed as being synonymous with social progress. Thus Jean-Etienne-Marie Portalis, the Catholic thinker and Minister of Religion under Napoleon I, linked the perfectionning of Man to the practice of sociability and in 1876 Jules Ferry argued that sociability (which he said was the scientific term for fraternity) combined the concepts of tolerance and charity, that fraternity was superior to all other dogmas, religions and philosophies.[2] As Zeldin has so aptly put it: 'whereas liberty had been the watchword of the previous century, fraternity was that of the utopians [of the nineteenth century]. Fraternity was expressed in religious terms as association or co-operation, as opposed to capitalist exploitation.'[3]

Somewhat unexpectedly, it was an aristocrat by no means wholly supportive of either democracy or associations, Alexis de Tocqueville, who produced one of the earliest positive assessments of the potential role of fraternal or voluntary associations within a democratic political system. After visiting America for two years in the 1830s, he observed that the Americans were 'a nation of joiners' and that 'Americans of all ages, all conditions and all dispositions constantly form associations.' De Tocqueville commented upon the connection between the proliferation of associations in the United States and the egalitarian, democratic character of American society. He concluded that, because individuals within a democracy are politically weak, the risk which they run of becoming totally powerless can only be effectively countered if they voluntarily associate, with individuals grouping in order to bolster one another.[4]

It was not, of course, solely due to de Tocqueville that, whereas liberty and freedom of the individual had been the key concerns of the eighteenth century in France, fraternity and freedom of association were to become those of the nineteenth century: mountains of intellectuals – including utopians, political theorists, politicians and religious thinkers – gave rise to ever-flowing streams of ideas for the reorganisation of society, doing so in a post-revolutionary world which acknowledged that change need not only be an aspiration, that it could be a realisation. Throughout the nineteenth century, as Robert Tombs has emphasised, 'religious, political and social prophets jostled each other to preach their creeds': there were fundamental disagreements concerning the very nature of society, over the interpretation of history, and over the relation of individuals with society. Tombs has recently provided a perceptive survey of this multifaceted search for a new order, identifying six differing social visions: Catholic, liberal, republican, socialist, Bonapartist and national.[5] Fundamental to the debate was how to reconcile the rights of individuals with those of the State. The liberal Benjamin Constant differentiated between 'the liberty of the ancients' (the right to participate fully in the political life of the community) and 'the liberty of the moderns' (the right to be left alone to pursue one's private affairs). The Revolution of 1789 had asserted 'the liberty of the ancients', emphasising the right and duty of every individual to be a patriotic citizen: the concept of fraternity constrained that of liberty and led individualism to be regarded with suspicion. Individualism thus came to be denounced by traditionalists, socialists and nationalists.[6]

The essentially traditionalist and conservative Catholic Church envisioned a social order based upon the paternal authority of God, the king, and the head of each family. Within the ordained social hierarchy there was, of course, a special place for the clergy and local *notables* (most often, landowners). This paternalistic social order was both sustained by and gave rise to a range of charitable societies. But in addition during the nineteenth century, the Catholic Church came to preach corporatism in opposition to secular individualism, adapting the concept of fraternity into its own authoritarian thinking and developing the concept of Social Catholicism.[7]

But in the aftermath of the tumultuous revolutionary period had come first of all the socialist utopians, seeking order and peace through a new Christianity or a new religion of humanity which 'would provide the spiritual cement to hold the society of the future together'.[8] Among these social philosophers was Henri Saint-Simon (1760–1825), usually seen as one of the founding fathers of socialism. While espousing social progress through economic expansion and technological application, through universal primary schooling and massive programmes of public works on, for example, road building, he also preached fraternity. He favoured a meritocracy – summarised in his slogan: 'To each according to his capacities, and to each capacity according to his works' – and preached fraternity – evidenced in his other slogan of the New Christianity: 'Love one another and help one another.' Although Saint-Simonianism never became a mass movement, remaining instead the creed of an intellectual elite, never developing into a prioritised programme, but retaining instead an *à la carte* menu of proposals, and although its practical achievements were few, its ideas and principles were to resonate throughout the nineteenth century as part of the discourse on the relations between the individual and the State.[9]

So, too, did the thoughts of Charles Fourier (1772–1837) who argued that social life should be based on the principle of co-operation and not on that of competition. He propounded the idea of small, self-supporting, voluntary communities, called *phalanstères*, whose 1,600 or so members would live in social amity through mutual co-operation. Fourierism offered one solution to the problem of reconciling class conflicts within society; it strongly preferred the principles of association and solidarity to those of *laissez-faire*, and it advocated a form of municipal socialism. Fourierist socialism spawned a vast literature and its ideas permeated social thinking throughout the nineteenth century; even if as a party it was not itself long-lived, many of its ideas were adopted by the socialists and radicals at the end of the nineteenth century. Eugène Sue's *Les misères des enfants trouvés* (1851) includes a description of a Fourierist commune established in the Sologne, a *pays* partly included within the department of Loir-et-Cher, which will be the focus of this present study.[10]

Complete socialisation and the founding of model communities was also advocated by Etienne Cabet (1788–1856): his was probably the most widely known socialist doctrine at the time of the Revolution of 1848. He advocated a return to the 'communism', as he argued, of Christ's own teaching and of early Christian practice. A non-revolutionary communist, Cabet favoured the establishing of co-operative associations.[11] Some of his ideas owed something to Babeuf, whose communism reflected the attachment of the poorest peasants to traditional, communal organisation of open-field systems of farming and rights of pasturing, gleaning and wood-collecting.[12] Another social utopian who insisted that for a stable society there had to be a religious foundation was Pierre Leroux (1809–65), probably the first Frenchman to use the term 'religion of humanity' and the first Frenchman to call himself a socialist, although he referred to himself as 'the fourth socialist', coming after Saint-Simon, Fourier and the Englishman Robert Owen, each of whom represented for Leroux respectively equality, liberty and practical fraternity. Leroux argued that the individual exists chiefly by virtue of his relationship to his society and, beyond even that, to mankind as a whole, deducing that man's true moral law lies in 'solidarity' and 'harmony' with his fellows.[13]

The impact of these and other social utopians has been admirably summarised by Zeldin: it was

not to create a series of coherent new [political] parties, but on the contrary to uproot tradition, to sow confusion, to stimulate hope, to construct dream-worlds . . . The attitudes they encouraged were perhaps more important, from the long-term historical point of view, than the precise details of their schemes, because men mixed up and recombined their ideas in a large variety of ways. What they achieved was to give a high status to ideals and theories . . . The tradition they established became one of the most powerful forces in French life. Their idealism, even if it was seldom implemented in practice, reflected widespread aspirations, and playing with idealism, verbally or more deeply, was a constant feature of [the nineteenth] century.[14]

The utopians were, then, important in promoting a discourse on fraternity during the first half of the nineteenth century and, although they were discredited as a nascent political movement by the failure of the Second Republic, their ideology also underpinned socialist and solidarist polemics during the second half of the century. But the socialist party created by Jules

Guesde (1845–1922) and consolidated by Jean Jaurès (1859–1914) was essentially an industrial workers' party. It endeavoured to win the support of the large number of peasant proprietors by promising both less taxation and exemption from nationalisation for all except the largest properties. In rural areas it was thus tempted to become a peasants' party rather than a socialist party, acknowledging the political power of the large number of small peasant proprietors and (with some significant regional exceptions) the small number of rural labourers. There was no serious attempt to win the peasants to the cause of socialism, merely an expedient endeavour to secure their electoral support.[15] The socialist party did not try earnestly to convert the large army of private property owners to the idea and practice of collective agriculture, to practical socialism or fraternity.

In relation to the discourse of fraternity, the most actively debated idea of the last decade of the nineteenth century and the first decade of the twentieth century was that of solidarism or solidarity, and the radical party – which was to be the largest single political party in France from 1902 until 1936 – came to adopt it as its central doctrine.[16] With its basic formula 'Every man his neighbour's debtor', solidarism represented a republican reinterpretation of the principles of the Revolution of 1789: it now came to be argued, as Zeldin has pointed out, that 'the individualism which the Revolution had consecrated was an evil and a delusion': the liberty of the individual which it proclaimed was an abstraction, for men were not independent beings without obligations and ties to others. The principle of *laissez-faire* came in the course of the century to be rejected by republicans, who came increasingly to argue that social progress had actively to be constructed. The use of Darwinian ideas about 'the survival of the fittest' which had been cited in support of *laissez-faire* because it led to an evolutionary progress came to be questioned, with some scientists now arguing that living organisms were made up of cells working together, so that the 'law of nature' was of co-operation not competition, which made it possible to assert that the 'law of society' was of solidarity not individualism. In the social sciences, Durkheim argued that the weakening of the traditional bonds of religion and family had combined with the growing division of labour and economic specialisation to undermine society and to produce moral chaos. Whereas the Revolution had asserted that change should be effected either by state intervention or by individual action, Durkheim now argued that 'a nation can be maintained only if, between the State and the individual, there is intercalated a whole series of secondary groups near enough to the individuals to attract them strongly in their sphere of action and drag them, in this way, into the general torrent of social life'.[17] Durkheim's advocacy of professional associations – precisely the groups which the Revolution had endeavoured to destroy – as mediators between individuals and the State was elaborated by political theorists, as it was by Eugène Fournière in his *L'individu, l'association et l'Etat* (1907). Fournière argued that democracy was the logical extension of eighteenth-century rationalism; that socialism was in the economic order of things a natural development from political democracy; and that the perfectionning of daily social life, as seen in the number and diversity of associations, was intimately linked to both of those phenomena and, like them, was both beneficial and ineluctable. For Fournière, association in daily life – like socialism in economic life and democracy in political life – was the ideal to be realised.[18]

Solidarism drew support from debates within the natural and social sciences, but it was in effect a reworking of the revolutionary concept of fraternity. It was given a firm theoretical base by Léon Bourgeois (1851–1925), with his doctrine of the 'social debt', which argued that individuals were not free but had both a moral duty and a positive obligation to repay their debt to society. This theory provided justification for a practical programme of social welfare, because the State was seen as being justified in compelling people to pay their social debts in income tax. In practical terms, solidarism required individuals to co-operate not in the production or consumption process, but more prosaically in insuring themselves against personal risks – those of illness, accident and unemployment – as a social duty. Thus solidarists were firm advocates of the development of voluntary mutual aid societies which, it was hoped, would provide for their members a wide range of social benefits without much cost to the State. In advocating mutualism, the solidarists were able both to present their doctrine in a practical and popular way to contemporaries and to draw upon deep wells of a tradition of mutual aid in French society, especially in its rural communities.

There was, then, a continuing but not a constant discourse of fraternity in nineteenth-century France. The idea of fraternity was complex, its practical possibilities were multiple. It came to be incorporated into the traditional, paternalistic, social order envisaged by the Catholic Church as well as into the new visions of social order presented by republicans, socialists and nationalists. The varied expressions which fraternity actually took in rural France now need to be considered: having reflected upon the rhetoric of fraternity, the time has now come to attend to the reality.

The practice of fraternal association

The legal framework of associations

Between theoretical discourse and empirical practice lay the intermediate legal structure within which authority gave 'official' recognition to 'popular' voluntary associations.[19] Before considering the historical geography of those associations, therefore, it is appropriate to consider the ways in which the discourse of fraternity was translated into a legal framework. The general picture is, of course, one of a gradually – but not continuously – increasing freedom of association during the 'long' nineteenth century from 1789 to 1914. Agulhon has suggested that, broadly speaking, associations were surveilled and controlled until the late 1870s, then tolerated for the rest of the century, and finally liberated in 1901.[20]

Given that the legal development of associations is only to be expected in a regime which recognises the legitimacy of bodies which are intermediate between the individual and the State, it is unsurprising that under the monarchical ancien régime the concept of voluntary associations had little standing: any such association certainly needed to have royal assent. The Revolution of 1789, adopting Rousseau's opposition to associations on the grounds that they fractured the unity of the State and constrained the liberty of the individual, was inimical to their development. The Declaration of Rights in 1789 did not proclaim the freedom of association, while laws passed in 1791 suppressed existing corporations and prohibited all kinds of professional association: there was considered to be no room within the French Constitution for such special interest groups. Although political societies were initially tolerated, gradually the central bodies became less favourably inclined towards them: articles 360–4 of the Constitution of 1795 prohibited the establishment of corporations and associations which were 'contrary to public order' and forbade any association from presenting a public petition or exercising political rights. Under the Consulate and the First Empire the legal grip on associations was tightened further. The Penal Code of 1810 provided a clear structure for the control of associations: no association with more than twenty members which aimed to meet daily or on regular, specified days for religious, political, literary or indeed any other purposes could be established without government approval and without agreeing to conditions laid down by the public authority.

The legal framework thus established, of tight control of local associations by the central authorities, persisted during the second and third decades of the nineteenth century. Following the Revolution of 1830 new tensions developed between the authorities and those political associations disillusioned by the monarchical character of the Revolution. But prosecutions brought under article 291 of the Penal Code of 1810 often resulted in acquittals of the accused by local juries, so that the legal vice was tightened further by a new law of 10 April 1834 which provided that thereafter the article would apply to associations of more than twenty members even if they were broken down into sections each with less than that number and even if they did not meet daily or on regular, specified days. Moreover, the penalties for breaking the law were stiffened and they were to be applicable not only to the officers of an association but also to its members; and charges would henceforth be heard not by local juries but by department courts (*tribuneaux correctionnels*). The legal grip of authorities upon associations was very tight.

With the Revolution of 1848 and the Second Republic, the legislative climate changed briefly in favour of associations. The Constitution of 4 November 1848 declared, for the first time, the principle of freedom of association: article 8 stated that citizens had a right to associate which could only be limited when its exercise infringed either the liberty of another individual or public order. This was thus a legally bounded freedom and when political clubs again began agitating a new law of 19 June 1849 authorised the government to prohibit clubs and other groupings and meetings which could be seen as a potential threat to public order. With the coming of the Second Empire, an authoritarian legal framework was firmly reimposed. The decree of 8 December 1851 provided for transportation to a penal colony of any person found guilty of belonging to an illegal (secret) society, and that of 25 March 1852 required all associations to seek prior approval for their formation. Only gradually was such a severe attitude towards associations relaxed and the government remained especially cautious about political societies, but in 1868 professional associations were again legally authorised.

From the early 1870s, with the beginnings of the Third Republic, the legislative climate for associations ameliorated both significantly and permanently, although it was to be another thirty years before the full freedom of association was legally acknowledged. A suspicious attitude towards political societies and most especially towards religious groupings persisted, so that only step-by-step were associations granted the legal right to exist. The law of 12 July 1875, relating to the freedom of teaching in higher education, declared that article 291 of the Penal Code of 1810 was not applicable to associations founded in connection with courses or institutions of higher education. More generally and so much more significantly, the law of 21–2 March 1884 authorised the formation and functioning of unions and other professional associations, even if they had more than twenty members. They were, however, still required to undergo screening, to seek prior approval from the authorities, and even in the 1890s there were still prosecutions of associations in breach of that aspect of the Penal Code of 1810. A liberal republican desire to introduce freedom of association in full measure was countered by an equally strong secular republican wish to control religious associations, and so it was not until 1 July 1901 that a law was passed which annulled article 291 (and associated articles) of the Penal Code and replaced the principle of prior authorisation by that of freedom of association. It stated: 'Les associations de personnes pourront se former librement et sans autorisation, ni déclaration préalable' ('Associations of individuals may be established freely without the need for authorisation or preceding notification'). From 1901 onwards, associations could be freely formed, their foundation needing to be reported to the authorities but no longer needing to be approved by them. The march towards freedom of association had taken a long road.

The empirical history of associations

The detailed history of associations in rural France has yet to be written. Surveys undertaken at the end of the 1980s of the literature on voluntary associations in France emphasised the relative paucity of historical studies and the focus of those in existence on the ancien régime and the early nineteenth century.[21] Since then there have been some further empirical studies (as later chapters of this book will make clear) but the basic position remains unchanged. In addition, a few tentative overviews have been essayed.[22]

Although the idea of association was alien to the underlying philosophy of the ancien régime, many groupings (both religious and corporate) did in practice exist on the eve of the Revolution: to the *confréries* of charity and the *corps* of some professions had come to be added some *sociétés, clubs* and *cercles*, many of which were directly modelled upon English institutions. They were essentially an urban and bourgeois phenomenon. Even the agricultural societies had a strongly bourgeois membership, but of people with landed interests even if they lived for much of the year in towns. Such societies continued to be encouraged during the revolutionary period, because as expressions of 'enlightened rationalism' they had an accepted role in the diffusion both of new knowledge and of best practice in the field of agriculture, even though they often combined an interest in agriculture with the other traditional concerns of such learned societies, such as archaeology, history, and the natural sciences. From the early nineteenth century there developed a greater degree of specialisation among these associations, with a clear difference emerging for example between literary *cercles* and agricultural societies.[23] The Revolution banned professional associations but in practice it tolerated, even for a while encouraged, political groups or *clubs*. The Jacobin popular revolutionary *clubs* or societies attracted a large membership (although their appeal and participation rates varied between town and country, and from region to region) and for a while they were important seats of political power and even of local administrative authority. But gradually the central bodies became less favourably inclined towards them. The Constitution of 1795 limited their activities and the Penal Code of 1810 imposed further constraints upon associations, with the need for prior approval for their formation and for their meetings being used by the authorities to tolerate only those considered to be supportive of the regime.

The law was applied inconsistently during the second and third decades of the nineteenth century, with associations experiencing varying degrees of toleration, but in general the combined forces of the civil and religious authorities have been seen as constituting a monolithic discouragement to minority groupings within society.[24] After 1815 and during the 1820s, secret societies were established by some groups (such as students and nostalgic army offers and under-employed non-commissioned officers) to whom the regime denied any means of expression other than such clandestine activity.[25] For a few years immediately after the Revolution of 1830 there was a greater degree of freedom in the practice of association, and even in the legalistically more repressive climate of the mid- and late 1830s some associations were in practice tolerated, including those electoral associations which were ultimately to become involved in the downfall of the monarchy and those journeymen's associations which existed as surviving forms of organisations in some trades least affected by industrialisation. Mutual aid associations (especially prevalent in Paris, Lyon, Bordeaux and Lille and primarily developed among artisans and miners) were often scarcely camouflaged forms of journeymen's associations. Self-governing associations, they collected dues that enabled them to help their individual members in case of illness or disability, but some of them also presented the more modern features of defensive societies of resistance and amounted to early forms of labour unions, which were of course illegal at the time.[26] For the most part, however, the late 1830s and 1840s saw political associations energetically persecuted and reduced to clandestine activity as secret societies, while politics to some extent also infiltrated associations not originally geared to them, such as the bourgeois *cercles* and Freemasons' lodges in provincial towns.[27]

For a brief period after the Revolution of 1848, the legal constraints upon association were presumed to be abolished and it was assumed that anyone could open a club or establish a society without formalities. Such euphoria was short-lived: new constraints upon the activities of societies introduced from mid-1849 meant that once again direct political association had to find expression in secret societies. Nonetheless, although formal associations were monitored, the concept of association had a much freer rein: fraternity was still promoted as a component of progress and found practical expression in the encouragement of mutual aid, both informally and institutionally. A collective mentality was being constructed, even if the formation of associations was strictly controlled.[28]

The best overviews of the history of associations have been provided by Maurice Agulhon. From the middle of the nineteenth century, even before the beginning of the Third Republic, there began to appear in various parts of France musical societies, gymnastic societies, rifle clubs, and volunteer fire brigades, although the exact geography of these societies remains unclear. With the Third Republic came societies related to schooling, such as 'friends' of a given school or type of schooling, and to military service, such as associations of 'old soldiers'. It was also towards the end of the nineteenth century that there emerged agricultural syndicates (although Agulhon – following Barral – believed that before 1914 they had hardly distanced themselves from syndicalism of the *notables*) and sports' societies (such as cycling clubs).[29]

Until fairly recently, much of the general history of associations in rural France has been written in terms of agricultural associations. At the national level, the Société des Agriculteurs de France was established in 1867: it was in effect a professional association of large landowners founded in order to co-ordinate the work of provincial agricultural societies. While it might have been agriculturally progressive, the Society was socially conservative and it was to become a powerful force for agricultural protectionism and against social change within rural France.[30] At the local and regional level, a very different range of agricultural associations developed some fifteen or so years later. The history of these agricultural co-operatives and syndicates has often been written in essentially economic terms, with their being seen as offering dis-counted supplies (because of the economies of scale they were able to achieve by placing bulk orders). Not only agricultural syndicates but also mutual insurance societies have been interpreted exclusively in economic terms by Désert and Specklin in their review of the reaction of French farmers to the agricultural crisis at the end of the nineteenth century,[31] and Zeldin concluded that it took a long time for agricultural co-operatives to develop, not doing so in significant numbers until the 1930s in response to the economic crisis of that period.[32] Similarly, Zeldin has argued that when wine producers, the *vignerons*, formed co-operatives they did so out of economic necessity, because 'they had to' in order to survive, but they did so 'in the spirit of solidarity and mutual-ity, not of co-operative labour', and each *vigneron* maintained his individual-ity on his small, owner-occupied, essentially independent holding.[33] In Zeldin's view the return to collective agriculture among the French peasantry as a whole during the nineteenth century and early twentieth century was very slight indeed.[34] But agricultural associations very clearly also had a significant social dimension: Michel Augé-Laribé claimed that the practice of association brought to the social fabric of farmers in Western Europe a change almost as fundamental as that wrought by the railway network in the economic sphere.[35] It has increasingly emerged that such associations, although explicitly eco-nomic in their intentions, also had implicit agendas which included political and other social objectives.[36]

Agricultural syndicates offered little benefit to rural dwellers without at least a parcel or two of land of their own to cultivate or on which to raise live-stock. For farm labourers, along with other sections of the non-landowning rural proletariat (like wood-cutters), the defence of their economic interests lay potentially in the formation of associations which had the characteristics of trade unions.[37]

During the 1890s and the early 1900s, solidarists placed their main hope on the development of voluntary mutual aid societies, a movement underpinned by Méline's law of 1898 on friendly societies (*sociétés de secours mutuels*).[38] It was hoped that such societies would provide a wide range of social services (such as employment exchanges, loans, medical attention, pensions and insur-ance) without much cost to the State. Such mutual societies emerged slowly during the early nineteenth century and their development received encouragement during the July Monarchy, which in 1837 allowed their formation provided that prior permission had been obtained from the authorities. Stimulated further by the Revolution of 1848, there were about 2,500 societies in existence by 1852 with about 250,000 members. By 1870 the membership of such societies had doubled but it has been argued that it was the solidarists who gave this movement an enormous boost. The law of 1898 gave these societies the same freedom as the law of 1884 had given to trade unions and also added financial privileges and promises of State subsidies. By 1902 total membership was more than one million, and by 1910 it was more than three million in almost 16,000 societies. Mutual aid societies proved to be a popular, practical expression of solidarism.[39]

The historical geography of voluntary associations in rural France remains very imperfectly known. Agulhon has noted that political historians have focused selectively upon secret societies, religious historians upon *confréries* or *congrégations*, and labour historians upon *compagnonnages* and mutual aid societies. In his view, other kinds of association have been relatively ignored. What is needed are more comprehensive studies of the entire range of voluntary associations which existed in particular places during the nineteenth century. In the meantime Agulhon has offered some general reflections upon the history of associations which led him to suggest that it was marked by their increasing number, their growing diversity and their developing liberalisation. But he readily acknowledged that many questions about associations would remain unanswered – including those relating to the geography of associations, to the differences between regions and between town and country – until much more empirical research had been undertaken.[40] Agulhon suggested, somewhat surprisingly, that such work should be undertaken without any preconceived ideas about the institutionalisation of social life. While his own work has indeed proceeded for the most part empirically, without reference to the general (theoretical) literature on voluntary associations, it must also be recognised that Agulhon situates such associations within the broader context of 'sociability', a concept which he has developed from and applied to his studies of everyday life in eighteenth- and nineteenth-century France. That concept now needs to be examined, ahead of a consideration of some of the broader issues relating to the understanding of voluntary associations.

The theorisation of voluntary associations

The concept of sociability according to Agulhon

Current approaches to the history of sociability were initiated by Maurice Agulhon almost thirty years ago in a number of fundamentally empirical studies which have in turn stimulated wider exploration of the field by other historians, not only in France but also in other European countries and in

Canada. Agulhon's early works have come to be regarded as classics: in 1966 he published a study of sociability in Provence at the end of the eighteenth century, focusing upon the role of religious groupings (*confréries, pénitants*) and of Freemasons, which was followed in 1977 by his analysis of the *cercles* as an index of the changing sociability of bourgeois France during the first four decades of the nineteenth century. Then in 1979, in his study of the department of Var during the first half of the nineteenth century, Agulhon traced the development of popular sociability, in particular examining the extent to which it was imitative of bourgeois sociability.[41]

On the basis of these three detailed, specific studies and of other similar studies which had started to appear in their wake, from the late 1970s Agulhon began to develop some general perspectives upon the history of sociability in France since the eighteenth century. He has offered some signposts 'towards a history of associations'; he has provided a *tour d'horizon* of associations in France since the late eighteenth century; he has probed the ways in which sociability may be viewed as an appropriate subject for historical study; and he has reflected upon the means by which further advances in the history of sociability might be made.[42]

Agulhon has been reluctant to provide a precise definition of 'sociability', perhaps because for him it is a broad concept which can embrace the whole range of interactive social activity which exists among individuals at scales intermediate between those of the family on the one hand and the State on the other. Similarly, Agulhon has been reluctant to provide a precise typology of 'sociability', perhaps because for him the concept embraces a very wide range of both informal and formal socialising, of spontaneous gatherings as well as institutionalised groupings. He has, however, suggested that there can be identified some 'traditional' and some 'modern' forms of sociability, while recognising that there were not only changes but also continuities in sociability during the nineteenth century. Traditional forms of sociability tended to be associated with work (such as bands of seasonal harvesters; of occasional get-togethers to undertake major tasks, such as constructing buildings; gatherings involving some collective labour, for example, at the oven, the mill, the forge and the washing place); with leisure (such as the *veillées*, gatherings of family and neighbours on winter evenings; the *débits de boissons;* the *jeunesses*, informal groupings of the young men of a village for a variety of fun and games; and the *chambrées* and *chambrettes*, social gatherings for drinking, playing cards, singing and generally passing the time); and with religion (such as the *confréries* of devotion and of charity). Whereas most 'traditional' sociability was informal, the nineteenth century witnessed the growth of formal sociability, of its institutionalisation into societies and associations (such as mutual aid societies, musical societies and agricultural syndicates). But Agulhon has not identified a clear historical pattern: he has not offered an historical typology of sociability because he has been at least as much aware, probably more aware, of the specificities as of the generalities, of the diversities rather than the simplicities of the forms of sociability. The fête, for example, had a multiplicity of forms and functions, and its character was far from constant throughout the eighteenth and nineteenth centuries. In short, Agulhon is acutely sensitive to regional contrasts in sociability, to its changing character through time, and to its variation in form from one social level to another.

Agulhon has been moving inductively 'towards a history of sociability' rather than deductively 'from a conception of sociability'. He has, however, been adopting more recently a broader, more conceptual if not exactly theoretical, approach to the history of sociability. He has, for example, sought to throw more light on the topic by examining the history of the idea of sociability, of the term 'sociability' itself, and thereby considered some of its psychological, philosophical and sociological meanings. The net effect of those explorations, however, has been to confirm him in his belief that sociability as a phenomenon is not only a legitimate object of study but that it varies in function, in space and in time. His most recent thinking on the problem presents a compelling case for both a history and a geography – in effect for an historical geography – of sociability. But he has also moved marginally towards a theorisation of sociability.[43]

New perspectives upon sociability

Studies of sociability by Agulhon have come to be viewed as a model and his approach and ideas have been extended not only to other parts of France but also, for example to Germany, to Switzerland, and to Québec.[44] In the process, a critique of his work has also begun to emerge which does not in any way belittle Agulhon's considerable achievement but instead builds upon it.

Studies of the history of sociability need, in the view of Jean-Claude Chamboredon, to confront more directly three sets of problems. First, there is a need to define the phenomenon more precisely, so that the relevant practices can be more readily identified; and Chamboredon acknowledges that although to restrict study to that of institutionalised sociability is to impose a somewhat arbitrary limitation, not to do so is to run the risk of dissolving the object of study into an infinite variety of manifestations of social life and in the process to deprive the concept of meaning and significance. Secondly, analysis of the functions of associations needs to be broadened away from seeing them as meeting a basic human need for sociability – products of a mythical *homo sociologicus* – to include a view of both informal and institutionalised sociability as an integral component of the history of politicisation and of the history of social control. Thirdly, there is the problem of the periodisation of the specific history of sociability, which might not be synchronous with that of other kinds of history, such as rural history more generally: on this point, Chamboredon argued that Agulhon and Bodiguel's overview of associations rested at least implicitly on the 'mythical history' of the delayed modernisation of rural France until after 1945. He also, more tellingly, argued the need even for studies of sociability, of groups, to recognise the key roles played by some individuals as significant agents within them.[45]

Both similar and different points have been made by Etienne François and Rolf Reichardt in their wide-ranging review of the forms of sociability in France between the mid-eighteenth century and the mid-nineteenth century. Dissatisfied both with historical usages of 'sociability' and with Agulhon's somewhat vague descriptions of the concept, they employed the term to include the concrete forms, the functions, structures and processes of socialisation and interaction across the whole field of social practice which are intermediate between the family on the one hand and the State and public authority on the other. They suggested that, despite their infinite variety, the different forms of sociability had (at least from the end of the eighteenth century) four essential characteristics in common: first, membership of a group resulted not from any social compulsion nor from a quest for material advantage, but instead from an ideal, general and abstract interest; secondly, a group was not socially exclusive but open to all (with the exception of occupational and residential groupings); thirdly, a group had no fixed social hierarchy and instead affirmed the principle of equality among its members; and fourth, members of a group shared common interests and conviviality. Popular sociability, they observed, tended to be less formalised than bourgeois sociability. Their review of the field led François and Reichardt to regret the mismatch between the success which the word 'sociability' had encountered in the literature and the paucity of studies extending the work of Agulhon or even making the phenomenon of sociability their central concern. The field, in their view, remains an open one.[46]

There have been a few endeavours to theorise the history of sociability in France, or at least to provide it with a more general framework. From her empirical, comparative study of three very different national associations during the second half of the nineteenth century – the Protestant Unions Chrétiennes de Jeunes Gens, the secular Ligue de l'Enseignement, and the Catholic Association Catholique de la Jeunesse Française – G. Poujol constructed an ideal-type model of the development of associations. She envisaged an initial grouping of people coming together on a voluntary basis, fired by an idea and functioning with very few rules: there follows a formative stage, which sees the development of an association that both seeks to ape an already established institution and receives support from another institution itself wanting to exercise some control over the fledgling association; in the next stage, a newborn association develops its own identity as a result of conflicts with external authorities opposed to its existence; then the association matures by transforming its rules to enable it to have effective relations with such external bodies and to select from it own members those most capable of pursuing those relations; then with new members joining the association the relations of the association with the external world are gradually modified but new ideas are discussed within a fundamentally unchanged structure.[47] Such a model may be applicable in part to other kinds of associations, but it has attracted little attention.

A more theoretical approach has been pursued by J. Ion. Rejecting existing functionalist typologies because they ignore the discourse in which associations developed and historicist typologies because they are too narrowly specific, Ion attempted to establish a more theoretical, more ambitious typology of associations which situated them within the *longue durée* of their discourse on the relations between the individual and society. Arguing that sociability should be recognised both as a form and as a process, Ion proposed three representations of the relations between the individual and the social to create a typology which had a diachronic dimension but which did not exclude the synchronic existence of different kinds of groups. In brief, his typology was as follows: 'type A' groupings are holistic, organicist, in which the grouping is imposed upon the individual as a fact (of history or of birth, or linked to an occupational, professional activity or to a geographical location); 'type B' groupings acknowledge a greater role for the individual who establishes or joins a group with a defined objective – these are groupings of individuals voluntarily united in the recognition that they have collective interests and a shared system of values (these are the formal societies and associations constructed on the basis of such interests and values); and 'type C' groupings are based on strong, concrete links among individuals with an identifiable reality (usually territorial, cultural or demographic) but a limited commitment, so that the individuals do not see themselves as voluntarily joining a cause but, much more practically, as accessing certain opportunities (such as those provided by sports' clubs). Ion considered that for a given group or association there could be a concurrence of forms or types, and that the essence of his typology lay not in the classification but in the analyses to which his model could lead, in its capacity to interpret contrasts, similarities and transformations.[48]

That might be considered an ambitious claim for an ambitious model, but at least by drawing upon more than just the empirical work of historians of sociability Ion has provided a new optic through which to view a particular phenomenon in the past. Until fairly recently, work by historians on sociability in France has drawn hardly at all upon work in other disciplines – somewhat paradoxically, given that the subject of their study is social interaction. While the need to situate the history of sociability within the history of ideas has come to be acknowledged, there is scant equivalent recognition either of the benefit potentially to be derived from an awareness of cognate work in other disciplines and in particular related work specifically on voluntary associations, the institutional form of sociability which is the subject matter of this present study.

Voluntary associations: some general concepts

Sociability in general and voluntary associations in particular have attracted the attention of a wide variety of scholars and their approaches and ideas deserve to be interrogated for the light which they might be made to throw upon the history of sociability and voluntary associations in France during the nineteenth century. For example, the anthropologist Robert T. Anderson reviewed the place of voluntary associations in history from the Neolithic to the modern period. While he recognised their diversity, Anderson nonetheless concluded that during the Industrial Revolution voluntary associations both contributed to social stability by providing social units intermediate between the individual and the community, and supported adaptive social change, facilitating the transition of individuals and societies to participation in the modern world. He argued that during this period the mode of operation of associations was also transformed, with the emergence of what he called 'rational-legal associations . . . possessing written statutes clearly defining the membership, participant obligations, leadership roles, and conditions of con-vocation'. They normally possessed 'a legally recognised corporate identity'. Such an association is 'rational in the sense that as a body it is geared to efficiency in making decisions and taking action, particularly as leaders are, in principle at least, impartially chosen by election of the most qualified to take office. It is legal in the sense that compliance in decisions and actions is sanc-tioned by the impersonal force of law.' In all these ways, according to Anderson, rational–legal associations represented historically 'a new kind of sodality'.[49]

The social scientist Arnold M. Rose developed a theory of the function of voluntary associations in contemporary social structure, arguing that such associations were important in supporting political democracy because they distributed power among a great many citizens, they provided a sense of satisfaction with modern democratic processes, and they provided a social mechanism for continually instituting social change.[50] Rose's theory was developed both in general terms and in relation to the United States, but he devoted a chapter of his book on theory and method in the social sciences to the research which he had conducted on voluntary associations in France. That work – in effect, a survey of the history of associations in France and of the legal framework structuring them – led Rose to conclude that there was only a weak tradition of voluntary association in nineteenth-century France, in part because governments feared associations as opposed to the power of the State and in part because many liberal thinkers considered associations as restrictive of the rights of individuals. Other reasons cited by Rose included the continuing concern of the Catholic tradition to encompass individuals within a community organised by the Church and its priests; the impact of a strong central government which performed many functions left in the United

States to local governments (which are closer to the people) or to the citizens themselves; and the persistence of the extended family.[51] Rose's view of the scarcity and weakness of voluntary associations in France both historically and recently was soon challenged, for example by Orvoell R. Gallagher on the basis of his work on such associations in a rural commune with about 800 inhabitants (where there were fifteen associations) and an urban community with about 50,000 inhabitants (where there were almost 300 associations). Gallagher argued that associations in France were and are more numerous and more significant than postulated by Rose, but he agreed that associations in France tended to be more expressive of the current interests of their members than instrumental in seeking social or other change. Indeed, Gallagher argued that French associations were primarily oriented towards protecting special interests and preventing change: they were defence mechanisms.[52]

In their discussion of the theoretical issues involved in the study of voluntary associations, N. Babchuk and C. K. Warriner identified three particular sets of concerns: first, that of the sociologist with the functions of voluntary associations within a social system; secondly, that of the social psychologist with voluntary associations as an environment of individual persons; and thirdly, that of the student of administration with the organisational processes of voluntary associations.[53] That list would no longer be regarded as being comprehensive but it remains indicative. Similarly, in his review of the sociological aspects of voluntary associations D. Sills usefully emphasised the distinction between their latent functions (those which their participants do not intend or recognise but which can be observed by an outside analyst) and their manifest functions (those which participants both intend and recognise). Sills also distinguished between the functions of voluntary associations for individuals and for society in general. He saw them both as training individuals in organisational skills and helping to integrate individuals into their social milieu, and as mediating between primary groups (notably the family) and the State, as offering a legitimate focus for the affirmation and expression of specific values, initiating social change and distributing power, integrating minority groups into the larger society, and governing in the sense of making decisions on policy and of providing services to individuals. He recognised that while there was some evidence in support of most of those hypotheses, by no means all of them were proved and in many instances there were contradictory findings from different studies.[54]

While there is no grand theory of voluntary associations into which specific cases can readily be mapped, there does exist a wealth of general ideas about them, some of which might be called upon to illuminate particular cases. There are, for example, useful ideas relating to the definition, classification and social (and historical) significance of associations.[55]

While the concept of sociability is somewhat vague, that of voluntary association is precise and so is much more analytical. Voluntary associations have been simply but clearly defined as 'organisations that people belong to, part-time and without pay'.[56] Voluntary organisations are non-profit making, non-governmental, private groups which individuals join by choice; members are not born into them, nor drafted into them, nor are they obliged to join them in order to make a living. Thus a voluntary association is a structured, formally organised, relatively permanent secondary grouping. The formal organisation is identified by the presence of administrative offices which are filled through some established procedure and the existence of qualifying criteria for membership.[57]

Attempts to understand the nature and roles of voluntary associations have often been initially founded upon exact descriptions of the forms and processes of association, upon structural and functional traits which have then been generalised into structural and functional typologies. Associations have been classified, for example, on the basis of their size, their internal political structure, their independence of or dependence upon outside control, their societal functions, their location, the class and other social characteristics of their members, and the intimacy of the contacts among their members. A fundamental functional distinction has been observed between 'expressive groups' (which exist in order to express or satisfy specific interests which members have in relation to themselves – such as sports' associations and other leisure societies) and 'instrumental groups' (which focus their activities upon the wider society in order to bring about a situation within a limited field of the social order which will be of benefit to their members – such as agricultural supply co-operatives). Such a binary classification does not, however, preclude the possibility that some associations will be both 'expressive' and 'instrumental'.[58]

Many sociologists have emphasised the distinction between the manifest functions of an association (as they appear to its members) and its latent functions (as they appear to observers) and this potentially provides a useful analytical tool when examining the historical role of associations, while also recognising both that different members of the same association might recognise it as having different functions and that the same association might come to have different or modified functions over time. Much more broadly, some early sociologists have tended to see voluntary associations as concomitants of modernisation, making their appearance in a society as it transformed from a folk or traditional social order into an industrial, urban social system. According to this view – drawing upon the work of Tonnies and Durkheim which distinguished between the Gemeinschaft or folk and the Gesellschaft or modern society – voluntary associations only become necessary as the ties of kinship, neighbourhood and religion break down and the individual, freed from traditional obligations, seeks new social relationships. But the assumption that voluntary associations were necessarily urban and modern in character has been questioned, not least by some anthropologists making room for their rural and traditional origins. Others have argued that while some associations might indeed have promoted the modernisation of a society, others have hindered or even endeavoured to counter that process, or at the very least have strengthened traditional social institutions.[59]

Until quite recently, much of the work on voluntary associations has been undertaken by sociologists and anthropologists, and there has been little work on the history of associations. The literature on the history of voluntary associations has grown considerably during the last twenty years or so, but it has been fundamentally empirical in character, showing scant awareness of relevant work in other disciplines. There has, for example, been a considerable interest in voluntary associations by political theorists but that body of literature has for the most part been ignored by historians. A key political role for voluntary associations has been argued by the pluralist school within political theory, which maintains that a democratic system requires a multitude of independent, voluntary, non-governmental associations to act as buffers between the individual and the State. Such associations are viewed by this school as preventing the arbitrary exercise of government power and as contributing to the maintenance of the polity by educating or socialising the citizenry, by enabling them to learn the basics of group and political action through participation in the governing of their private organisations. The growth of voluntary associations might thus be seen as an indicator of the development of a democratic state. The assumptions of pluralism have never been completely accepted, for example by those who argue that private groups can become inimical to the public interest and that they tend to develop their own non-democratic and bureaucratic structures as they grow older. But the strongest challenge to pluralism has come from the elitist school which interprets power structures in terms of key decision makers, of a power elite, the history makers. According to this view voluntary associations might occupy the middle but not the top rungs of political power; moreover, the most powerful individuals are often seen as belonging to associations which they use effectively as vehicles for their own policies. The two, polar views have been reconciled in elitist democratic theory which acknowledges the existence of elites, the role of individual leadership, and the minimal concern of the average member of an association with power; according to this theory, democracy is delivered essentially in terms of the procedures by which officials of an association are chosen, their selection by its members, so that the association is not literally self-governing. There are, instead, competing elites who have to submit themselves to the electoral process. Within this context, voluntary associations are seen as being of especial significance because they socialise the activists who come to constitute the elite, the agents who come to control the structure: voluntary associations thus play a pivotal role in the process of the structuration of society.[60]

The practice of voluntary association, of *fraternité*, reflects the principle of the freedom of the individual, of *liberté*. Such associations are created and developed to promote diverse objectives, ranging from idealism to material-ism, but they share in common a wish on the part of individuals to have some control over the making of their own histories and to prevent the political authorities from having a monopolistic hold on that process. For that reason, associations tend to be regarded with caution, even suspicion, by political authorities which regard them as being potential threats which therefore need to be controlled. To some extent, therefore, the condition of voluntary associations within a society may be an indicator of the status of democracy within that society. It is worth noting that in France the freedom of associa-tion was recognised as a constitutional principle only as recently as 1971.

As Debbasch and Bourdon have pointed out, freedom of association is powered by a deep-seated ideology which regards it as one of the basic free-doms of democracy but there are at the same time strong checks upon the uncontrolled growth of voluntary associations. They have summarised the justifications for freedom of association in the following terms: (a) associa-tions mediate between citizens and their governments, protecting them against possible abuses of political power; (b) associations constitute a form of participation in power, both by mediating between the wishes of the adminis-tered and the services of the State and by having some power devolved to them; (c) the freedom of association supports other freedoms, enabling other-wise powerless individuals to exercise their other rights; (d) associations are a factor promoting social innovation – whereas the apparatus of the State adapts only slowly to social needs, associations can respond more rapidly; (e) associations are factors in the formation of citizens – within them, individu-als develop and motivate themselves for collective causes, they undergo a social apprenticeship, becoming effective agents within society; (f) the freedom of association is both a means of defence and of attack – associations can promote altruism, resolve problems, obtain material benefits for their members, providing collective support for an individual's cause. Associations are, then, potentially powerful catalysts of social change. That potential does, however, have acting upon it some significant brakes, for reasons identified by Debbasch and Bourdon as follows: (a) associations are seen as competitors by the authorities (especially by political and religious authorities), as challenges to their power, so that the authorities endeavour to impose controls upon associations; (b) associations are seen as bringing into play democratic free-doms which might in turn release revolutionary forces, so that the State uses those perceived dangers as reasons for controlling associations; (c) associa-tions, being non-profit making, are seen to be able to accumulate property and thus become powerful, a potential threat responded to by limiting their legal powers; (d) associations carry within themselves the risk of sclerosis, of coming to oppose change rather than to promote it – associations, like all institutions, carry the risk of ossifying and encounter the temptation to build monopolies by regrouping associations and federating them, so that the 'freedom' of association becomes a travesty; and (e) associations often see within themselves the triumph of special interests, of pressure groups, over the general good.[61] Voluntary associations in general can be very diverse institutions and any particular association can itself undergo considerable transformation. To view voluntary associations simply as part of the 'modernising' process might be to distort their role, given the plurality of their forms and functions, of their origins and developments.

Towards an historical geography of voluntary associations in France during the nineteenth century

Voluntary associations, then, constitute an organised and institutionalised form of sociability. Although *la vie associative* itself can be expected to have great variety, and to be dynamic rather than static in character, it may be viewed as a precisely identifiable form of *sociabilité* and so is amenable to close examination. The foregoing review of the general field leads to the suggestion that what is now required is a theoretically-informed, systematic study of the history of voluntary associations as part of the broader phenomenon of sociability, a detailed analysis of a segment of the whole. It also leads to the view that such studies need to be conducted synthetically, in relation to particular places or regions – it is striking that many non-geographical scholars have argued the case for more explicitly geographical studies of the history of rural *mentalités* and sociability in general and of voluntary associations in particular.[62] In their 1987 review of studies of the forms of sociability in France from the mid-eighteenth to the mid-nineteenth century, François and Reichardt concluded that although the concept of sociability had attracted considerable attention there were still regrettably few historical analyses of the phenomenon and that most of them were in effect limited studies of one particular form or another of sociability. They argued that what was needed were comprehensive, systematic studies of the forms of sociability within a well-defined area, such as a village or a *quartier*, and for those studies to be much more than histories of particular institutions and their social structures, to be broader analyses of their social context and social significance.[63] Similar but independent thinking has provided the rationale for the detailed, analytical study presented here of the historical geography of rural voluntary associations in one department of France during the nineteenth century.[64]

No such study has hitherto been attempted. Probably the nearest comparable study is that by Pierre Goujon of associations and *la vie associative* in the department of Saône-et-Loire.[65] Goujon's study, although briefer and more descriptive than the one offered here, will serve in due course as a useful point of comparison with this historical geography of voluntary associations in Loir-et-Cher.

3

Loir-et-Cher during the nineteenth century: period, place and people

In his travels through France on the eve of the Revolution of 1789, an English topographer and agronomist, Arthur Young, identified a basic contrast between on the one hand northern and eastern France, with its nucleated villages and open-field systems of farming, and on the other hand southern and western France, with its hamlets, dispersed farmsteads and enclosed fields. The frontier zone between these two settlement and agrarian systems extended across France roughly from Le Havre on the Channel coast to Geneva on the Swiss border.[1] In 1934 Roger Dion, drawing upon Young's observations, published a map of France depicting the frontier between open-field *pays* and enclosed field *pays* crossing what is now the department of Loir-et-Cher between Vendôme and Blois and then following the Val de Loire north to Orléans.[2] The contrast between the 'two Frances' was subsequently elaborated in some detail, for example in 1836 by Adolphe d'Angeville in his pioneer statistical survey of France, which noted that people to the north of a line running from Saint-Malo to Geneva were, for example, taller, more literate, better fed and better housed, but also more likely to be illegitimate and to commit suicide, than people to the south of that line.[3] Just over a century later, the geographer Albert Demangeon observed that a line joining Geneva, Besançon, Dijon, Montereau, Orléans, Blois, Chartres and Rouen more or less separated two territories: to the north-east lay that dominated by nucleated villages, while that to the south-west was dominated by hamlets and isolated farms.[4] In fact, many observers (both historical and modern) have commented upon the duality of French 'civilisations' on either side of that frontier, with the northern and eastern zone 'modernising' sooner and more profoundly than the southern and western zone.[5] Loir-et-Cher was selected for this present study partly because, as a department, it straddled that frontier between the two Frances and could be expected to provide a position from which to view them both. But Loir-et-Cher was principally chosen because it was here that France's first agricultural syndicate was founded. This particular frontier had certainly produced what was to be a distinguished pioneer in the field of voluntary associations.

The department and *pays* of Loir-et-Cher

One of the first initiatives of the Revolution was to devise a spatial strategy for the democratisation and dispersal of political power, to formulate a rational system of administration through a highly geometrical and egalitarian division of French political space: eighty-three departments were constructed as the apex of a nested, spatial hierarchy of *arrondissements*, cantons and communes designed to facilitate local administration (the principal town of each department was to be no more than a day's horse-ride from any of its communes), to accentuate the control by national government (ministers in Paris were in direct contact with the prefect of each department), and to create a new mental map and set of geopieties, with the historic provinces of the ancien régime being erased from both public and private memory and replaced by the new names and boundaries of the Revolutionary departments.[6]

The department of Loir-et-Cher was created in 1790 principally on the territories of the former *baillages* of Blois, Vendôme and Romorantin. Covering 6,424 km², which was very slightly larger than the average for the new departments in France as a whole, Loir-et-Cher was divided into three *arrondissements*: in the east, that of Romorantin covered 2,124 km² and comprised six cantons; in the centre, Blois had 2,580 km² and ten cantons; and in the west Vendôme had 1,720 km² and eight cantons (figure 3.1). The cantons were themselves divided into communes (figures 3.2, 3.3 and 3.4): in 1848 there were 296, with considerable variations in their sizes reflecting both their physical and their cultural geographies (the largest being in the physically poorer and less-settled areas, like the Sologne, and the smallest being in better endowed, more settled and better connected areas like the Beauce plateau and the valleys of the Loir, Loire and Cher rivers). Such an arrangement provided a clear and potentially effective administrative link between central and local authorities, running from Paris through the prefecture at Blois (photograph 3.1) to the sub-prefectures at Romorantin and Vendôme out to the cantonal centres and so down to the communes (and for many purposes the prefects and sub-prefects communicated directly with the mayors of communes, omitting the cantonal tier). This unifying administrative system was superimposed upon a diverse topography, upon differing localities.[7]

In general terms, Loir-et-Cher is characterised by only moderate relief, with gentle undulations and some extensive near-level surfaces (figure 3.5). But the apparently monotonous relief of the department as a whole disguises the real diversity of its constituent *pays*.[8] The department is centred on the wide valley of the Loire, with the river itself close to the fairly steeply rising valley slopes on its right bank and its alluvial plain and terraces stretching out more gently on the left bank. The Val de Loire has, of course, provided the major epic story of embanking and river basin management in France from the Middle Ages through to the present day, and it constituted a major corridor of communication in the department throughout that period.[9]

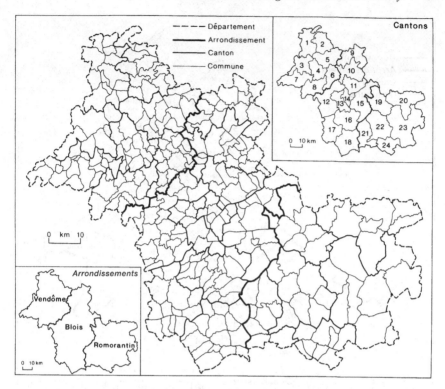

Fig. 3.1 The communes, cantons and *arrondissements* of the department of
Loir-et-Cher
Key to cantons: 1 Mondoubleau; 2 Droué; 3 Savigny-sur-Braye; 4 Vendôme;
5 Morée; 6 Selommes; 7 Montoire-sur-le-Loir; 8 Saint-Amand-Longpré; 9 Ouzouer-
le-Marché; 10 Marchenoir; 11 Mer; 12 Herbault; 13 Blois-Ouest; 14 Blois-Est;
15 Bracieux; 16 Contres; 17 Montrichard; 18 Saint-Aignan; 19 Neung-sur-Beuvron;
20 Lamotte-Beuvron; 21 Selles-sur-Cher; 22 Romorantin; 23 Salbris; 24 Mennetou-
sur-Cher

To the south and east of the Val de Loire lay the Sologne, a *pays* of gently
broken topography, of numerous (mainly man-made) ponds (*étangs*), and of
slow-flowing streams like the Sauldre, the Beuvron and the Cosson. The
Sologne comprised sands and clays which had not favoured much agricultural
development by the early nineteenth century: the beds of clay were often thick
and impermeable; where the clays were overlain by thin beds of sands the soils
were also frequently waterlogged; and where the beds of sands were thick they
did not retain moisture, becoming very dry and podsolised. In the larger part
of the Sologne, in the Grande Sologne, heath lands and woodlands alternated
with often-poor farmland on soils awaiting improvement. In the Petite
Sologne, to the west of a line running roughly from Blois on the Loire to Saint-
Aignan on the Cher, soils were both more varied (being not only sands and

Fig. 3.2 The communes and cantons of the *arrondissement* of Blois

Fig. 3.3 The communes and cantons of the *arrondissement* of Romorantin

clays but also derived in places from *limon*, from marls (*faluns*) and from chalk) and more favourable to agriculture.

Between the Val de Loire and the valley of the river Loir lay *pays* which were very different from the Sologne. To the north and east of a line running roughly from Vendôme to Blois was the south-western extremity of the Paris Basin's *pays* of Beauce, a plateau formed from almost horizontal beds of limestone, usually covered to varying thicknesses with rich *limon* soil. But within Loir-et-Cher even this *pays* was not homogeneous, because in the north Beauce proper (sometimes referred to as the Grande Beauce) was separated from the Petite Beauce[10] by the Forest of Marchenoir, situated on an outlier of clay-with-flints, while in the south variation was provided by the valley of

Fig 3.4 The communes and cantons of the *arrondissement* of Vendôme

the river Cisse, dry in its upper parts but flowing as it left Beauce to cross the Gâtine tourangelle in order to enter the Val de Loire. The Gâtine tourangelle, lying to the west of the Vendôme–Blois line, was a gently dissected terrain formed on the clay-with-flints which largely covered the chalk solid geology of this area. Similar but higher and more accentuated terrain lay to the north of the valley of the Loir in the *pays* of Perche.

In general, within Loir-et-Cher these major topographic divisions of Perche, Beauce and the Sologne were emphasised by the alignments of its two principal rivers, the Loir and the Loire. In detail, there were also other significant, more local nuances reflecting both edaphic and topographic variations over short distances. In particular, the valley of the Cher and even the

Photograph 3.1 Blois, from the south, looking across the Loire towards the Cathedral
Source: 3 Fi 7797 photo, Archives Départementales de Loir-et-Cher

smaller river valleys like those of the Sauldre and the Cisse had their own personalities, as did the larger forested areas like those of Marchenoir.[11] All of these *pays* had some meaning for those living in the department in the nineteenth century, although they were not likely to have been equally well known to all of its people. *Pays* were, after all, cultural constructions and not simply productions of their physical geographies. A *pays* was, in the vivid description of Paul Vidal de la Blache, 'a medal struck in the image of a people'.[12]

Population and settlement patterns

Loir-et-Cher – like much of France – was not densely populated during the nineteenth century, nor – again, like much of France – did it experience the dramatic population growth seen in many parts of Europe in that period. In 1806 it had a recorded population of 212,453; in 1911 of 271,231, an overall increase of 27.7 per cent. The population had increased slowly for much of the nineteenth century, to reach a peak of 280,392 in 1891, and it thereafter declined.[13]

A detailed reconstruction of the demography of Loir-et-Cher in 1851 has been provided by Dupeux.[14] The department's 261,892 inhabitants were very unevenly distributed (figure 3.6). Most striking was the low density of population (usually of less than 20 per km$^{2)}$ in the Grande Sologne and the relatively

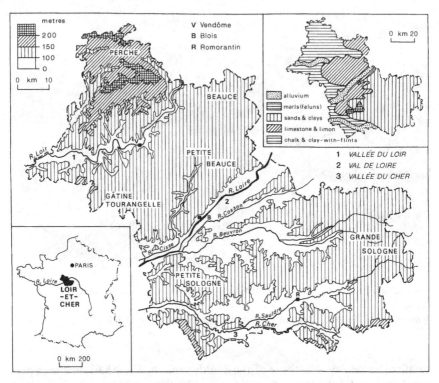

Fig. 3.5 The location, physical geography and *pays* of Loir-et-Cher

high densities (usually of more than 60 per km^2 and often of more than 100 per km^2) in the valleys of the Loir, Loire and Cher. Elsewhere in the department, in the Petite Sologne, in Gâtine tourangelle, in Beauce and in Perche, population densities were for the most part of the order of 20–40 per km^2 (although a few communes containing cantonal centres – like Mondoubleau – were more populous).

The 'urban' communes (defined officially by the census as those in which the principal settlement had a population of at least 2,000) were concentrated in the main valleys: in the Cher valley they were Selles-sur-Cher (4,500), Saint-Aignan (3,400), and Montrichard (2,800); in the Loir valley they were Montoire-sur-le-Loir (3,200) and Vendôme (9,325); and in the Loire valley they were Onzain (2,100), Mer (4,200) and Blois (17,729). Beyond these valleys there were only three other 'urban' centres: Romorantin (7,962) was capital of the Sologne and of its *arrondissement*, and it also had close contacts with the Cher valley; the other two, Savigny (2,900) and Contres (2,575), were the *chefs-lieux* of their cantons. Overall, the 'urban' population of Loir-et-Cher in 1851 accounted for only about one-fifth of the total.[15]

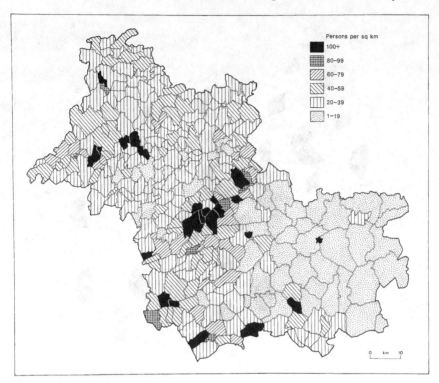

Fig. 3.6 Population densities in Loir-et-Cher, by communes, 1851
Source: AD 201M 35

The rural population itself exhibited a considerable variety of settlement pattern and of degree of concentration. Within the main valleys, villages were often elongated along the roads and the roads between the villages were themselves often dotted with houses, so that these communes had an 'urban' feel to them. Large nucleations were also characteristic of Beauce, where it was common for the population of a commune to be concentrated into the village and perhaps one or two large hamlets. By contrast, the Grande Sologne, Gâtine tourangelle and Perche were characterised by populations living in dispersed, isolated farmsteads. In the Petite Sologne settlement was a mixture of isolated farms, dispersed hamlets and some quite large villages.[16] There was an underlying stability to these patterns throughout the nineteenth century, which were largely unaffected by the demographic changes of the period (figure 3.7).

Overwhelmingly, the population of Loir-et-Cher in the middle of the nineteenth century was dependent upon agriculture. In 1851, 67 per cent of the economically active population was engaged in agriculture, whereas

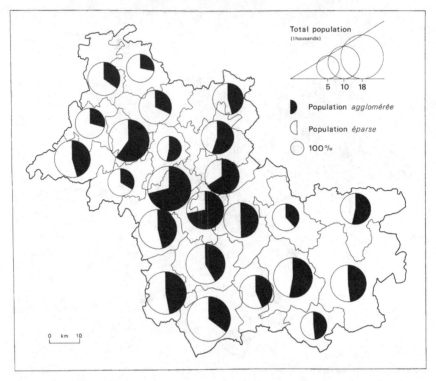

Fig. 3.7 Population concentration in Loir-et-Cher, by cantons, 1891
Source: AD 201M 143–57

manufacturing engaged only 2 per cent, mainly in textiles. Commercial and artisanal work (mainly in the building, clothing, and food trades, and in transport services) involved 18 per cent of the economically active population, the liberal professions 8 per cent, and domestic service 5 per cent.[17] In midcentury, then, the largest town in Loir-et-Cher had just under 18,000 inhabitants; three-quarters of the population of the department was 'rural'; and two-thirds of its active population was employed in agriculture (photographs 3.2 and 3.3).

During the second half of the nineteenth century the population of Loiret-Cher grew slowly to a peak in 1891 and then declined. The population had grown from natural increase, most especially as a result of the fall in infant mortality, but its decline stemmed both from emigration and from a fall in the birth-rate. Whereas between 1821 and 1865 the department had a migration deficit (i.e. an excess of emigrants over immigrants) of 200 people, between 1866 and 1890 the figure was 12,600 and between 1891 and 1910 it was 15,100.[18] This reflected a general rural exodus not only to the towns of the

Photograph 3.2 Peasant, horse and plough on the edge of the nucleated village of Mulsans, on the Petite Beauce
Source: private collection, Jacques Leroy

department but also beyond it during the second half of the nineteenth century, and also a specific adjustment from the 1890s onwards to the agricultural crisis and to the devastations of the phylloxera in the department's vineyards.[19] Widening marriage fields indicate that the population was becoming more mobile: between the early 1870s and the early 1920s there was almost a doubling in the proportion of marriages in Loir-et-Cher in which one partner came from within and one from without the department.[20]

The 1911 census recorded 219,692 French people living in Loir-et-Cher who had been born in the department and 47,683 who had been born elsewhere. But 81,248 French people who had been born in Loir-et-Cher were also recorded in that year living in other departments: most were then living in northern and western France, with about two-fifths of them in the Paris region. The department's population decline also reflected a fall in its birth-rate, which was more significant than the simultaneous fall in the death-rate: the birth-rate fell from 27.3 per 1,000 persons in the period 1840–60 to only 18.8 in the period 1901–10 (a decline of almost one-third), while in the same periods the death-rate fell from 22.8 to 17.4 (a decline of one-quarter).[21]

This overall loss of population was not uniformly distributed within the department. Different *pays* saw their maximum populations being reached, and their losses of population being experienced, at different times. Broadly, the population of Perche peaked before 1851; the populations of Beauce and

979 — **En Sologne** - Le Laboureur

L'Aube a blanchi le Ciel. Déjà le laboureur
Creuse au flanc du vallon le sillon producteur
Où germera, l'été, la belle moisson blonde
Le Produit de sa sueur, le pain de tout le monde.
MARTIAL FOURNIT

Photograph 3.3 Ploughing in the Sologne
Source: 3 Fi 6436 photo, Archives Départementales de Loir-et-Cher
The poem may be translated as:
>   Dawn has blanched the sky. Already the ploughman
>   is creating on the valley side a fertile furrow
>   in which will germinate in summer the beautiful golden
>   harvest, the product of his sweat, bread for all.

the Val de Loire reached their maxima in the 1850s and 1860s; those of Gâtine tourangelle and the Petite Sologne in the 1880s; and those of the Grande Sologne in the early 1900s.[22] Between 1851 and 1911 population densities fell in many communes of the Cher valley and the Petite Sologne, and in most communes of the Val de Loire and of all the *pays* to the north of the valley; by contrast, most communes of the Grande Sologne increased their populations during this period. Generally speaking, the most densely populated areas of the department in the mid-nineteenth century (and especially the valleys) had seen substantial population reductions by the early twentieth century, while the areas which had been least densely settled saw their populations increase. Dupeux has argued that the former areas were at that time being integrated into a wider, market economy, whereas the latter remained more isolated, within closed economies.[23] The continuing growth of the population of the Grande Sologne in this period is interpreted as a sign of its persistent 'backwardness' at a time when the other *pays* were being caught-up in the process of 'modernisation'.

If, J. H. Clapham once argued, the pace of industrialisation – or what some might call 'modernisation' – can best be measured in general terms by that of urbanisation,[24] then it was proceeding only slowly in Loir-et-Cher during the second half of the nineteenth century. Between 1851 and 1911 the 'urban' population increased only slightly from about one-fifth to about one-quarter of the total.[25] Most of the increase had been absorbed by Blois, which grew by more than 6,000 from 17,749 to 23,933; some of the increase was due to the growth of two cantonal centres in the Sologne (Lamotte-Beuvron was now 2,702 and Salbris 3,036). But the other major towns had seen only slight increases: Vendôme grew by less than 400 to 9,707, and Romorantin by less than 100 to 8,102. Many of the communes with 'urban' cantonal centres lost some of their populations between 1851 and 1911: even in the Loir valley Montoire-sur-le-Loir's declined by 200 (to 2,970), in the Val de Loire Onzain's fell by 150 (to 2,249) and Mer's by 600 (to 3,578), and in the Cher valley Selles-sur-Cher's dropped by 470 (to 4,074) and Saint-Aignan's by 400 (to 2,992). Other 'urban' communes, such as the cantonal centres of Contres (2,243) and Montrichard (2,769), and the few other, exceptional 'urban' communes like Cour-Cheverny (2,241) and Pontlevoy (2,267), more-or-less maintained the sizes of their populations as they had been in the mid-nineteenth century. Generally speaking, the decline of Loir-et-Cher's small towns of between 2,000 and 4,000 people can be traced to the 1870s and in some cases even earlier.[26]

The pre-eminence of Blois within the settlement system of the department, already apparent when the department was created in 1790, became even clearer during the course of the nineteenth century and by 1911 Blois accounted for some 9 per cent of the department's population. The central location of Blois within Loir-et-Cher geographically and its central role within it politically and administratively ensured that it also had an essential part to play in the 'modernisation' of Loir-et-Cher, in the transformation of the department's society, economy and culture. It is now appropriate to examine that transformation and it will be convenient to do so initially in terms of some of the main catalysts of change operating within rural France during the nineteenth century.

Processes of change

In his thesis about the transformation of peasants into Frenchmen, Weber argued that 'between 1880 and 1910 fundamental changes took place [in rural France] on at least three fronts. Roads and railroads brought hitherto remote and inaccessible regions into easy contact with the markets and lifeways of the modern world. Schooling taught hitherto indifferent millions the language of the dominant culture, and its values as well, among them patriotism. And military service drove those lessons home.' [27] We might disagree with Weber both about the precise period and the particular processes of change in relation to Loir-et-Cher, but there can be no doubt that underlying the changing geography of France as a whole in the nineteenth century were the annihilation of distance (or at least a substantial time–space compression) and the elimination of illiteracy (with its resultant cultural compression). The vastly accelerated circulation of an increasing volume of people, commodities, capital and ideas significantly differentiated the Loir-et-Cher of 1914 from that of 1815, transforming the lifestyles and attitudes of many of its inhabitants.

Communications

Improving the networks of communications in Loir-et-Cher was a constant concern throughout the nineteenth century, but the most significant developments took place from mid-century onwards. By 1848 the six *routes nationales*, totalling 306 km, and the sixteen *routes départementales*, totalling 450 km, provided an improved, well-maintained infrastructure, but one which only effectively served the main towns (most especially Blois, upon which three of the national and five of the departmental roads converged) and some of the principal *bourgs* of the department. Away from these main roads much movement was difficult even in summer and often impossible in winter. From the mid-1830s communes were given direct responsibility for their own *chemins* and, once they had come to accept it and to appreciate its significance from the early 1850s onwards, expenditure on local roads was often the largest item of a commune's budget during the second half of the century. In addition, during the 1870s and 1880s, some of the increasingly used local roads were reclassified and became a national or departmental responsibility. During the 1850s and 1860s special attention was given – partly stimulated by Napoleon III's direct interest – to improving the economy of the Sologne by creating a network of *routes agricoles*: more than 300 km of such roads were constructed in Loir-et-Cher's portion of the Sologne, breaking down the isolation of many of its communes. By the early 1880s improvements to rural roads had brought hitherto remote and relatively isolated communes into much closer contact with the more dynamic, better connected, more precociously 'modernising' parts of the department.[28] Even so, dog-drawn carts persisted in some parts as one form of road transport for both people and goods until the twentieth century.[29]

One very distinctive transportation improvement in the Sologne deserves special mention because of its exceptional, if essentially symbolic, character. The Canal de la Sauldre had an agricultural purpose: to transport marl cheaply into the central communes of the Sologne as a means of improving the fertility of their soils. The first section was started in 1852 and completed in 1859 and the second opened in 1869, when the full canal extended 43 km between Launay (in the adjacent department of Cher) and Lamotte-Beuvron. In the 1870s a fleet of a dozen barges was transporting 25,000–35,000 metric tons of marl annually but the operation was not financially viable in the long-term: the canal's construction and use was heavily subsidised by the State, there was virtually no product transported in the opposite direction to the carrying of marl, and the costs of transhipping the marl and transporting it over-land by road or rail were high (so high that it is doubtful whether by 1880 the marls were much used beyond a band more than two or three kilometres wide on either side of the canal). Although the Canal de la Sauldre undoubtedly contributed to the agricultural improvement of some communes of the Grande Sologne, its actual achievements fell far short of the aspirations of its promoters.[30]

Water-borne transport in Loir-et-Cher was not restricted to this exceptional canal because its rivers, and especially the Loire, had historically played an important role. But their contribution to communications was, of course, restricted (as it had always had been) to those communes and *pays* with access to navigable water and in the nineteenth century the rivers of the department came to be overshadowed by the improved roads and the new railways. During the 1840s commodity traffic on the Loire was in the region of 200,000 metric tons annually, but it declined rapidly in the 1850s to only a little more than a quarter of that amount and virtually disappeared in the 1880s. The railways in particular were able to offer not necessarily a cheaper but a much more flexible and reliable means of transportation for both merchandise and passengers than could boats on the Loire.[31]

Although Loir-et-Cher was one of the first departments of France to benefit from a railway link with Paris, it thereafter developed its network of primary, secondary and tertiary lines relatively slowly, not completing it until after 1914 (figure 3.8).[32] Its two first lines were part of the national network radiating out from Paris: that from Orléans to Tours along the Loire valley was opened in 1846 with (in Loir-et-Cher) stations at Mer, Ménars, Blois, Chouzy-sur-Cisse and Onzain; and in the east of the department the line traversing the Grande Sologne from Orléans to Vierzon and Châteauroux (in the department of Cher) was opened in 1847 with (in Loir-et-Cher) stations at Lamotte-Beuvron, Nouan-le-Fuzelier, Salbris and Theillay. Thus even before 1848, thanks to its geographical position, Loir-et-Cher had acquired 101 km of main-line railway. But it then had to wait almost twenty years before seeing the completion of its primary network with the opening in 1866 of the line from Paris to Tours via the Loir valley with (in Loir-et-Cher) stations at Saint-Jean-Froidmental, Saint-Hilaire-la-Gravelle (photograph 3.4), Fréteval, Pezou, Vendôme, Saint-Amand-de-Vendôme and Villechauve. This was followed in 1867 by the line from Tours to Vierzon via the Cher valley with (in Loir-et-Cher) stations at Chissay, Montrichard, Thézée, Saint-Aignan, Châtillon-sur-Cher, Selles-sur-Cher, Gièvres, Villefranche-sur-Cher and Mennetou-sur-Cher, with a branch–line being opened in 1872 from Villefranche-sur-Cher to Romorantin.

Thus, at the beginning of the Third Republic, the department's primary

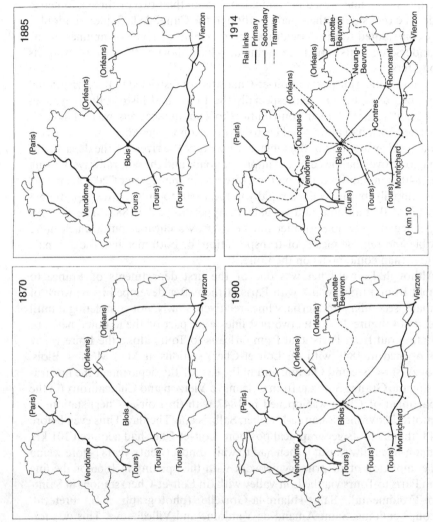

Fig. 3.8 The railway and tramway network of Loir-et-Cher, 1870–1914

Sources: Annuaires de Loir-et-Cher 1870–1914 and Richard Turner (personal communication)

Photograph 3.4 The railway at Saint-Hilaire-la-Gravelle, on the line between Paris and Tours opened in 1866
Source: 3 Fi 8407 photo, Archives Départementales de Loir-et-Cher

railway lines were completed as an integral part of the national network. No more railways were constructed in Loir-et-Cher in the 1870s, so that still in 1880 Blois and its two sub-prefectures of Vendôme and Romorantin were linked by railway to Paris but not yet to each other. The primary network emphasised the pre-existing significance of the major river valleys and of the *routes nationales* in the department's communications. By 1880 the railways had enhanced links between some parts of Loir-et-Cher and places beyond it (and especially Paris) much more than with other localities in the department.

From the mid-1860s there were debates about projects for a secondary network within the department but it was not until 1881 that a line linked Blois and Vendôme and not until 1883 that a line linked Blois and Romorantin. The late 1880s and 1890s saw the opening of five other secondary lines with routes partly within the department: the lines from Vendôme to Pont-de-Braye (opened in 1881, with stations in Loir-et-Cher at Thoré-la-Rochette, Saint-Rimay, Montoire and Trôo); from Brou to Bessé-sur-Braye, opened in 1885 and connecting Droué, Boursay, Mondoubleau, Sargé-sur-Braye and Savigny-sur-Bray; from Tours to Sargé-sur-Braye, opened in 1893, linking Authon, Prunay-Cassereau, Lavardin, Montoire and Savigny-sur-Braye; from Thorigny to Courtelin, opened in 1900 with stations in Loir-et-Cher at La Fontenelle, Arville, Le Gault-du-Perche, Saint-Avit and Le Plessis-Dorin; and the line from Château-du-Loir to Saint-Calais, opened in 1900 and crossing a small section of the department in the north-west.

Photograph 3.5 The tramway at Bracieux, in the Sologne, on the line between Blois
and Lamotte-Beuvron opened in 1888
Source: 3 Fi 5573 photo, Archives Départementales de Loir-et-Cher

To these secondary lines – linked to the national system and so partially
integrated with it, but essentially serving the department – were added from
the late 1880s a tertiary network of narrow-gauge local lines which often followed the routes of existing roads and on which travelled the steam-powered
trains (*tramways à vapeurs*) (photographs 3.5 and 3.6). Such lines were opened
from Blois to Lamotte-Beuvron (1888), from Blois to Ouzouer-le-Marché
(1888), from Blois to Saint-Aignan (1899), from Blois to Montrichard (1900),
from Oucques to Vendôme (1900), from Vendôme to Mondoubleau (1906),
from Blois to Château-Renault in Indre-et-Loire (1907), from Vendôme to
Droué (1908), from Ligny-le-Ribault in Loiret to Neung-sur-Beuvron (1905)
and from Neung-sur-Beuvron to Romorantin (1906). In addition the line from
Le Blanc in Cher to Argent in Indre crossed the Grande Sologne in the south-
east of Loir-et-Cher, between Pierrefitte-sur-Sauldre and Gièvres (1902). On
these essentially rural lines, intended to serve small, isolated localities, there
were many stops and stations but facilities were minimal. Stations were often
simple wooden, rather than brick or stone, constructions. To these rural steam
tramway lines were added in 1910 five urban electric tramways in Blois itself
and then in 1913 there was opened an electric *tramway* line along the left bank
of the Loire from Amboise (in Indre-et-Loire) to Cléry (in Loiret), running
through the whole of the section of the valley in Loir-et-Cher.[33]
    From the early 1880s onwards, then, constructing the secondary and ter-

Photograph 3.6 The tramway at Maves, in the Petite Beauce, on the line between Blois and Ouzouer-le-Marché opened in 1888
Source: private collection, the author

tiary railway lines had increasingly focused the railway network as a whole upon Blois, whose population grew markedly between then and 1914. But the combined improvements and developments to both road and railway systems meant that from the 1880s the smallest and most remote villages became linked by improved roads to their nearby markets (photograph 3.7) and through their nearest railway station or halt brought into contact with the rest of the department and even more distant places. Even the most isolated localities had become integrated into the national space-economy by 1914 and for many localities in Loir-et-Cher that process had begun in earnest in the 1840s and 1850s.

Primary schools

Expenditure on local roads and primary schools together accounted for the major part of the budgets of most rural communes in Loir-et-Cher for much of the nineteenth century, an unambiguous statement of the importance which both they and the central authorities attached to the improving role of communications in the broadest sense.[34]

The Revolution of 1789 had affirmed the rights of all citizens to free schooling and the obligation of the authorities to provide it, but it was unable to deliver the means whereby such principles could be put into practice and the

Photograph 3.7 The pig market at Romorantin
Source: 3 Fi 6056 photo, Archives Départementales de Loir-et-Cher

provision of schooling remained very uneven both geographically and socially.[35] An *ordonnance* of 1816 delegated to each commune the responsibility to provide elementary schooling but it was not until 1833 that each commune with 200 or more inhabitants was required, by the *loi Guizot,* to have at least either a private school or a public primary school funded mainly from local taxation and from parental fees (with free schooling only for children from the poorest families, as designated by the commune's council), supplemented by grants from the department and the State. A commune with fewer than 200 inhabitants was required to co-operate with a neighbouring commune in setting up a school. The 1833 law also required each department to establish a teachers' training college for men: a college for women teachers was not to be required until 1879.[36]

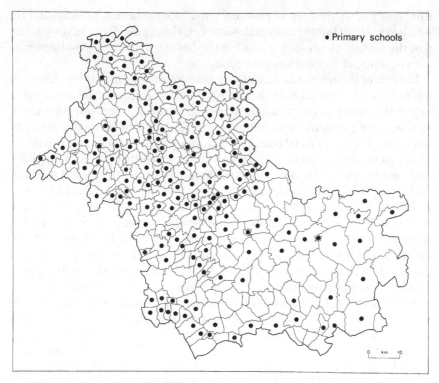

Fig. 3.9 Primary schools in Loir-et-Cher, by communes, 1833
Source: AN F^{17}115 no. 60,099

The Guizot law also required that a survey be conducted of primary schools throughout France and that survey provides, with all of its limitations, a basic datum from which to assess the development of schooling and its impact during the nineteenth century. Almost two-thirds (63%) of the communes of Loir-et-Cher already had a primary school in 1833: that was also the proportion in communes of the *arrondissement* of Vendôme but in that of Blois it was higher (70%) while in that of Romorantin it was markedly lower (43%) (figure 3.9). Some communes had not established a school because they had too few inhabitants to justify one (as was the case at Roches and Villeneuve-Frouville) and others because their proximity to their cantonal *chef-lieu* or to another, large village enabled their children to attend its school (as at Courmenon and Coulanges). Some communes too small to have their own schools were also unable to send their children to schools in neighbouring communes because of their inaccessibility: an inspector of schools reported that children from Saint-Loup and from Maray, in the Cher valley, could theoretically have gone to the school in Mennetou-sur-Cher but were discouraged from doing so by the need to cross the river. More generally, the inspectors – who had to travel throughout the department, visiting all of its schools – noted that the paucity of primary schools in the Sologne reflected the indifference, even hostility, of its inhabitants to education.[37]

Important though the *loi Guizot* was in terms of educational legislation, it has to be acknowledged both that primary schools existed in the department before 1833 and that the principles underpinning the law were not always, not fully and not promptly implemented in practice.[38] Many commune councils were fearful of the financial burdens which new schools would create and many of the *curés* were suspicious of an institution which they saw as potentially undermining their own authority; and there was a scarcity of good teachers and appropriate premises. For all of these reasons, the impact of the *loi Guizot* in Loir-et-Cher was neither immediate nor ubiquitous. In 1838, the prefect reported that there remained fifty communes – 17 per cent of the total – without a public primary school, most of them in the Sologne; by 1848 the figure had fallen to twenty, but they were communes with fewer than 200 inhabitants so that by then, some fifteen years after the *loi Guizot*, the infrastructure for a system of primary schooling was in place throughout Loir-et-Cher.[39]

Its effectiveness, however, was far from uniform. Primary schooling was not made compulsory until 1882 by the *loi Ferry* (which also made it free and secular). Before then there persisted significant variations in matters such as school attendance, the quality of the teachers, and the schools' premises. Attendance varied between the genders, with the seasons and among localities. In 1839 almost half (45%) of school-age children attended for at least part of the year, the figure for boys (51%) being higher than that for girls (40%). Only some 23 per cent of boys of school age and some 19 per cent of girls were reported by the inspectors as attending school regularly and completing a full course of instruction for five or six years (most children who did go to school attended irregularly and only for two to four years). Attendance was much higher during the winter months, from October to March, than during the summer: children of school age in rural areas were always expected to help out on the farm, for example guarding and feeding livestock, and during the summer months the demand for their labour was even greater. There were also variations from place to place. In general, attendance rates were higher and showed less seasonality in the towns and in larger villages than in the more remote and smaller communes. There are also identifiable differences among the *pays* and rural communes. In 1839 most boys of school age in Perche and in the Sologne were not attending school but in the other *pays* most were for at least part of the year; by contrast, most girls were not attending school in most of the department, for it was only in parts of the Vendômois (in the cantons of Vendôme, Selommes, Montoire-sur-le-Loir and Saint-Amand-Longpré) and of the Blésois (in the cantons of Blois, Mer and Marchenoir) that most girls of school age were doing so (figure 3.10). These contrasts in

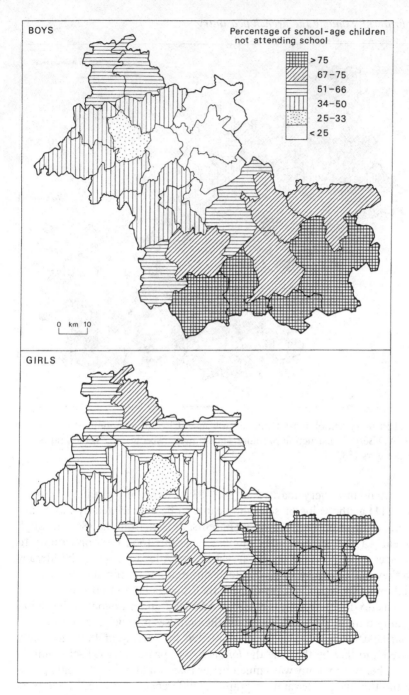

BOYS

Percentage of school-age children
not attending school

> 75
67 – 75
51 – 66
34 – 50
25 – 33
< 25

0   km   10

GIRLS

Fig. 3.10 Primary school non-attendance by boys and girls in Loir-et-Cher, by
cantons, 1839
Source: AD Série T Instruction primaire: statistiques, inspection des écoles,
rapports des inspecteurs 1833–49

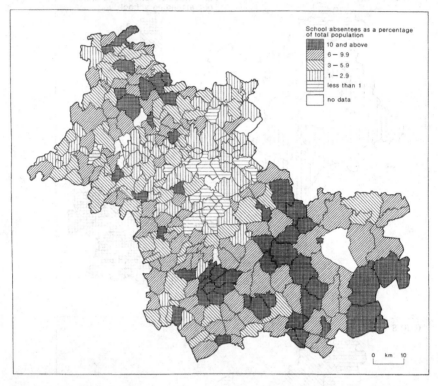

Fig. 3.11 Primary school non-attendance in Loir-et-Cher, by communes, 1848–49
Source: AD Série T Instruction primaire: statistiques, inspection des écoles, rapports des inspecteurs 1833–49

school attendance, very marked in the 1830s, were only gradually eroded (figure 3.11) as the value of schooling came to be appreciated by more children and, more importantly, by more parents. Indeed, from the 1860s through to the early 1900s an increasing number of adults took the opportunity to attend special winter-time classes in school in order to acquire the literacy which they had not themselves obtained through not having attended school as children and which now they wanted their own children to possess.[40]

There is no doubt that, in the 1830s and 1840s, and probably in the 1850s, the quality both of the teachers and of the premises in which they taught was very variable.[41] There is also little doubt that the quality of both gradually improved and that by the time the *loi Ferry* was passed in 1882 schooling in Loir-et-Cher was in many ways much better than it had been fifty years previously and that the gender and geographical differences of the 1830s had been virtually eliminated. By the early 1850s, just over 50 per cent of children were receiving some schooling; by the early 1860s the figure had risen to 80 per cent

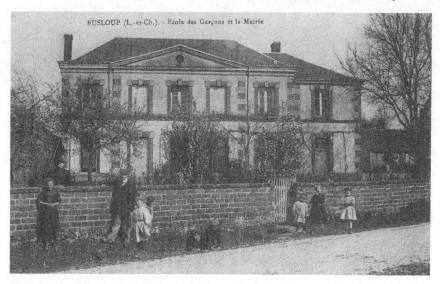

BUSLOUP (L.-et-Ch.). - Ecole des Garçons et la Mairie

Photograph 3.8 The boys' school and the *mairie* of Busloup, in the valley of the Loir
Source: 3 Fi 8280 photo, Archives Départementales de Loir-et-Cher

and by the early 1880s to almost 90 per cent.[42] New school buildings, constructed in great numbers during the 1870s and 1880s, stood in the rural landscape as icons of a new rational, national and increasingly secular and
republican ideology (photographs 3.8 and 3.9).[43] Much more remains to be
discovered, particularly about the content of the teaching (about the knowledge and skills being imparted) and about its impact upon the new generations
of Loir-et-Cheriens.[44] Such a subject deserves its own book.

In a general way, primary schools clearly contributed increasingly to a widening of the intellectual, social and geographical horizons of the rural populations. They specifically contributed to the fundamental cultural transition
from an oral to a written culture, without of course being solely responsible
for it because literacy was taught both in the home and at the workplace as
well as in schools and a gradual improvement in literacy rates can de detected
well before the public system of primary schooling was established.[45]
Nonetheless, the growth of literacy was associated with primary schooling
and the annihilation of illiteracy during the nineteenth century was a change
as significant in its social impact as the annihilation of distance had been in
the economic sphere. In conjunction, these two processes laid the foundations
for the 'modernisation' of Loir-et-Cher as they did for many other regions of
France at this time. They were both closely associated, for example, with the
development of a national and provincial press, with both the supply and
demand of an expanding reading public.[46]

Photograph 3.9 Schoolboys and their school at Suèvres, in the Loire valley
Source: private collection, Jacques Leroy

For the department as a whole, the literacy of military conscripts increased from 26 per cent in 1827 to 95 per cent by 1890 (figure 3.12).[47] Of the three *arrondissements*, Blois was the most and Romorantin the least advanced; Vendôme increased its literacy rate faster than the department as a whole. The growth in literacy at faster than average rates in Blois and Vendôme *arrondissements* significantly widened the gap between those two on the one hand and that of Romorantin on the other during the 1850s and 1860s and only from the 1870s did the gap begin to narrow. The percentage gap between the 'most-literate' and the 'least-literate' canton was 36 in 1827, widening to 63 in 1860 and narrowing to 16 by 1890. Of course, given that the conscripts were aged twenty, the significant periods in terms of the actual growth of literacy may reasonably be assumed to have been about ten years earlier than these dates suggest. This would mean that near-total literacy – at least for boys – had been achieved in Loir-et-Cher by 1880, before schooling became compulsory. Local variations in conscript literacy were clear at the cantonal level in the 1820s, with highest rates in the Val de Loire, Beauce and the lower valley of the Loir and the lowest rates in Perche, the Sologne and the Cher valley (figure 3.13). By 1860, with differences in conscript literacy rates among the *pays* having been accentuated, the same broad pattern existed with the Val de Loire, Beauce and the Loir valley having the highest rates but among the 'laggard' *pays* Perche, the Petite Sologne and the Cher valley had made significantly more progress than had the Grande Sologne, where fewer than

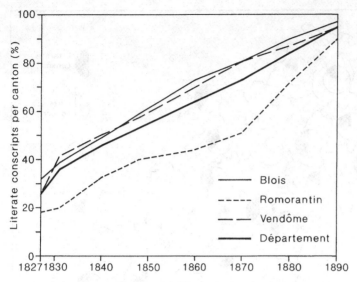

Fig. 3.12 The growth of conscripts' literacy in Loir-et-Cher, for the department and its *arrondissements*, 1827–90
Source: AD Série R and Baker (1992b) 211

half of conscripts were able to read and write (figure 3.14). By 1890, with near-total conscript literacy having been achieved throughout most of the department, only in parts of Perche (in the adjacent cantons of Mondoubleau and Savigny-sur-Braye) and of the Grande Sologne (in the adjacent cantons of Neung-sur-Beuvron, Lamotte-Beuvron and Salbris) did significant pockets of conscript illiteracy persist (figure 3.15).

Military service provides one optic through which to view the growth in literacy, but it has also been attributed a direct role in the transformation of French rural society.

Military service

Systematic military conscription in France was introduced in 1798. The precise form that it took varied somewhat during the course of the nineteenth century, but in broad terms every canton was responsible for supplying a certain quota of recruits. Lots were drawn annually and the youths drawing numbers higher than the required contingent were exempt. The size of the quota varied from time to time, as did the categories of those who were exempted – such as priests, married men, only or elder/eldest sons of widows or of large families, and well-educated men. Those who did draw a number requiring them to serve were able to pay for a substitute to serve in their stead,

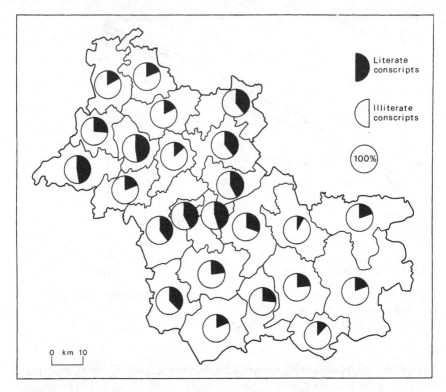

Fig. 3.13 Conscripts' literacy in Loir-et-Cher, by cantons, 1827
Source: AD Série R Listes départementales de recrutement

if they could afford to do so. The length of service also varied according to military needs: it was six years after 1818, eight years after 1824, seven years between 1855 and 1868, then five years until 1889. The relatively long term of service meant in turn that a relatively small proportion of the relevant cohort of young men (the call-up related to those in their twentieth year) were conscripted (it was usually 10 per cent or less of the age group). Until the 1850s, at least one-quarter of those registered as conscripts were substitutes – either poor young lads viewing military service as a way of earning some money or veterans who intended to re-enlist in any case and who viewed substitution as a way of earning a bonus in doing so. Substitution was abolished in 1873. From 1889 the term of service was reduced to three years and all those categories of men who had previously been exempted were now required to serve for one year. In 1905 military service was made compulsory for all for two years and in 1913 the term was extended to three years.

It has been persuasively argued in principle by Weber that such military service throughout the nineteenth century produced a considerable 'institu-

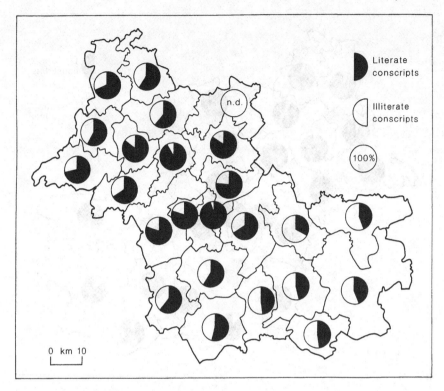

Fig. 3.14 Conscripts' literacy in Loir-et-Cher, by cantons, 1860
Source: AD Série R Listes départementales de recrutement

tionalised migration and kneading together' and that it was a crucial part of the process whereby peasants were transformed into Frenchmen. Both the idea and the practice of conscription, it has been argued, brought home in each commune the existence of France as a nation. Young men from the communes were required to present themselves before the military selection board at their local cantonal centre and those recruited were then sent to the barracks of the department, usually in the principal town, which many of them had rarely or even never visited (photographs 3.10 and 3.11). They would then be required to complete their service anywhere in France or its colonies, or even fighting for France in foreign countries.

Military service was, for a variety of reasons and unsurprisingly, unpopular throughout the French countryside for much of the nineteenth century, but it was, in its way, an education for those who experienced it. This was so both directly in the sense that it encouraged reading, writing and counting and promoted French as the national language, and indirectly in that it uprooted young men from their own *pays* and showed them other 'countries', where

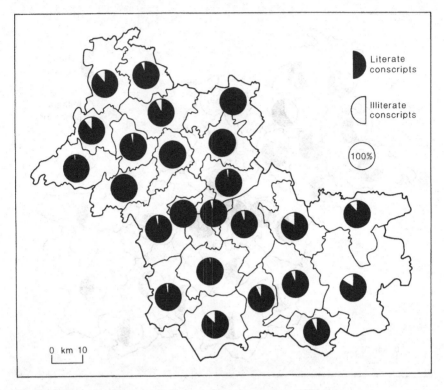

Fig. 3.15 Conscripts' literacy in Loir-et-Cher, by cantons, 1890
Source: AD Série R Listes départementales de recrutement

things were done differently, both in France and in many cases abroad. Military service also provided an experience of living standards which, although Spartan, were far above those of many at home. Military service gave young men new ideas and experiences, new knowledge and expectations, widening their horizons. To their feelings for their *pays* was developed, especially from the 1870s onwards, a loyalty to their *patrie*.

Although perhaps only two or three young men in each commune were conscripted and removed in this way each year from their localities, over the decades of the nineteenth century as a whole this process must have had a cumulative impact, both directly upon the conscripts themselves and indirectly upon the populations of their own communes for whom the recruits from the locality were surely a talking-point both when they were away and when they returned for visits, or for good, and to whom they brought their new knowledge and reported their experiences, thereby becoming agents of change – of 'modernisation' – within their communities. Of course, having been so educated, not all conscripts returned to their original communes. Weber has

Photograph 3.10 The 'call-up' of conscripts in 1912 at Mondoubleau, in Perche
Source: private collection, Marcel Landier

Photograph 3.11 The military barracks at Blois, to which those selected for
conscription were sent from throughout the department
Source: 3 Fi 4986 photo, Archives Départementales de Loir-et-Cher

claimed that 'a good proportion of peasant soldiers chose not to return to their villages', deciding instead to live and work elsewhere, especially from mid-century onwards when job opportunities multiplied in the manufacturing and service sectors of the national economy. Weber suggested that between a third and a half of conscripts did not return to their original communes when discharged.[48]

In sum, Weber concluded that the army acted in conjunction with the communications and schools as 'an agency for emigration, acculturation, and in the final analysis, civilisation'. It must be said, however, that in relation to military service at least Weber's argument was based more on common sense and anecdotal evidence than on any hard analysis. For example, his claim about the failure of a high proportion of conscripts to return to their homes upon discharge was based upon a newspaper report about non-returning conscripts from one commune over the ten-year period ending in 1896 and upon data relating to a total of fifty-one conscripts in another commune over the same period.[49] Nonetheless, in the absence of argument and evidence to the contrary, Weber's general thesis about the general role of military service as providing an educational experience may be accepted.

The history and geography of military service in Loir-et-Cher during the nineteenth century has yet to be written – to date it has attracted remarkably little attention. The precise role of military service in the department remains unclear and can so far only be glimpsed in general terms and by assuming that it was not significantly different from the general French experience. Documentation on the subject, in both the departmental and the national, military record offices, is vast and awaits scholarly, systematic analysis. These records do make clear, for example, the exceptional, 'backward' condition of the Sologne by comparison with other *pays* of the department: conscripts from communes in the Sologne were, during the first half of the nineteenth century, not only, as has been seen, less literate but also smaller in stature and more were rejected as being unsuitable for military service on the grounds of their physical condition than were conscripts from elsewhere in the department. The difference – which may be inferred as reflecting the Sologne's lower living standards – diminished during the second half of the nineteenth century when conditions in the area significantly improved.[50] As far as any link between conscription and migration is concerned, however, analysis of the conscript classes from two cantons (Mer and Neung-sur-Beuvron) for 1856 and 1891 has shown that there was not a statistically significant difference between those young men recruited and those not recruited in terms of their propensity to migrate away from their communes. More important was the link between occupation and migration, with young men with skills (such as stonemasons, carpenters and clerks) being more likely to migrate than were those more tied by bonds of land ownership (such as *cultivateurs* and *vignerons*).[51] But the military records would repay closer examination than they have so far received.

There can be no doubt, however, that one military episode – the Franco-Prussian War – had particular significance within Loir-et-Cher. The French Army of the Loire had bases in the department in the Sologne and in Beauce. On 10 November 1870 it recaptured Orléans, which had been taken by the Prussians on 27 September. But the fall of Chartres and of Châteaudun to the Germans gave them access to Loir-et-Cher in the north, via Mondoubleau. On 4 December Orléans was again lost to German forces which then moved on 10 December along the Val de Loire and into Loir-et-Cher at Josnes in the Beauce, and then from 14 to 16 December into the Loir valley taking Fréteval and Vendôme. French forces evacuated from Blois on 18 December. A line of German forces in the north of the department, stretching from Mondoubleau to Droué, then advanced to Vendôme, occupying it and the surrounding region for three weeks before moving westwards into the department of Sarthe. Thus from early December until early January the north of Loir-et-Cher was a theatre of war and then until March the department was under German occupation. From late September 1870 the National Guard had been mobilised, involving the call-up of all 21-year-olds, unmarried males and childless widowers. Each commune throughout the department had to form a company of the National Guard, each canton a batallion, and each *arrondissement* a legion: thus was constituted the department's brigade. Only the legion of Blois *arrondissement* had been properly established by the time the Prussians invaded the department and even it had only 2,000 rifles for its 3,177 men. A considerable part of the cost of the mobilisation fell directly upon all of the department's communes and the war itself inflicted heavy casualties and physical destruction on some of them. In addition, communes in the occupied areas were required to provide the Prussian forces with food, supplies and money and, after the French defeat, all communes were required to contribute towards the cost of post-war reparations to Germany.[52]

The burden imposed by the war on the communes of Loir-et-Cher was immediate and must have been known to all of its inhabitants, whatever the extent of their own direct involvement in these extreme *événements*. Afterwards, the effort had to be made to adjust once again to a less dramatic but also dynamic daily life, to the *conjonctures* of economic, social and cultural change.

The character of change

It is not my purpose to provide in this book a new synthesis of the history and geography of Loir-et-Cher during the nineteenth century. It will be my intention instead to offer an original analysis of the historical geography of its rural voluntary associations in that period. As setting for that analysis, however, a brief survey of the changing rural society and economy, culture and *mentalités*, of Loir-et-Cher in the nineteenth century is necessary, drawing and elaborating upon the classic, magisterial monograph of Georges Dupeux (1962) and the briefer and more popular study by Jean Vassort (1985). Between 1815 and 1914 rural Loir-et-Cher saw some fundamental changes, of which the two most notable were probably the transformation of its agricultural methods and productivity, and the development of political democracy and of republicanism.[53]

Society and economy

In the mid-nineteenth century the nobility owned about one-quarter of the land and the bourgeoisie about one-third. Many of the nobility possessed not only rural estates and *châteaux* but also urban properties and *hôtels*. Many of the families of the nobility were interlinked through marriage and it was by marriage that some individuals ascended the social ladder. Despite the upheavals of the Revolution of 1789, and with the relaxed conditions after the Revolution of 1830, rural society was still dominated by the *notables* during the first half of the nineteenth century: status within society was closely associated with land ownership. Owners of large estates figured prominently in the activities of the Society of Agriculture, whose meetings – frequently held in Blois, where it was based – were part of the social scene of the department in addition to their explicit concern to improve the condition of its agriculture.

Status was attached to land, but it could also be attached to wealth and to occupation. The better-off merchants and millers were well regarded within rural society, as were doctors and lawyers. The mass of the rural population – the peasantry – were dominated by these smaller but much more powerful classes. Some peasants were owner-occupiers, as was the case usually among the *vignerons* of the valleys and often on the farms of the Beauce, while elsewhere in the department tenanted farms were the norm. But the picture is complex and many peasants owning and/or renting land were also labourers or domestics on other farms for at least some of the time. In addition, of course, there were many landless labourers. As well as that diverse population directly engaged in agriculture there was also the usual range of people employed in craft and commercial activities related to agriculture, and – in areas like the Forest of Marchenoir and the Grande Sologne – wood-cutters and charcoal burners.

Agriculture was the principal economic activity of Loir-et-Cher throughout the nineteenth century. In 1851 it was employing about two-thirds of the active male population. By then the small farms on the Beauce plateau and in the valleys contrasted with the large holdings in the Sologne. Agricultural improvement was encouraged both by the public authorities, anxious to counter the food shortages and its attendant problems of inflation and social unrest, and by enlightened, progressive landowners keen both to improve the economic productivity of their own properties and to serve as models for the diffusion of knowledge about agricultural innovations. Improvements there were, but they were not always unopposed and the routine of peasant life was not readily broken down. Crop yields and livestock numbers did increase during the first half of the century, but even in 1852 it was still the case that a quarter of the cultivated land lay fallow each year.

There was much agricultural improvement and change during the second half of the century and by the end of it the centuries-old concern with subsistence food shortages – still, given the crises of 1846–8, in recent memory – had been rendered obsolete.[54] In 1914 agriculture was still employing more than four-fifths of the active population but the amount of fallow land had been reduced from 90,000 ha at mid-century to less than 60,000 ha; the cultivation of artificial meadows and of potatoes, the application of chemical fertilisers and the mechanisation of harvesting as well of cultivation were key components of the modernisation of farming which contributed to an almost doubling of cereal production, concentrated by 1914 mainly on wheat to the virtual disappearance of maslin and buckwheat. In the case of livestock production, while the number of sheep declined by about two-thirds (essentially because of the reduction in the acreage of fallow in the Beauce and of *landes* in the Sologne), milk production increased by 86 per cent from 1852 to 1892.

Of the *pays* of Loir-et-Cher, it was the Sologne which saw the most remarkable transformation during the nineteenth century. Improvement here was stimulated by some enterprising landowners (including Napoleon III, who had farms at Vouzon and Lamotte-Beuvron), by the creation in 1859 of a regional agricultural committee (the Comité Central Agricole de la Sologne), by State subsidies and by considerable improvements to the communications' networks through and on the boundaries of the Sologne. Especially during the 1850s, 1860s and 1870s much heath land was reclaimed and degenerate woodland cleared and brought into cultivation, and much existing farm land was improved through marling and draining. While the acreage under arable cultivation and under vineyards was significantly expanding, so too was that under woodland as the poorer soils of the Sologne were afforested: between 1852 and 1914 its woodland area almost doubled. When the impact of the phylloxera crisis was exacerbated by that of the general agricultural crisis towards the end of the century, the high tide of cultivation in the Sologne receded. The region was to become more important for its forestry and for *la chasse* (building upon the growing fashionability of its *châteaux* and hunting lodges during the Second Empire to such an extent that from 1880 onwards rents for using the Sologne's land and lakes for shooting and fishing were higher than were rents and incomes from farm land).[55]

The phylloxera dramatically affected the entire department, increasingly invading its vineyards from the mid-1860s onwards, just at the time when they were being expanded in response to the growing commercial demand for wine. The area of vineyards expanded from 25,000 ha in 1850 to 45,000 ha in 1889, but had retracted again to 28,000 by 1913. The formation of syndicates to combat the phylloxera will be considered separately in due course.[56] In brief, the solution which was eventually and generally adopted was that of replanting the vineyards with grafted American vine-stocks in the late 1890s and early 1900s. Not all owners of vineyards could afford the costs of such an expensive solution, even though it was subsidised by the State, and so many vineyards were either converted to other forms of production (such as vegetables and fruit, and especially the growing of asparagus, in the Val de Loire and in the Petite Sologne) or completely abandoned.[57]

Notwithstanding the many changes to the rural society and economy during the nineteenth century, the broad contrasts among the *pays* remained as clear in 1914 as they had been in 1850 and in 1815. The Sologne was characterised by large estates, owned now by both notables and the wealthier bourgeoisie. In the Beauce large tenanted farms predominated; in Perche, average-sized properties; in the main valleys were to be found the smaller farms of the *vignerons*.

Culture and mentalités

Rural society was dominated by notables, by men in the liberal professions and by priests. But these were far from being a united group and conflicts among them were often played out on the commune councils, which they tended to run for much of the century. For the first half of the century, the social distance between the notable and the mass of the population within a commune was often mediated through the priest and resolved in a form of paternalism (this was especially the case in the Sologne).

Religion
The Church's role clearly declined during the nineteenth century. Both materially and ideologically, the Church had come under direct attack during the closing revolutionary decade of the eighteenth century and, although it recovered to varying degrees in the nineteenth century, its influence was generally diminished. The process was no doubt both gradual and discontinuous, but it was inexorable. Religious practice and religiosity in Loir-et-Cher cannot be monitored precisely during the nineteenth century but their gradual decline is, paradoxically, evidenced in both the persistence of popular superstition (especially in the Sologne and in the Vendômois) and in the emergence of scientific rationalism (especially among teachers in the primary schools).

During the 1820s and 1830s perhaps only about one-third to one-half of the rural population took Easter communion, with far more women doing so than men. Nonetheless, religion still touched directly upon the lives of most people. The priest was involved in critical family events, such as baptisms, first communions, marriages, the last rites, and burials, all of which were occasions

Photograph 3.12 The church at Mulsans, in the Petite Beauce,
with its dominating eleventh-century bell-tower
Source: private collection, the author

for family reunions. He was there, also, at community gatherings such as the
celebrations and festivals of the liturgical calendar, as well as at fairs, markets
and even the *veillées* during the winter. When parishioners were anxious about
their own health or about the condition of their crops and livestock, the priest
would be called upon for his blessing and the protection which it was believed
it could assure. The Church provided theatre and spectacle – with its pilgrimages and processions, its bell-ringing, music and singing, its sermons and its
costumes. The church building was usually still the dominant cultural structure within the local rural landscape (photograph 3.12) and roads and fields
were punctuated with wayside crosses. The Church was, it seems, still
omnipresent. But its influence was clearly declining and much religious

practice seems to have been superficial. Other activities – like dancing and bowls and drinking in the *cabarets* – offered an alternative attraction, a different kind of theatre. The constraints of the Church came increasingly to be questioned, as did the moral authority of priests and the financial burdens which they imposed upon local communities. The tendency towards de-Christianisation was more marked among men than among women.[58]

In 1840, the diocese of Blois (which corresponded with the department of Loir-et-Cher created in 1790) recorded that 99.7 per cent of its population was Catholic but in reality its bishop was aware that there was a general ignorance of basic religious truths, religious practices were being abandoned, and there was indifference or even hostility to religion. An attempt was made in the 1840s to reinvigorate the celebration of Saints' days throughout the department, with special services, processions and pilgrimages, given that more and more people were working on Sundays rather than attending devotions.[59] In the early 1850s, only about one-third of the populations of communes in the Sologne presented themselves for Easter communion and in those in the valleys of the Loire and Loir the figure was only about one-tenth. By then the Beauce also seems to have become indifferent to the Catholic Church. The few Church schools existing in the department were located in its three principal towns, in some cantonal centres and in some communes of the Loire and Loir valleys: throughout most of rural Loir-et-Cher there were no Church schools.[60]

From the 1830s more and more opportunities were provided for more secular forms of celebration and sociability. For example, new roads gave rise to festivities associated with the public opening ceremonies and more communes sought to institute fairs or markets which themselves became secular festivals, with entertainments such as dancing bears and opportunities for drinking and gaming. The *Annuaire* of 1848 lists 245 such commune fairs or *assemblées* in Loir-et-Cher. Similar fêtes were promoted by the department's agricultural societies or *comices,* and the end of the school year was also often marked by some form of public celebration, usually involving children reciting, singing or acting a play. The new republican regime of 1848 also gave rise to celebrations – like those linked to the planting of 'liberty trees' – distanced from the Church.[61]

During the 1850s and 1860s the Church in Loir-et-Cher endeavoured to retain a reasonably vital public image: active *congrégations féminines* were established at Blois and at Vendôme in that period and the pilgrimage of Notre-Dame-des-Aydes was well supported. Also, during the second half of the nineteenth century about fifty churches were built or refurbished and social Catholicism was a much-discussed idea. Nonetheless, the religious indifference detectable in mid-century (and even earlier) gained ground. The Church's role in primary school teaching declined; from 1882 teaching in state schools was entirely secular. Anti-clerical feeling was quite strong in Loir-et-

Cher: proposals to rid schools of the influence of the allegedly reactionary clergy were well-supported in the department, where petitions of the Ligue d'Enseignement for lay education attracted greater than national-average support. The experience of the diocese of Blois cannot have been very different from that of the adjacent diocese of Orléans, where – as Christianne Marcilhacy has shown – the bishop's energetic campaign to revive Christianity during the 1850s, 1860s and 1870s might have had some short-term success but failed in the long term, undermined by the development of schools, of a local press and of *cabarets*.[62] The Church probably reached its apogee around 1860 and thereafter 'popular' religion declined, even in the Sologne.[63] The depth of anti-clerical feeling which developed within Loir-et-Cher was perhaps symbolised in February 1906 when, to enforce the compilation of an inventory of Church property (required as part of the process of separating Church and State), troops broke down the doors of the cathedral in Blois – recalling the physical attacks upon churches and the conversion of many of them (temporarily) into Temples of Reason in the early 1790s. Perhaps more realistically, the decline in the role of the Church was to be seen in the dramatic fall in the number of Easter communicants in various localities of the department between 1890 and 1910, from about 15–35 per cent to about 10–25 per cent: in practice, the reduction might not have been quite that dramatic, because attendance before the separation in 1906 of Church and State might have been more superficial, less committed, than it was afterwards.[64]

For much of the nineteenth century, of course, virtually all of the population of Loir-et-Cher would have been regarded officially – and not only by the Church – as being Catholic. The Census of Population taken in 1872 recorded the religion of each individual and suggested that virtually the whole of the population, and certainly as much as 99 per cent, was indeed Catholic. But two reservations have to be made about that figure. First, it clearly relates to what people were prepared to claim as their faith rather than what they practised as their religion. Secondly, for the *arrondissement* of Blois, the data is extant on a commune basis and this reveals the existence of significant numbers of non-Catholics (mainly Protestants, but including a few Jews and some declaring themselves to be of no religion) in a set of communes extending from the town and neighbourhoods of Blois up the Val de Loire to Mer, then on to the Beauce plateau around the eastern end of the Forest of Marchenoir (figure 3.16).[65]

In the context of the time, there were significant proportions of non-Catholics (i.e 1% or more) recorded in Mont-près-Chambord (4.1%) and Saint-Claude-de-Diray (4.7%), on the left bank of the Loire, opposite Blois; in Blois itself (1.8%) and at Villebarou (1.6%), to the north of the town, on the edge of the Beauce plateau; but the largest concentrations were to be found upstream from Blois, at Ménars (1.7%) and at Mer (8.8%), and in the group of communes to the north of Mer on the Beauce plateau (Josnes 15.2%;

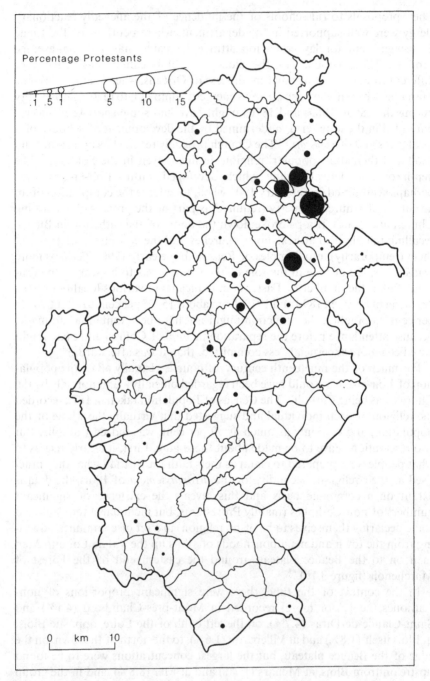

Fig. 3.16 Protestants in the *arrondissement* of Blois, by communes, 1872
Source: AD 201M 110

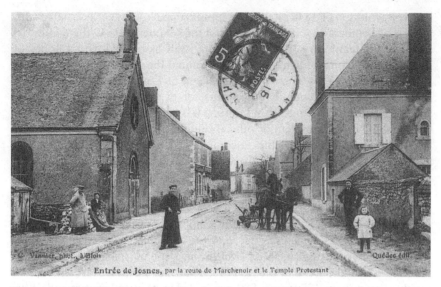

Photograph 3.13 The Protestant chapel on the outskirts of Josnes, in the canton of Mer
Source: 3 Fi 8189 photo, Archives Départementales de Loir-et-Cher

Lorges 8.6%; Briou 5.7%). There had been a Protestant community in the small town of Mer and neighbouring communes for a very considerable time. Following the Revolution of 1789 and its initial, violent but – given its prosecution of *liberté* – contradictory attack upon the Church, the right to freedom of worship was recognised in February 1795 and not only Catholics but also long-suppressed Protestants were again able to practice their religions. From 1819 the Protestant community at Mer, which was then said to number about 300 or about 8 per cent of the population, acquired a French Methodist pastor for its *Eglise réformée*: Armand de Kerpezdron, a Breton who was incarcerated by the English on a prison-ship on the River Medway during the Napoleonic Wars, was to minister to the Protestants at Mer for thirty-five years, until 1854. The maintenance and development of Protestantism in Mer and in the nearby communes of the Beauce plateau and the Val de Loire owed much to the endeavours of that remarkable man, building though he was on earlier foundations (photograph 3.13).[66]

Politics
During the first half of the nineteenth century, Loir-et-Cher as a whole moved politically away from the right and gradually towards the centre and left. Especially among the *vignerons* of the valleys there was a marked republican spirit, manifested most explicitly in the *unions plébéiennes* created during the July Monarchy by the notary Jean-Jacques Delorme of Saint-Aignan in the Cher valley.[67]

The Revolution of 1848 introduced universal male suffrage, associated with the rise of democratic socialists. But from the summer of 1849 there followed a marked (if temporary) shift to the right, itself associated with a policy of repression which immediately (if briefly) limited the activities of republicans whose hopes were shattered by the *coup d'état* of Louis-Napoléon. While the short-lived Second Republic was a time for experimenting with universal suffrage, the longer-lasting Second Empire was a period of political manipulation, making use of a system of official candidates and a variety of administrative pressures to influence the outcome of elections. Following the political amnesty of 1859, republicanism once again extended its power base, both socially (by gaining more and more support among traders and artisans, as well as among the *vignerons* and liberal professions) and geographically (by becoming significant in Beauce as well as in the Cher, Loire and Loir valleys).

The leftward drift of the department politically was clear in the elections of 1871, 1876 and 1881, which saw the department becoming increasingly republican: as Vassort has put it, those elections at last confirmed in Loir-et-Cher the promise of the Revolution of 1789. That movement towards moderate republicanism in the 1870s and 1880s was aided by politically aware and active members of the liberal professions and abetted by a widely held fear (promoted by those on the right) of social unrest or (promoted by the left) of a return to the conditions of the ancien régime.

From the mid-1880s, moderate republicans in the department popularised their views, not least by establishing their newspaper, *Le Progrès de Loir-et-Cher*, in 1884. The 1890s saw political debates and even some strikes, often promoted by people coming into the department from elsewhere (and the strikes were usually restricted to the working-class populations in the main towns). During the early 1900s can be detected the emergence of socialism linked to more local militants (often secondary school teachers, artisans, shopkeepers and *vignerons* but rarely industrial workers): by 1914 the socialists had established twenty local branches in the department but none of their candidates in that year gained sufficient votes to be elected.

But the growth of democracy in rural Loir-et-Cher during the nineteenth century involved much more than exercising the right to vote in national elections: it also, and perhaps more importantly, involved voting in local, municipal elections and even the possibility of standing as a candidate for election to the commune council. The functioning and composition of commune councils became increasingly democratic during the nineteenth century. Broadly speaking, they functioned at three levels. First, they were empowered to take executive decisions about some strictly local and short-term matters (such as the administration of property owned by the commune, and the implementation of leases and the collection of rents on such properties). Secondly, they debated other local issues which had medium- or long-term implications (such as the commune's annual budget, which included the

Photograph 3.14 The late nineteenth-century *mairie* and, to its left, post office obscuring from view the early medieval church at Suèvres, in the Val de Loire
Source: private collection, Jacques Leroy

levying of a local tax, the purchasing, exchanging and selling of property by the commune, and its own construction projects – for example, for roads, public squares, schools and a *mairie* (council office) (photograph 3.14): but any actions which a council wished to take in such matters had to be approved by the central administration, represented by the prefect). Thirdly, a commune council was able to express its views on issues initiated elsewhere but which might affect its own commune (such as development plans for departmental or national roads and railways, and the state and repair of religious properties in the commune). The commune council was essentially a debating chamber because it was the mayor's responsibility to be the commune's executive officer and simultaneously to act as the local agent for central authority. During the first half of the nineteenth century, mayors and their deputies were appointed by the prefect from among council members who were themselves elected by an electoral corps comprised of the commune's principal tax-payers. Within rural communes, each council comprised a dozen or so members (the precise number being a function of the size of the commune's population). In 1848 local democracy was significantly extended with the introduction of universal male suffrage and with mayors being elected by the councils (instead of being appointed by the prefect) in communes with less than 6,000 inhabitants (which, in effect, meant most communes and certainly all rural communes). Thus throughout the nineteenth century, local commune councils were both taking their own initiatives and responding to those coming from the central administration, and in that process mayors often played pivotal roles.[68]

The decisions of a council had a direct bearing upon the lives of those living in its commune and upon the commune's landscape,[69] and a growing awareness of that role and of the right to participate in it developed among rural populations during the nineteenth century. For example, Silver has shown how in the Vendômois between 1852 and 1885 peasants wanted increasingly to influence, even to control, communal affairs and mobilised themselves politically in order to challenge within rural communes both the traditional authority of notables and of priests and the new, post-revolutionary authority of prefects. This politicisation of the peasantry was sustained by the right to vote for municipal councillors and to present petitions to the council. Appreciation of the significance of local politics then developed into a recognition of the extent to which they in turn were integrated into national politics: in the mid-1870s, 'Vendômois peasants voted for republicans nationally when they recognised the affinity between their local problems and the republican national ideology'.[70] Furthermore, the composition of commune councils changed during the second half of the nineteenth century, becoming less dominated by the large landowners and the liberal professions and more reflective of the general character of a commune's population. In addition, the membership of councils turned over more rapidly, breaking down the earlier tendency for councils to be dominated over long periods by particular individuals, families or social groups.[71]

In sum, the years of the *fête impériale* (1852–71) in Loir-et-Cher marked a period of economic expansion, better living standards, social improvement and democratic development. The rural community as a whole seemed to be flourishing and within it the *vignerons* appeared to be especially buoyant, but even enterprising day-labourers were able to acquire a parcel or two of land. During the 1880s and increasingly during the 1890s, the situation changed dramatically: the phylloxera crisis and the agricultural depression, singly or together, affected most rural communities whose income levels generally declined for almost twenty years after 1885. The social optimism of the 1850s and 1860s came to be replaced by a pessimism in the 1880s and 1890s: the rural exodus initiated in those earlier years in some localities came to be more general throughout the department and to deepen its impact.

Towards an historical geography of voluntary associations in rural Loir-et-Cher during the nineteenth century

There has not yet been a systematic study of sociability in Loir-et-Cher during the nineteenth century. Only fragments can be pieced together incidentally from other existing studies. Similarly, if less surprisingly, there has not yet been a systematic study of the history and geography of voluntary associations in the department during the period. This present study attempts to address this latter question, at least insofar as rural Loir-et-Cher is concerned. Before doing so, it remains finally to pull together such work as has been undertaken to-date on these two, related themes.

Rural life in Loir-et-Cher during the first half of the nineteenth century still for the most part faced traditional threats and followed traditional rhythms, with a clear seasonality imposed by both agricultural and liturgical cycles. Sociability has been described by Vassort as 'considerable', taking the traditional forms of organised fairs and *assemblées* (village gatherings) and of spontaneous *charivaris*.[72] Even at the end of the century, much informal sociability still revolved around markets and fairs, *veillées* (social evenings), Church, community and family celebrations, and the call-up of conscripts, and collective activities such as harvesting and well-digging. It was also occasionally enlivened by an itinerant circus or performing bear.[73]

As far as formal sociability is concerned, the existence of some voluntary associations in rural Loir-et-Cher has been noted *en passant* but, with perhaps one exception, has not been seriously examined. The Revolution of 1789 saw the formation of *sociétés populaires* (societies of patriots) in the towns, encouraged by and linked with correspondents in Paris. These tended to have among their members especially lawyers and teachers, those who (according to Vassort) knew how to go about constructing a new world: they were impatient to do so and, in order to give a sense of immediate achievement to their long-term projects, they had recourse to symbols (such as secular fêtes, liberty trees, red bonnets, substituting *citoyen* for *monsieur*, and cleansing some street names and even commune names of religious associations). Such societies did not, however, catch the rural imagination and their impact was largely urban as well as being short-lived, although it might be that they sowed the seeds for the later flowering of republicanism [74] and they might well have served as a model for the somewhat similar *unions plébéiennes* (popular associations) which developed among the *vignerons* of the Cher valley during the July Monarchy.[75] At Blois a Société des Amis de la Constitution was founded in early 1791 and existed until November 1792. Its aim was to promote respect not only for the Constitution of 1791 but also for the Declaration of the Rights of Man (26 August 1789). It was affiliated to the Club des Jacobins in Paris, and itself gave rise to some local societies in turn affiliated to it, such as those at Saint-Dyé, Villebarou and Chambord.[76] At Montoire-sur-le-Loir a Société Populaire, Républicaine, Sabotière, Révolutionnaire des Sans-culottes had a short-lived but lively pro-revolutionary and anti-clerical existence between November 1793 and July 1794.[77] The late eighteenth century also saw the beginnings of a very different association. The Société d'Agriculture de Loir-et-Cher was initially founded in 1799, but then foundered in 1801 only to be re-established in 1805. The existence of this society, comprised principally of notables and with its base in Blois, is well known but much else about it remains to be discovered. It tends to be assumed that its role was that of promoting not only agricultural improvement but also contact among the social elite of the department as a whole, but especially of the *arrondissement* of Blois, since that was the role of many such learned societies throughout France in the first half of the nineteenth century.[78] In the other *arrondissements* of Romorantin and Vendôme a similar role, if in a somewhat lower social key, was played by their *comices agricoles* (agricultural committees), comprised principally of large landowners and set up officially by the central authorities in order to promote the diffusion of knowledge about best practice in farming through annual agricultural fairs featuring exhibitions, demonstrations, competitions and exhortations. The Society of Agriculture and the *comices agricoles* were, it is clear, fundamentally urban-based and socially selective institutions, but their history has yet to be written.[79]

Similarly, the contribution of agricultural associations to the modernisation of farming in Loir-et-Cher at the end of the nineteenth century has been simply asserted and reasserted but not seriously assessed or reassessed.[80] It is remarkable that the best account remains that provided by Roger Dion in his classic regional monograph on the Val de Loire: he observed in 1934 that the development of professional agricultural associations was then, together with the use of railways and the cultivation of vegetables as field crops, one of the dominant characteristics of the rural economy in all parts of the Val de Loire where small holdings predominated. Dion specifically noted the significant historical role played in the development of agricultural syndicalism by the Syndicat des Agriculteurs de Loir-et-Cher since its foundation in 1883. Its membership was particularly strong in Beauce and in the *pays* of small viticultural properties in the Val de Loire and its neighbouring plateaux between Blois and Beaugency (up the valley, towards Orléans), as well as in the viticultural districts of the Petite Sologne, and of the Cher and Loir valleys. Dion was particularly struck by what he saw as the precocious awakening of the spirit of agricultural association in that section of the Loire valley: he 'searched in vain' the whole length of the Val de Loire, between Decize and Nantes, for an agricultural association founded earlier than the threshing-machine syndicate established in 1878 at Saint-Claude-de-Diray, just across the river from Blois.[81] Within the very general scope of his historical and regional synthesis, Dion's treatment of agricultural syndicalism was necessarily brief and, as we shall see, it failed to capture the historical depth, geographical spread and functional diversity of the spirit of association in the Val de Loire and its neighbouring *pays*. In her survey of the viticultural *pays* around Blois, Monique Touvet noted their reorientation towards the production of fruit and vegetables during the early-twentieth century and the creation of some agricultural marketing co-operatives between 1918 and 1939 as a result of significant *initiatives paysannes*, but she made no reference to the existence of co-operatives or syndicates in the region before then.[82]

Beyond these associations, there have been some references in the literature to a few others but there has been no systematic study. For example, in his study of Freemasons in the Val de Loire, Jacques Fénéant noted that their Lodges were essentially urban associations, although some were established in larger rural centres: in Loir-et-Cher there were Lodges at some time during the nineteenth century not only at Blois and Vendôme but also at Montrichard and Saint-Aignan, two such centres in the valley of the Cher, a hearth of republicanism within the department.[83] Similarly, André Prudhomme has thrown some light on the secret republican societies which operated during the 1850s, mainly in Blois, in some of the more populous centres of the Loire valley and also of the Cher and Loir valleys, their membership being predominantly craftsmen and traders and some farmers or viticulturalists.[84] The existence of a handful of religious *confréries* (confraternities) – charitable and devotional – in some rural communes in the early nineteenth century has been noted by Maurice Gobillon.[85] In a study of social Catholicism in Loir-et-Cher between 1875 and 1902, Olivier Martin has very usefully discussed those associations established by the Catholic Church in Loir-et-Cher (principally in its towns and larger cantonal centres) in the course of nineteenth century. Relatively few in number and variously called *cercles, conférences* and *sociétés*, these Catholic associations promoted diverse activities related mainly to worship and charitable activities, and they were of course in general intended to maintain, even preferably increase, membership of the Church. From the 1880s, under the impulse of social Catholicism, there also developed some associations more directly associated with the workplace, aimed at urban craftsmen and traders on the one hand and the farming community on the other. Such associations were, of course, swimming against the anti-clerical tide and their impact has to be assessed accordingly.[86] The Church's endeavour to lead people towards a spiritual security via the material and social security provided by mutual aid societies has been noted by Elisabeth Muller and by Annick Notter.[87] Looking elsewhere, André Prudhomme has written a brief local history of the volunteer *corps de sapeurs-pompiers* (fire-fighting brigades) in Loir-et-Cher from 1762 until 1914. He has provided considerable factual material about each association as well as about the legislative framework within which they operated, but the brigades have not been assessed within their broader historical and geographical contexts.[88]

With the aid of these earlier studies, it now becomes possible to take some further steps towards an historical geography of voluntary associations in rural Loir-et-Cher during the nineteenth century. In this book I will address those rural voluntary associations which were functionally connected to the world of work and which were related to the question of peasants as survivors. I will examine the timing, the spacing and the functions of the principal voluntary associations which were developed in Loir-et-Cher during the nineteenth century, examining them both as practical expressions of the principle of fraternity and as a 'modern', pragmatic means by which peasants might cope with some persistent risks which threatened their existence. The following five chapters examine some voluntary associations developed to provide peasants with new ways of managing risks to themselves, their families, their properties and their livelihoods.

4

Insurance societies

'L'Union fait la force, aidons-nous les uns les autres'
Slogan of La Mutuelle de Droué, Société d'assurance contre la mortalité des animaux
des races bovine et chevaline (1899)

The declining influence of the Church in Loir-et-Cher during the nineteenth century has often been commented upon,[1] but the developing role of insurance societies has scarcely been noticed, especially in relation to its rural areas. The history and geography of insurance in the department is a vast research topic in its own right and no attempt will be made here to fill that particular lacuna. But one aspect – that of the role of livestock insurance societies – is of especial significance not only to the historical geography of insurance but also more generally to that of voluntary associations, and so it will be considered in some detail. Before proceeding down that avenue, however, a general prospect of agricultural insurance in the department during the nineteenth century will provide a base from which to undertake the exploration.

Agricultural insurance companies

The idea of agricultural insurance was being actively considered in the department during the opening decades of the nineteenth century. A circular of 24 January 1810 from the Ministry of the Interior to the prefect drew attention to the view of the Conseil d'Etat about the developing role of *compagnies d'assurance mutuelle* in countering the damage done to crops by hail-storms and to livestock by epidemics. The usefulness of such associations had come to be appreciated in several departments and the government saw them as contributing to agricultural prosperity: the concern of the Conseil d'Etat was that the establishment of such associations should be properly controlled by the authorities. The prefect wrote on 20 March 1810 to the president of the Society of Agriculture of Loir-et-Cher, inviting the Society to consider the question of such crop and livestock insurance societies, but he

101

had also to send a reminder on 27 November before the Society replied. On 12 December 1810 the secretary of the Society wrote to the prefect, saying that the Society's committee had in fact considered the matter in 1807 and had then told the prefect that establishing such societies in Loir-et-Cher would pose some difficulties: firstly, because hail was not as great a problem in the department as it was in some other regions of France where insurance against hail damage was then being practised; secondly, because there was the dual problem of having to assess both the value of the insured property and of the damage done. The secretary concluded by saying that the Society was pre-occupied with the problem of how to combat damage to vines inflicted by *vrebec*, an insect causing leaf-curl.[2]

The Society's negative response serves to emphasise both the range of natural hazards faced by the farming community in the early nineteenth century and the fact that the Society – a robust promoter of agricultural innovation and denigrator of routine – was not always itself immediately receptive to new ideas and practices which in due course were to prove to be of considerable benefit to the farming community.[3] That episode also demon-strates how agricultural practices in Loir-et-Cher were potentially open to influences from well beyond it even in the early nineteenth century. The idea of formal insurance came from outside and putting it effectively into practice was also to involve companies based outside of the department. A list com-piled by the prefect for the Ministry of the Interior in October 1858 indicates that there were then twenty-five insurance companies operating in the depart-ment, with only one of them based in the department itself and the rest in Paris, except for two based at Le Mans.[4] A somewhat similar list compiled for December 1908 records more than 120 companies operating in the depart-ment but based outside it, predominantly (more than three-quarters) in Paris but also in some of France's major provincial cities like Bordeaux, Lyon and Toulouse, as well as in substantial cities closer to Blois, like Orléans, Chartres, Dreux, Le Mans and Tours.[5] These companies offered insurance of crops against hail damage, of property against fire and flood, and of livestock against death or serious accidental injury. The last of these merit closest examination here.

Livestock insurance companies

At the end of 1833 the prefect received from Paris a copy of the statutes of a company then being founded to provide insurance for horses and cattle dying from epidemics or diseases in eighteen departments, mainly in the Paris Basin and including Loir-et-Cher. In their preamble to the draft statutes, the pro-moters of the society set out its rationale: they argued that the farming com-munity was everyday exposed to enormous losses occasioned either by hail damage or by the death of livestock; that throughout France hail and fire insurance societies had been successfully organised and warmly welcomed; and that benefits of association, although often little understood, were one of the incontrovertible truths, because through association losses shared in common (especially if by a large number of individuals and over a large geographical area) were much lighter and so without serious impact upon either families or communities. But the company was to be open to all those with horses or cattle they wished to insure, such as those running postal and transport services. The company would require animals to be insured for five years; it would require a minimum insurance valuation of 200 fr. for horses, donkeys and mules, and of 100 fr. for cattle (although several owners could group their animals for this purpose and have them insured in just one of their names); insurance valuations were always to be to the nearest 10 francs. The company's operations were to be effected primarily through its agent and its veterinary doctor located in the *chef-lieu* of each *arrondissement*. An annual premium was to be paid for each animal insured, which would be at most 2.5 per cent, or 4.5 per cent or 5 per cent of the total value of the livestock insured by a member, according to the kind of animals being insured. In addition, each member would have to pay on joining the company a variable entry fee based upon the value of the livestock being insured. Those fees would be used to enable the payment of immediate but provisional indemnities to those suffering losses; the final and definite indemnity, and the specific premium to be paid each year, would not be determined until the end of each year. If the income in any given year was insufficient to meet in full the cost of the members' losses, then they would be met proportionately. The closely printed draft statutes of this proposed company ran to seven (A4–sized) pages, but it is not clear that it actually came into existence and operated in Loir-et-Cher.

During the 1840s a number of companies (sometimes called societies) similar to the one just described were set up and endeavoured to be active in Loir-et-Cher, as can be determined by the materials which they sent to the prefect. They included La Ligérienne Tourangelle, based in Tours, and L'Agricole and La Providence, both based in Paris. L'Agricole had been authorised in September 1840 and La Ligérienne Tourangelle was established during 1842. Although the prefect appears not to have any views about these two, the growing number of such societies appears to have led him to become more critical of their utility to the community of Loir-et-Cher. On 24 June 1843 the prefect wrote to the president of La Providence both to thank him for having been sent a copy of the company's statutes and to say that he would refrain from commenting upon them, because Loir-et-Cher already possessed such a company which was sufficient for the department's needs. Similarly, on 15 December 1843 the prefect informed the Minister of Agriculture that, although he had no objection to the statutes of another livestock insurance company being created in Paris, La Bucéphale, with the intention of operating in part in Loir-et-Cher, the company would have no purpose because there already existed in the department a society, La Palès, with the same aim which met the needs of the region. Again, on 28 February 1848 the prefect told the Minister that although another Paris-based livestock insurance company, La Thémis, then being established and wishing to have permission to operate in Loir-et-Cher, would have a general agricultural utility, it would have no specific purpose for the department because it already had a society, La Palès, which was sufficient for its needs.

Similar reticence about the need for more than one such company or society in the department was expressed by the prefect when considering in 1846 an initiative coming from within Loir-et-Cher itself. In April 1846 the deputy mayor of Vouzon informed the sub-prefect at Romorantin that someone in the commune wanted to set up a livestock insurance society for several cantons in the Sologne, notably those of Neung-sur-Beuvron, Salbris and Lamotte-Beuvron and was requesting advice about what formalities needed to be followed. On 12 July 1846 the sub-prefect submitted to the prefect a request to authorise the livestock insurance company, La Confiance, which had been established in some cantons of the *arrondissement*, and enclosed a copy of its statutes. Only after being reminded by the sub-prefect again about this matter did the prefect on 23 July forward the request to the Ministry of Agriculture, supporting it himself and making the point that the company had 135 insured members, but also remarking that there already existed in the department a society of the same kind, La Palès, based at Mer (in the *arrondissement* of Blois). The Ministry replied on 1 October 1846 that it was unable to consider the draft statutes unless it also had a full list of names of those wishing to join the company and a statement of the total value of the livestock to be insured. Also the Minister wanted the statutes to be modified, to accord with legal requirements, and he suggested that those founding the company at Vouzon should consult the statutes of La Palès, which had been authorised by ordinance on 28 December 1843. On 29 January 1847 the prefect forwarded to the Ministry the revised statutes of La Confiance and sought authorisation for the company which, he said, already had sixty-five members. The statutes indicated that the company, based at Vouzon, would operate in five *arrondissements* (those of Blois and Romorantin in Loir-et-Cher and of Orléans, Gien, Sancerre and Bourges in adjoining departments). The Ministry's initial reply, dated 20 May 1847, required some further modifications to the statutes, and the process of modifying the statutes to meet with the Ministry's approval – particularly in relation to the administrative and financial arrangements – continued in a series of exchanges of letters for a further two years: the company was finally authorised by decree of the Conseil d'Etat on 9 August 1849. Even then it could not be activated immediately, because the only list existing of those wishing to join the company had been sent to the Ministry, which eventually returned it to the prefect in mid-November 1849. The printed statutes of this company, which run to seven pages, describe in seventy articles its aims and functioning in great detail: the company insured horses and cattle; it had precisely defined and carefully monitored procedures for all of its operations; it had annual maximum premiums of 6 per cent for cattle and 8 per cent for horses, with the actual premium to be paid being determined at the end of each year in relation to the losses incurred by members.[6]

It is difficult to know for how long these livestock insurance companies existed. None are included in the list of insurance companies certified by the prefect as existing in October 1858, while that for January 1909 records twenty-six such companies operating in Loir-et-Cher but with their headquarters (*sièges*) outside it. On the other hand, as will be shown, there were by the latter date numerous local livestock insurance societies in Loir-et-Cher with their *sièges* in the department.

The impression gained from this brief, and as yet partial, look at insurance is that while there was undoubtedly a significant development in the number and range of insurance companies operating in Loir-et-Cher during the nineteenth century they were generally run from outside of the department and few of them were likely to have attracted many members from within its rural society in general or from among its farming community in particular. Most of the companies and societies identified so far would, it is not difficult to imagine, have had little appeal to the peasantry. They were largely impersonal groupings of individuals not known to each other; they were run by administrators largely unknown to their members; the remoteness of their headquarters from the daily life worlds of people living in Loir-et-Cher meant that few who joined could expect to attend general meetings or to participate in the running of the companies or societies; not only participation but even simple membership assumed a level of literacy and numeracy which was certainly not common throughout Loir-et-Cher before mid-century. As far as the crop and livestock insurance companies based outside the department were concerned, they assumed a high level of trust among members in relation both to establishing crop and livestock valuations initially and subsequently to determining the cost of losses sustained by individuals. Moreover, all of these companies and societies assumed both an ability and a willingness to pay fees and premiums in cash. For all of these reasons, it is not surprising that such insurance companies and societies appear to have made little impact upon rural Loir-et-Cher. They were, it might be argued, too far removed from the daily experiences of its peasantry. Nonetheless, the new secular insurance – fundamentally an urban, modern phenomenon – was far from being ignored in rural Loir-et-Cher where, it will be argued, local communities sought new forms of fraternal association which were adaptations of traditional notions and practices of locally based co-operation, of mutual aid based upon simple labour exchanges rather than upon complex cash payments. This will be argued in due course in relation to the development of local mutual aid societies and fire-fighting corps, which will be seen as forms of personal and fire insurance. But it will be examined first in relation to the most explicit case, that of local livestock insurance societies.

Livestock insurance societies

Data

For most of the nineteenth century local livestock insurance societies developed within a legal framework designed for other purposes; they emerged along the margins of the legal system. Not until the 1890s was there an explicit endeavour to employ in relation to their formation and functioning some simplified formalities derived from the law of 21 March 1884 relating to the authorisation of *syndicats professionnels*. Then from 1898 the Ministry of Agriculture began systematically awarding grants to local livestock insurance societies. It is, therefore, only from the late 1890s – more than half a century after such societies began to be founded – that information about them was systematically and regularly collected by the prefect for submission to the Ministry. From 1898 there were compiled annual lists of the societies, providing in relation to each of them information about the kind of livestock insured, the sort of premium on which it was based, the number of members in the association, its annual income, the total value of the insured livestock, the total cost of claims, the total of indemnities paid to members, and the percentage relationship between indemnities and claims. In addition to this summary data, the prefect maintained a dossier on each individual society which principally included correspondence (or copies of correspondence) with the Ministry, with the society's officers, with the mayor of the commune in which it was based, and with the department's own professor of agriculture. They also often included copies of the society's statutes and rules. The surviving records are, of course, not complete and the data are by no means consistent through time and over space: but from this mass of unpublished (and, during the period when the research was conducted, uncatalogued) material it is possible to reconstruct the historical geography of local livestock insurance societies in Loir-et-Cher. The picture will not be complete and accurate in every detail, given the fragmentary nature of the sources, voluminous though they are. But the general outlines will serve to demonstrate the considerable significance of these local voluntary associations.[7]

Historical development

Determining the date when a particular livestock insurance society was founded is not easy, because there was in many cases a time lag between the informal establishment of a society in a locality and its formal recognition by the central authorities. During that interim period, some nascent societies awaited prefectoral authorisation while others functioned as though it had already been given (or, in many cases, unaware that such authorisation was even necessary). Such ambiguities were only slowly eroded during the course of the nineteenth century and it was not until 1898, when the State started to grant subsidies to these societies, that a precise mechanism for monitoring them was established. While this uncertainty has to be borne in mind when considering specific societies, it is of less seriousness in relation to their overall historical development and geographical distribution.

As has been seen, the idea of livestock insurance was current in Loir-et-Cher by the 1830s and 1840s, when some insurance companies began to operate in the department: but they did so usually from central offices outside of the department and each company operated over a very large geographical area. This same period also saw, however, the emergence of livestock insurance societies which were based locally and which functioned within restricted localities. There were at least four such societies founded in Loir-et-Cher during the 1830s and 1840s, with a further four being established during the 1850s (figure 4.1). That relatively slow growth in the absolute number of livestock insurance societies was then followed by a rapid increase in the 1860s: more precisely, the fourteen-year period from 1858 to 1872 saw a three-fold expansion in their number, from 7 to 22. The next thirty years saw steady growth, to just over 60 by 1902. There then followed a second period of rapid expansion until at least 1912, when their number exceeded 160. There can be no doubt that the 1860s and the first decade of the 1900s were particularly significant periods in the formation of local livestock insurance societies, nor that in absolute terms the latter period was the more important: 61 per cent of those societies founded by 1912 had been established since 1902, but that florescence drew upon the roots of a movement which went deep into the nineteenth century. Explanations for the particular significance of the 1860s and of the early 1900s in the historical development of local livestock insurance societies will need to be sought in due course, after a consideration of their geography.

Geographical distribution

The changing geography of local livestock insurance societies in Loir-et-Cher was quite remarkably patterned, with a primary hearth area in the Val de Loire and, much later, a secondary one in the Loir valley, from which centres such societies subsequently spread into other *pays* of the department. There was a marked geographical concentration of these societies until the end of the 1890s, and only in the early years of the twentieth century did their distribution become more dispersed. In the early phase of the development the societies were focused upon two sets of communes in the Val de Loire on both of its banks upstream from Blois. In 1860 there were six societies in the three

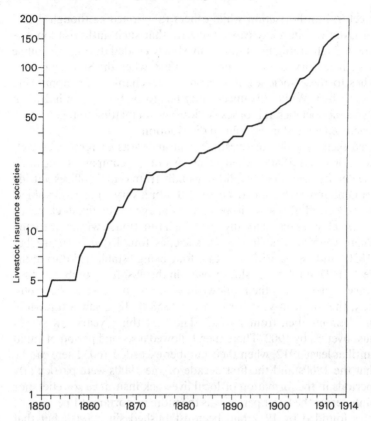

Fig. 4.1 Livestock insurance societies in Loir-et-Cher, 1850–1914
Sources: AD 7M 360–8

adjacent communes of Mer, Courbouzon and Séris, and another two in the
adjacent communes of Montlivault and Maslives (figure 4.2); by 1870 the
former set had been joined by a society at Cour-sur-Loire and the latter had
expanded into being a group of nine societies in seven communes (figure 4.3).
In addition, there was then one at Saint-Laurent-des-Bois, on the northern
edge of the Forest of Marchenoir, and another at Naveil, adjacent to
Vendôme, in the Loir valley. The pattern established by 1870 was intensified
but its distributional characteristics remained basically unchanged during the
1870s and 1880s, although the cluster on the left bank of the Loire did during
that period begin to expand a little southwards into la Petite Sologne (figures
4.4 and 4.5). That fundamental pattern was still clearly recognisable in 1900
(figure 4.6), when there remained a primary concentration of livestock insurance societies in the Val de Loire and the northern district of la Petite Sologne,
with secondary clusters around Saint-Laurent-des-Bois and around Vendôme

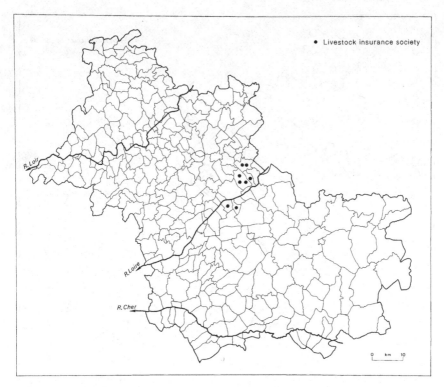

Fig. 4.2 Livestock insurance societies in Loir-et-Cher in 1860
Sources: AD 7M 360–8

and just one society in Perche, at Droué. In 1900 there were 58 societies; by 1910 the figure had rocketed to 145 and they had become much more widely distributed throughout the department, although there were still remarkably few in communes of the Cher valley, the Grande Sologne, the Gâtine tourangelle and Perche (figure 4.7).

The paucity of these associations in Perche is initially somewhat surprising, given its reputed emphasis upon livestock rearing, not least of the renowned percheron horses.[8] But closer consideration indicates that there was little, if any, correlation between the localities in which horses and cattle were found in significant numbers and those in which livestock insurance societies developed first and to the greatest extent. Making use of the 1852 Enquête agricole, Dupeux calculated cattle densities per square kilometre: while it was indeed the case that densities were moderate to high (15–19 per km^2) in the cantons of Blois, Mer and Marchenoir in which livestock insurance societies were established relatively earlier, cattle densities were as high or higher (15–24 per km^2) in cantons in the north-west of the department (those of Saint-Amand-

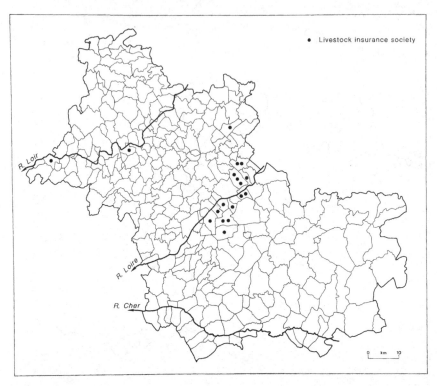

Fig. 4.3 Livestock insurance societies in Loir-et-Cher in 1870
Sources: AD 7M 360–8

Longpré, Montoire, Savigny-sur-Braye and Mondoubleau) where few societies were created and those that were only developed after 1900.[9]

A further picture of the significance of livestock within the department may be derived from the Enquête agricole of 1862.[10] The enumerations of livestock in different categories reveal regional variations in emphasis. More than 2,000 horses were recorded in each of the cantons of Droué, Mondoubleau, Savigny-sur-Bray and Montoire, in the north (Perche), and in the canton of Blois (no doubt influenced by the town of Blois itself, which then had almost 20,000 people). There were, by contrast, fewer than 1,000 horses in each of a group of cantons in the south (Neung-sur-Beuvron, Lamotte-Beuvron, Romorantin, Mennetou-sur-Cher and Selles-sur-Cher), mainly in the Grande Sologne where most of the department's donkeys and mules were to be found. The regional variation in the numbers of dairy cows in 1862 was not as great as that of horses: it ranged from 1,775 in Neung-sur-Beuvron to 4,559 in Montoire-sur-le-Loir; there were more than 2,500 in almost every canton except those of the Grande Sologne where instead were to be found the great-

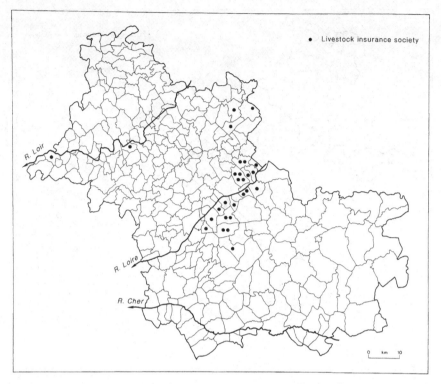

Fig. 4.4 Livestock insurance societies in Loir-et-Cher in 1880
Sources: AD 7M 360–8

est numbers of beef cattle (more than 200 in each of the cantons of Neung-sur-Beuvron, Lamotte-Beuvron, Romorantin, Salbris, Mennetou-sur-Cher and Selles-sur-Cher).

Absolute numbers of livestock and even cattle densities per square kilo-metre provide, of course, only crude measures of the economic and social significance of livestock. A somewhat less crude measure has been derived from the 1862 Enquête agricole, which records for each canton not only live-stock numbers but also the number of working farms. For the two separate categories of livestock covered by the insurance associations (horses, donkeys and mules; cattle), and also for both together, have been calculated the ratios of livestock per farm for each canton (figure 4.8).[11] Farms of the cantons of Blois, Mer and Marchenoir had relatively low livestock ratios but witnessed the precocious development of insurance societies; farms in the cantons of Bracieux and Contres had moderate livestock ratios and it was indeed into communes of those cantons that the movement to establish insurance societies spread; but many cantons with moderate to high livestock ratios on their

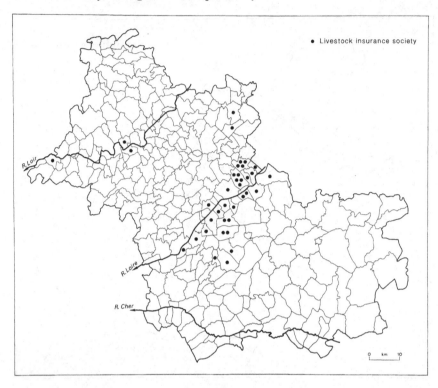

Fig. 4.5 Livestock insurance societies in Loir-et-Cher in 1890
Sources: AD 7M 360–8

farms hardly participated in that movement (such was especially the case in the north-west, in the cantons of Mondoubleau and Droué, and most especially in the east, in the cantons of Neung-sur-Beuvron, Lamotte-Beuvron, Romorantin, Salbris and Mennetou-sur-Cher). Quite simply, the geography of livestock insurance societies was not congruent with that of livestock production. Other explanations need to be sought.

Both the timing and the spacing of local livestock societies were distinctively patterned: the outline, descriptive, historical geography so far provided now needs to be interpreted in terms of the formation and functioning of the societies themselves and ultimately in terms of their broader cultural context.

Formation and functioning to c.1870

Most farms in Loir-et-Cher in 1862 had one or two horses, or a horse and a donkey or mule, and many had a few dairy or beef cattle. The latter were, of

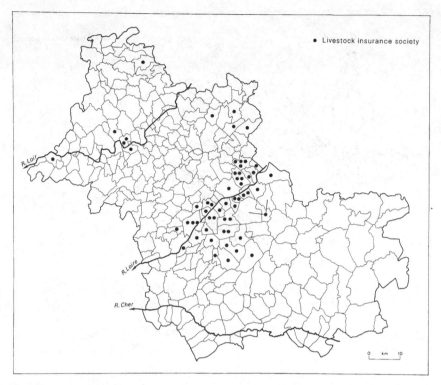

Fig. 4.6 Livestock insurance societies in Loir-et-Cher in 1900
Sources: AD 7M 360–8

course, valued for their products (mainly milk and meat, but also for others, like manure and skins), while the former were especially prized as draught animals and were essential to many farming operations. According to the 1862 Enquête agricole, the use of horses for ploughing had by then become the norm throughout the department, although both horses and cattle were thus employed in two cantons of the Sologne region (Mennetou-sur-Cher and Selles-sur-Cher) and also, more surprisingly, in the canton of Bracieux, on the border of the Sologne and the Val de Loire.[12] For varied reasons, therefore, horses and cattle were highly valued by farmers and represented a considerable capital investment, second only to that of their land if they were owner-occupiers and of foremost importance if they were not themselves landowners. Around 1900 a horse was valued at about 800 fr. and a dairy cow at about 350 fr.: the cost of a horse thus amounted to more than the annual expenses of a family of five persons, while the cost of a cow amounted to about half such expenses.[13] The protection and welfare of valuable livestock was unsurprisingly a major concern for peasant farmers (photographs 4.1 and

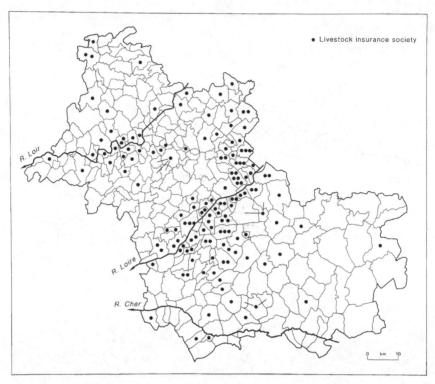

Fig. 4.7 Livestock insurance societies in Loir-et-Cher in 1910
Sources: AD 7M 360–8

4.2). Traditionally, they relied upon superstition and religion; gradually but increasingly during the nineteenth century they turned to science and reason, to veterinary medicine and insurance.

Information about the first handful of local livestock insurance societies to be established in Loir-et-Cher, those created before 1860, is sparse and much of it is derived from later documentation containing retrospective references whose accuracy cannot readily be checked. But the statutes and other contemporary information relating to a number of societies established around 1870 indicate how such societies were then being formed and how they were intended to function. It will be instructive to take a look at a few of those in particular before considering the characteristics and development of societies in general from around 1870 onwards.

A Société d'assurance mutuelle contre la mortalité des vaches was established at Saint-Claude-de-Diray, on the left bank of the Loire just upstream from Blois, on 31 December 1869. Its aim was to insure cows against death by accident or natural causes, against serious accident which rendered the animal

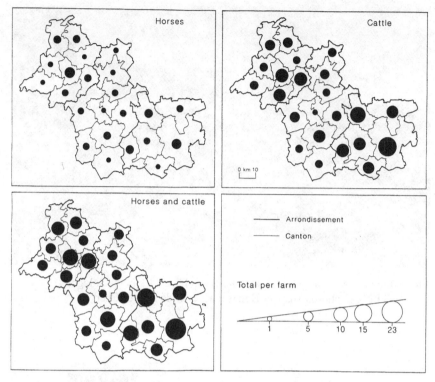

Fig. 4.8 Horses and cattle per farm in Loir-et-Cher, by cantons, 1862
Source: AN F[11] 2705 Enquête agricole de 1862 (Loir-et-Cher)

useless, and against compulsory slaughter required by the authorities during
an epidemic. Anyone keeping cows in the commune could join the society and
members were committed to belonging to it for one year at a time, with the
last Sunday of December being the last day of the society's year. Each new
member was required to pay a joining fee of 50 centimes, creating a fund which
met the society's limited administrative expenses. The society was administered by a committee comprised of a president, a vice-president, a secretary
and a treasurer, and by a variable number (two for each designated *quartier*)
of local agents (*experts*). All of these administrators were elected by the
membership to serve for three years, and were re-electable for unlimited
periods of three years. None were paid, except the secretary who was given an
honorarium of two francs for his work on each occasion an insured animal
died. The four officers ran the society's main business, organised its meetings,
and kept its records and accounts; its *experts* monitored its interests on a day-
to-day and local level. It was the responsibility of the *experts* to check
that insured animals were carefully looked after and to report any contrary

Photograph 4.1 A farming family, farmhouse and livestock in the Sologne
Source: 3 Fi 5882 photo, Archives Départementales de Loir-et-Cher

Photograph 4.2 A *paysanne* with her valuable cow, worth in 1900 about half of the
annual expenses of a family of five
Source: private collection, the author

indications to the president so that the committee could investigate and, if necessary, warn or even expel a member for not giving his animals appropriate care and attention. To be put on to the register for insurance, an animal had to be approved by two *experts* (but they were not allowed to admit on to the register any heifer aged less than one year): insurance cover for cows of new members would not commence until ten days after the *experts* had approved the animal, but the delay was only for one day in the case of cows of newly insured animals of existing members. If the *experts* declined to approve an animal for cover because of its poor condition, they had to inform the president and secretary, who, with the agreement of the *experts,* could pronounce the member's suspension from the society until the animal was in a condition to be approved.

Members of the society at Saint-Claude-de-Diray did not pay a fixed insurance premium: instead, they were required to make payments when a claim by a member was recognised. When an animal insured with the society became ill, the owner had to summon the two *experts* of his *quartier*; if they agreed, with the owner, that it was necessary to call a veterinary doctor, then the owner would be issued with a certificate to that effect, because the costs of two such certified vet's visits would be met by the society if the animal were to die and the certificate could be presented to the treasurer for reimbursement. Whenever possible, the decision to call the vet was to be taken after consulting several of the neighbouring members of the society. Whenever the *experts* and the neighbouring members agreed that there was no hope of saving the animal and that it would die before a vet arrived, they had the authority to order the animal to be killed on the spot, in order to preserve the quality of its meat, and the same procedure was followed in relation to animals found dead at their grazing stakes in the fields. In either case, the secretary had then to be alerted because it was his responsibility to fix a time when the meat would be distributed among the society's members. It was the responsibility of the owner of the dead animal to inform the *experts* about the agreed time, and in turn their role to inform the society's members. It was also the responsibility of the owner to obtain from the vet a certificate attesting that the meat was safe to eat.

Rules about the distribution of the meat were very precise. Only the four quarters of the carcass were to be weighed and distributed: the rest of the carcass remained the property of the animal's owner. The butcher was instructed to cut the four quarters into as many joints of meat as there were animals insured with the society; each piece of meat was given a number and a corresponding number placed in a box; at the pre-arranged time, the secretary (assisted by the dead animal's owner and by the *expert* who lived furthest away) would hold a lottery; each member of the society would draw a number from the box and then claim and immediately pay the Secretary for the similarly numbered piece of meat, the price to be paid per kilo having been agreed annually at the general meeting at a level intended to provide an indemnity which reimbursed the dead animal's owner as much as possible for his loss. To cover the costs of administering the incident, the sum of five centimes was added to the amount to be paid by each member for his piece of meat. It was expressly stipulated in the society's statutes that the lottery should not last more than four hours and that, at the end of it, the secretary would immediately pay the proceeds to the afflicted member. The principle of mutuality upon which such an association was based was thus made public in its practice. A different procedure operated when the vet declared that the meat was not fit to eat. In such cases, the secretary had the carcass weighed and then sent to each of the *experts* a list of the members in his locality and the sum that each was required to pay to meet the cost of the insurance claim, calculated in proportion to the number of cows each member had insured with the society; the *experts* had to collect the moneys due from their local members and hand them to the secretary within four days of the animal's death, and the secretary was obliged to pay the indemnity to the owner on the following day. The society at Saint-Claude had no fixed capital insured and no fixed or regular premiums: it simply paid an indemnity related to the weight of the dead animal and raised the sum needed by requiring its members to pay for portions of meat (or just equivalent payments if the meat was deemed not to be edible). This society was still operating in this manner at the outbreak of the Great War: in 1898 it had 140 members, with a total of 173 cows insured, but the number of members declined to about 100 by 1914.

Similar, but at least in one significant respect different, rules governed the Assurance mutuelle contre la mortalité des vaches created at Avaray, on the right bank of the Loire, just upstream from Mer, in November 1872: its manuscript statutes were signed by seventy founding members. The society was open to all of the residents of the commune (except, revealingly, tenant farmers practising *la grande culture* – indicating that this was a mutual association intended for small farmers) and to other owners of cows pastured in the commune. The society divided the commune into five *quartiers* (two focused upon the *bourg,* two upon the hamlet of Tertre and one upon that of La Place): each *quartier* had two *commissaires,* the ten of whom (together with a secretary) constituted the association's *conseil d'administration.* The administrative costs of the society were very low, being restricted to the expenses incurred in acquiring registers for maintaining lists and accounts, and the honorarium (of five centimes per insured cow and five centimes on each occasion when a cow died) paid to the secretary for keeping the society's records. The registers recorded the names of members and the identifying mark of each cow. The responsibilities of each of the unpaid *commissaires* in relation to his *quartier* were considerable and included: agreeing the valuations of cows being registered with the association; collecting the payments to be made by members whenever a beast died; verifying the requests for compensation submitted by members; summoning members to the society's annual general meeting and providing that meeting with a report of their activities during the year. Each member was required to pay an initial, fixed premium (*prime*) of two francs per insured cow (which was considered to represent half of one per cent of the average value of a cow) into the society's reserve fund (*fond social*) and would receive a fixed indemnity of 250 fr. on the properly certified death of each insured animal, with payment being made within a week of the catastrophe. As soon as a member's cow fell ill, the member was obliged to inform his *commisssaires,* whose duty it then was to check that the animal was receiving appropriate attention from a vet (and to call in a vet of their own choice if they wished). All of the veterinary costs had to be met by the member (although he could appeal to the association's *conseil d'administration* if costs mounted because of an animal's prolonged illness) and, when an insured animal did die, the owner was allowed to retain its skin by way of compensation for the veterinary costs. When a cow, as result of accident or illness, was declared by a vet to be irrecoverable and its carcass fit to eat, it was killed and the meat sold on behalf of the association (*au profit de l'assurance*). The cash thus received was used to pay the indemnity to the owner; any shortfall between the moneys received from the sale and the indemnity of 250 fr. was met by requiring each member to pay, within a week of the animal's death, a special subscription (*cotisation*) calculated in proportion to the number of animals each member had insured with the society. This association was still operating in this way in 1903, some thirty years after its creation, when it had about fifty members, but it was no longer in existence by 1908.

Judging both from the statutes of societies drawn up at the time of their foundation and from revised versions compiled around 1900 (as required by new legislation), it is clear that all of the societies established by about 1870 functioned very similarly to those at Saint-Claude and at Avaray, with those paying indemnities by requiring their members to purchase meat portions matched in number by those selling the meat openly and requiring their members only to make cash contributions if the money so raised was insufficient to pay the indemnity. The approximately twenty societies established by 1870 had increased to just over fifty by 1898, when significant legislation created a new operational framework for local livestock insurance associations. It is time to consider the basic characteristics of those societies before examining that new legislation and its impact.

The basic characteristics of societies until c. 1900

The essence of the early livestock insurance societies lay in their grouping relatively small numbers of individuals, furthering their mutual interests in a manner which relied considerably upon self-regulation and minimally upon cash transactions. Their deceptively simple role and operation can be shown, however, to have been quite complex, sophisticated and effective in practice.

Insurance cover

At least four out of five of the societies founded by the end of 1897 insured cattle (*race bovine*), predominantly dairy cows; almost all of the others were for horses, donkeys and mules (*race chevaline*), and only two certainly covered both. The insurance generally provided was against the loss incurred by an owner as a result of the death of an animal from natural causes or by accident, or as a result of a serious accident to an animal which meant it could no longer serve the purpose for which it had been acquired.

In some societies (such as Cour-Cheverny, est. 1877; Cheverny, est. 1888; and Cormeray, est. 1888) the cover was extended to include the loss incurred when an animal was compulsorily slaughtered in accordance with public health regulations during an outbreak of some contagious disease. Some societies included cover for animals sold to a butcher and subsequently found to be unsuitable for human consumption (such was the case with the society at Blois, est. 1893). A few societies (such as Candé, est. 1887) paid in full or in part the cost of a vet called to deliver a calf, in full if the calf were to die within one month but only one-half if it survived longer.

Of course, insurance was not available for all animals, whatever their condition: societies could refuse to register those animals presented by an owner for cover but deemed by its officers to be ill or to have some condition which might lead to its death. Some societies (such as Nouan-sur-Loire, est. 1872) specifically excluded from their cover the death of, or serious injury to, an animal as a result of a fire if the owner also had fire insurance, while others (such as Mazangé, est. 1885) excluded such eventualities entirely.

Importantly, most societies retained the right to reduce or even suspend entirely the payment of indemnities if the losses suffered by members in any one year were especially heavy as a result of an epidemic.

Geographical coverage and membership

Most societies restricted their membership to farmers living in a particular commune: they were based upon a geographically identifiable set of individuals who were all known personally to each other. This geographical limitation, itself a means of ensuring the effective operation of an association and especially the policing of its regulations, meant that societies did not have large numbers of members. In 1898, when data began to be collected systematically on these societies, their memberships ranged from 21 in the cow insurance association at Saint-Denis-sur-Loire (est. 1891) to 156 in a similar society at Mont (est. 1872); at that time 80 was the mean membership.

Some societies even had territories which were less than an entire commune, merely covering one or more settlements within it. The earliest society for which there is evidence in Loir-et-Cher, La Mutuelle de Saint-Marc, was created in 1833 to serve cow owners in the two hamlets of Saint-Marc and Montcellereaux in the commune of Mer; by the end of the nineteenth century it had widened its membership and was open to anyone in the commune and even to those in neighbouring hamlets outside the commune. In 1879 a horse insurance society was created for owners in those same two hamlets. Also within Mer, the hamlet of Aulnay had established two societies in 1843, one for cattle, one for cows. At nearby Séris, a horse insurance society was established in 1852 and a cattle insurance society in 1858 for owners living in the *bourg* of the commune; then in 1883 two similar societies were founded for those living in the hamlet of Lussay. Although one cow insurance society was established at Cour-Cheverny in 1877, its membership was concentrated in the southern and eastern portions of the commune and a second society was founded in 1893 to cover the commune's northern and western sections. The three hamlets of Granges, Villierfins and Villejoint in the commune of Blois established a cow insurance society in 1893; that of Bas-Rivière in the same commune founded a similar society in 1898. On the other bank of the Loire, a cow insurance society had been established in the commune of Vineuil in 1864 but another was set up in 1896 for owners living in the hamlet of Noëls.

While it was the norm before 1898 for a society to limit its membership to owners from a single commune or even from a portion of a commune, there were some exceptions. Some societies (such as those at Avaray, est. 1872; at Binas, est. 1880; and at Villermain, est. 1894) were open to livestock owners from outside its commune if within it they regularly grazed their animals. A few societies were open not only to livestock owners in the commune in which the association was based but also to those in neighbouring communes. Such was the case with the cow insurance society established at Naveil in 1868 and with the horse insurance society founded at Saint-Dyé-sur-Loire in 1897.

Functioning and financing

Most societies were administered by a committee, usually comprised of a small, central group of officers (a president, a secretary, and a treasurer) and a larger group of local agents (*commissaires* or *experts*). All of these were elected by the membership at the society's annual general meeting. While the committee ran the society on a day-to-day basis, the membership as whole took strategic decisions at its annual meeting – decisions about the society's statutes, about its financial arrangements, about the composition of the committee. As such, these societies provided their members with an opportunity to acquire experience of democratic principles put into practice locally and in a context which affected them directly. Unlike many of the other voluntary associations to be considered in due course, the livestock insurance societies did not have two categories of members, honorary and ordinary. In the livestock insurance societies all of the members were of the same standing and all were active participants in the affairs of the societies. These were fraternal, mutual societies of (usually small) farmers and were run by themselves: for example, the Société de secours mutuels contre la mortalité des bestiaux founded at Mazangé in 1895 had grown by 1905 to embrace eighty-five members insuring 217 cows (an average per member of 2.5 cows with an average value of 304 fr.): its president had 2 cows insured with the society, its secretary 6 and its treasurer 3.

The affairs of most societies were very carefully regulated. The average society, with about eighty members and perhaps twice as many insured animals, required a good deal of policing by its local agents. Its credibility with its members depended in part upon mutual trust and in part upon effective, and visibly effective, monitoring of all of its activities. Its local agents had a crucial role to play in that process. Most societies operated, as has been observed, over just one commune, but even those societies usually broke their territories down into a number of smaller areas (*sections* or *quartiers*), for each of which two or more agents were given responsibility. Their duties varied from one association to another but they generally included: checking applications for admission to membership of the society; carrying out valuations of livestock to be insured; ensuring that insured animals were properly registered with the society and marked with its insignia; verifying that insured animals were given due care and attention by their owners, and reporting cases of malpractice to the secretary for investigation; attending reportedly sick animals and arranging, if necessary, for a vet to be summoned; reporting an insured animal's death, and thus claim for indemnity, to the secretary; and collecting payments from members. Given that animals could be added to, and withdrawn from, the insurance register throughout the year, for example because of purchases and sales, there was much for the local agents to do. Only rarely (as at Muides, est. 1865, and Cellettes, est. 1893) did a society's statutes explicitly state that the agents must know how to read and write, but it is difficult to envisage the efficient functioning of a society unless that were the case.

It was the role of agents to see that the detailed regulations of the society were being followed by its members. Many societies forbade their members to give a sick animal anything which might endanger the quality of its meat; many expressly stated that owners were only allowed to give powders and smuts (of cereals; *suie*) to their sick beasts.

The cash income of societies was derived from fees payable by members on joining, which were usually viewed as a reserve fund from which some administrative expenses might be met, and from irregular or regular, variable or fixed, premiums or subscriptions from which indemnities would be paid to members who had suffered losses. On being admitted to membership, an owner would normally be required to pay a fee of between fifty centimes and three francs for each animal insured. Members were normally required to engage for at least one year. A small income came from fines imposed upon members for non-attendance at gatherings to distribute meat and for late payment of sums due for a member's share of the meat.

The administrative expenditure was usually small, resulting from the need for registers in which to maintain a record of a society's members, of the live-stock insured with it, and of its accounts, together with honoraria of a few francs paid to its secretary, upon whom usually fell the main burden of running a society. Some societies also paid honoraria to their presidents and vice-presidents if they required them to assist the secretary in organising a share-out of meat. Many societies paid the costs of two visits by a vet in the event of the animal's death, but the animal's owner was expected to pay for any medicines administered.

The main expenditure of a society was the payment of indemnities, which varied between 50 and 100 per cent of the estimated value of an animal. Most societies paid less than 100 per cent in order to encourage their members to look after their animals properly. Some societies simplified their business by having all animals insured for a fixed sum: for example, that at Lestiou (est. 1874) valued all its cows at 200 fr. each and paid indemnities which were usually about two-thirds of that amount. The necessary funds were raised in almost half of the societies established before 1898 by requiring members to purchase a share of the meat from the deceased animal at a price which was usually slightly lower than the local butcher's price (or to pay a cash equiva-lent if the meat was officially deemed to be unsuitable for consumption). In slightly fewer of the societies, the carcass was sold on the open market and only if the funds so raised amounted to less than the indemnity to be paid were their members required to make a supplementary cash payment (unless, of course, the carcass was inedible, in which case they were required to make cash contributions to enable the indemnity to be paid). Payments due when the meat was inedible were calculated in relation to the number (not the value) of the animals each member had insured with the association.

Thus, in four out of five societies established before 1898, the regular cash burden upon their members was neither great nor regular, and when, because funds had to be raised to meet a claim by one of their number, a payment was demanded by the association the real payments due from individuals were kept low because of the contributions made to the overall fund-raising from meat sales. With little surplus money in the peasant economy and with credit not yet available from rural banks, such a method of financing insurance no doubt appealed to the many small farmers of Loir-et-Cher.

It was, nonetheless, a method which imposed some financial burden upon them without warning from time to time, whenever a society's member sub-mitted a claim. One way of reducing that residual uncertainty was for a society to require its members to pay a minimum annual premium, based upon a per-centage of the value of his insured livestock, and then to levy a supplementary premium (but not in excess of an agreed maximum) at the end of the year if necessitated by the level of indemnities to be paid. Societies working on this more modern, cash-based, system developed very gradually in Loir-et-Cher from the early 1880s, and fewer than ten of those established before 1898 functioned in that way. The first to do so, it seems, was the Assurance mutuelle contre la mortalité des vaches founded at Binas in 1880: each member had to pay a minimum of 2 per cent of the value of his insured cows; the indemnities paid were 85 per cent of the losses; if necessary, a supplementary premium could be levied but the total premium in any one year was not to exceed 10 per cent. If the total losses incurred by members could not be indemnified to 85 per cent from a premium levied at 10 per cent, then the indemnity was lowered. It seems that an animal that had died remained the property of the owner. But if an animal had an incurable disease which did not make its meat inedible, the society could require the beast to be slaughtered: the skin of the animal remained the property of the owner, and was not taken into account by the society, but if the owner sold the carcass the price received for it was deducted from the indemnity paid by the society; if the slaughtering and sale of meat were undertaken by the society, then the indemnity was raised to 90 per cent. This society had 89 members in 1898 and 100 by 1905 (maintaining that number until the outbreak of the Great War).

The variable premium of between 2 and 10 per cent allowed for in the statutes of the society at Binas had a much wider range than it in fact employed and a much wider range than allowed for in the statutes of other similarly financed societies. Other societies were run on the assumption that an annual premium of between 1 and 3 per cent of the value of the insured livestock was appropriate and between 1897 and 1913 the society at Binas was managing very effectively on annual premiums within that range, consistently paying indemnities of 80 per cent.

For much of the nineteenth century, then, livestock insurance societies developed as local, participatory voluntary associations: they drew upon the principle of fraternity and in practice provided a relatively simple form of mutual aid. They were grounded within their localities and communities. They offered a means whereby small farmers could protect one of their major capital assets against risk, doing so for the most part without requiring regular or substantial cash payments. What is also remarkable is that these associations developed spontaneously, as grass-roots organisations, and that they did so as popular societies almost outside of, or at least on the margins of, the law. Not until the end of the nineteenth century did they become a matter of serious public, official discussion and, ultimately, promotion.

The great debate of the 1890s

The agricultural crisis in France during the last two decades of the nineteenth century was accompanied by a great debate about how best to counter it and to protect the large proportion of the population still dependent upon agriculture.[14] That debate embraced a wide range of problems and policies, including those relating to the special needs of agriculture in the field of insurance. In part, of course, the debate was conducted nationally but it was also actively prosecuted regionally and locally: for present purposes it seems most appropriate to follow it in the agricultural press and in the administrative records of Loir-et-Cher itself.

During 1895, for example, there appeared a series of articles in *L'Agriculture Pratique de Loir-et-Cher*, the journal of the department's major agricultural syndicate, which had at that time about 4,000 members (and whose wider role will be considered in detail in chapter 8). The issue of 19 February 1895 included a short article by the Marquis de Chauvelin entitled 'L'assurance mutuelle'. Noting that the government had appointed a committee to examine the possibility of establishing a state system of agricultural insurance, in order to reduce the cost of insurance by foregoing the profits made by insurance companies, the Marquis de Chauvelin argued that for some time already the problem had been effectively tackled by mutual insurance societies which had the three advantages of cheap premiums, absolute guarantees against risks, and the attribution of any profits to those insured. He saw these *mutuelles* as in practice being the earliest form of agricultural syndicate and he expressed surprise that the considerable interest in the formation of agricultural syndicates in France at that time had not spilled over into the creation of insurance associations.[15]

In the same issue of *L'Agriculture Pratique* and in that of 28 February 1895, a practising vet writing anonymously argued that a local livestock insurance *mutuelle,* based on a commune, was potentially of particular benefit to small farmers because it provided insurance which was both cheaper and more secure than that offered by more widely based societies or companies operating on fixed premiums. From first-hand knowledge of twenty local mutual associations and an historic survey of their activities over two decades, the vet claimed that the annual, average, livestock losses incurred in *la petite culture* were about 2 per cent of the total value, whereas companies charged fixed premiums which ranged between 3.5 and 7 per cent of the capital insured. In local mutual insurance societies, by contrast, the average premium was 2 per cent or even less, because losses were kept to a minimum by effective local monitoring of the care and attention given to insured animals. Another advantage of the local society was that it paid an indemnity almost immediately after the death of an insured animal. These two articles then described in detail how such a society functioned. The vet emphasised that in order to ensure that owners should maintain an interest in looking after their animals, and to ensure that they would not make a profit from an insurance claim, two conditions had to be met: the estimated value of an animal for insurance purposes had to be less than its real value, and losses were not to be indemnified fully but only by 90 per cent, leaving 10 per cent to the association to accumulate as a reserve fund for use when an epidemic caused unusually heavy losses for its members.[16]

The anonymous vet's advocacy of the commune-based society was challenged by the Marquis de Chauvelin, on the grounds that such a society would have to have a variable premium because it would have few members and would cover a small area, so that even a minor epidemic affecting its livestock would result in considerable costs to its members because of the higher premium thus necessitated. He argued that a 'good' association would be one in which there were sufficient members to permit all indemnities to be fully paid within known premium limits, but an 'excellent' society would have sufficient reserves to make variable (i.e. higher) premiums unnecessary. Livestock insurance, the Marquis argued, had to be based on a wide geographical territory and that it was accordingly impossible to imagine it working in a commune-based society.[17] The anonymous vet responded that some such societies had existed for twenty years, some even for forty years, and that their persistence indicated that their members were satisfied with them. He saw no objection to variable premiums (thereby accepting that such a system of insurance reduced but did not eliminate uncertainty), but he recognised that contagious illnesses presented local societies with considerable difficulties, so he argued that they should exclude such risks from their cover (thereby again accepting that such societies reduced but did not eliminate risk).[18] The Marquis in turn responded that the need was for a system of livestock insurance which covered all risks, including epidemics, and which did so with a fixed premium: such a system required a wider spatial territory than a single commune and he argued for societies based instead upon groupings of communes.[19]

The fundamentals of the debate were to occupy the department's administration, in the form of its professor of agriculture and members of its Conseil Général, for some considerable time from 1898 onwards, prompted by central government. On 15 April 1898 the Ministry of Agriculture sent a circular to prefects announcing that subsidies were to be made available to *les sociétés d'assurances mutuelles agricoles*. The Ministry had come to be increasingly dissatisfied with the long-standing system of compensating individual farmers who had experienced exceptional agricultural losses, because that system had many deficiencies (the sums awarded were only a small fraction of the losses sustained, the payments were often made after considerable delays because of the administrative formalities involved, and they frequently were counter-productive by creating jealousies and complaints within farming communities which became a source of embarrassment for both local and central authorities). While the compensation system would nonetheless continue for some time, the funds hitherto allocated for that purpose would increasingly be used to subsidise mutual insurance societies. The Ministry urged prefects to encourage farmers to join existing local livestock insurance societies and to form new ones where none already existed. The prefects were to be helped by their departmental professors of agriculture and the Minister suggested that prefects should also encourage any agricultural syndicates existing in their departments to set up *caisses d'assurances mutuelles* for their members. Given that it was not the Minister's intention to replace private responsibility and capital by state funding, subsidies would only be awarded to associations levying appropriate charges upon their members and paying appropriately high indemnities. Any request for a subsidy submitted by the prefect on behalf of a society had, therefore, to be accompanied by a copy of its statutes and accounts. In addition, the Ministry would now be requiring from the prefects returns of information about the societies in their departments on 1 January each year.[20] Similar ministerial circulars encouraging the formation of livestock insurance societies were sent out on 4 February 1899 and 18 January 1900.

The prefect of Loir-et-Cher responded to the April 1898 circular from the Ministry of Agriculture with no great sense of urgency: it was not until 20 September 1898 that he revealed its contents to the sub-prefects and mayors of the department. At its meeting on 25 August 1898, however, the Conseil Général had agreed that the prefect should examine the idea of setting up a *caisse départementale d'assurances contre la mortalité du bétail* and the matter was then taken up by M. Vezin, the department's professor of agriculture. He produced a lengthy report, dated 10 April 1899, on the whole question of agricultural livestock insurance. Surveying three systems of insuring, he dismissed two and strongly favoured one. He had a low opinion of insurance companies, because their high administrative costs and their need to make profits for their shareholders meant that the cost of insuring with them was high in relation to the benefits they offered, and he similarly disapproved of the *caisses départementales d'assurances* because of their disproportionately high administrative costs, which he argued could amount to as much as one-third of the premiums charged. Instead, he strongly favoured mutual insurance societies because they provided their members with maximum benefits at minimum costs. Acknowledging both that effective insurance provision depended upon grouping together a large number of individuals and that administrative and policing costs were lower the smaller the geographical area covered by the grouping, M. Vezin argued that the best way of resolving that dilemma was to encourage both the growth of local mutual societies and their linkage to a central reinsurance fund, by forming a union of the local societies. He was of the view such a union would best be established by the Syndicat des Agriculteurs de Loir-et-Cher, with the help of grants from the State and the department.

At its session in April 1899, the Conseil Général was informed by the prefect that only two departments in France had set up *caisses départementales* for livestock, that Loir-et-Cher's professor of agriculture favoured the promotion of local mutual insurance societies linked into a union which provided reinsurance, and that the department already had forty local societies which were operating satisfactorily. At the August session that year, he additionally reported that the Syndicat des Agriculteurs de Loir-et-Cher was examining the idea of establishing a *caisse de réassurance*. The issue continued to be debated for some years and M. Vezin again reported in great detail to the prefect in February 1902, with a printed copy of that report being distributed to each member of the Conseil Général.

M. Vezin recalled in his report that in January 1900 the Syndicat des Agriculteurs de Loir-et-Cher had decided to encourage the formation of local livestock insurance societies in the department but reported that a campaign by its director had not been very productive, with only two such new societies having been formed during the previous eighteen months. He argued that the whole question needed thorough re-examination, to find the best method of insurance which would provide for the reimbursement of the maximum proportion possible of the losses sustained, would distribute the costs of insurance equitably, and would have minimum administrative costs. He argued that some societies (not necessarily those in Loir-et-Cher) had not been successful because on some occasions they had only paid out indemnities of between 30 and 50 per cent and on other occasions had required premiums of between 5 and 11 per cent. Some societies had been salvaged by state subsidies, but given the limited budget for that purpose and the growing number of insurance associations, it was vital for them to be self-funding. M. Vezin therefore undertook a thorough review of the different methods of providing livestock insurance. He acknowledged that the provision of livestock insurance was in essence a geographical problem, so he considered in turn basing a society upon a commune, a canton, an *arrondissement* and a department, and the advantages and limitations of each which he identified may be summarised as follows:

> commune societies: easy to undertake livestock valuations and to monitor the care given to animals; possible to limit losses by deciding to slaughter an animal thought to have a contagious disease; easy to carry out accurate and fair evaluations of losses, eliminating (or at least significantly reducing) speculation and fraud; self-regulation of activities by members; low administrative costs; low amount of capital insured; premiums had to be variable, to enable unusually high losses in some years to be indemnified; exceptionally high losses might require indemnities to be suspended at a time when losses being suffered by members are greatest; limitation to a commune increased the risk of high losses as a result of contagious diseases.

> cantonal societies: risks more widely spread geographically, it being unlikely that members distributed throughout a canton would all sustain the same level of losses; the capital sum insured would be of sufficient size to nor-

mally permit a balance between losses to be met and premiums to be paid, so that premium variability would not be great; administrative costs of an agent needed to deal with the association's business throughout the communes of the canton (recruiting members, collecting premiums, allocating indemnities, etc.); policing of the association's rules more difficult than in an association based on a commune; still liable exceptionally to sustain serious losses through epidemics or contagious diseases, requiring the payment of higher premiums or of lower indemnities, thereby disappointing and disillusioning members.

societies based on *arrondissements* or departments: would have higher administrative costs (for example, of surveillance and of accounting) than a society based upon a canton; would provide greater protection in times of heavy losses resulting from an epidemic than that offered by a society operating over a smaller area; even so, such protection was not absolutely guaranteed, because epidemics could have an impact over an entire department.

On the basis of this analysis, Vezin concluded that the soundest economical arrangement would be provided by a society based upon a single commune, or upon two or three communes, but for such a society to provide satisfactory insurance (i.e. to have a low and hardly variable premium combined with a high and invariable indemnity, of say 80 per cent) it would have to be linked to other similar societies, either through a *caisse de réassurance départementale* or through an even geographically wider federation of local societies. Vezin, unimpressed by the performance of such *caisses* elsewhere in France because with insufficient reserves they had not been able to guarantee the payment of indemnities, favoured the idea of federating local societies. A union so formed spread risk among a large number of members over a wide geographical area, distributing losses among all of the federated societies. Vezin went to La Rochelle to visit the offices of such a federation established in 1898, L'Union Féderale des Associations Cantonales et Communales de France, and he subsequently engaged in correspondence with M. Héronneaux, its director. Vezin himself was clearly keen to promote local societies based upon single communes but Héronneaux argued that it was impossible to admit such narrowly based societies into his federation, because they covered too small a territory and had too little capital insured, so that they would bring too much risk into the federation. For a society to be admitted to his federation, Héronneaux argued that it would need to have about 1,500 to 2,000 animals, or a capital value insured of about 300,000 francs. He therefore advised Vezin either to see that societies based upon a single commune or upon two or three communes were established throughout the whole of Loir-et-Cher or to group all of the societies existing within a canton or *arrondissement*. Vezin concluded his massive report by saying that three systems of insurance deserved attention of the Conseil Général: firstly, that based on single-commune societies, a simple and economical system but with the serious limitation of variable premiums and indemnities; secondly, that based on commune societies backed up by a *caisse de réassurance départementale*, a system which had some of the advantages of the former but which could not provide adequate guarantees to cover losses in calamitous years; and third, a cantonal federation of commune-based societies, a system which possessed the advantages of both of the other systems without their limitations. Vezin urged the Conseil Général to provide both moral and financial support for the promotion of livestock insurance societies in Loir-et-Cher where annual livestock losses, he claimed, often amounted to 1.5 million francs.[21]

In his campaign to promote livestock insurance societies, Vezin was supported not only by the Conseil Général but also by the two other professors of agriculture, based one each at Romorantin and Vendôme, and by the Syndicat des Agriculteurs de Loir-et-Cher. Model statutes were prepared both by the department's professor of agriculture and by that Syndicat, as aids to those interested in forming such an association. The campaign, conducted for more than ten years in public lectures and in a variety of publications, gradually came to have a significant impact but it was not until 1910 that Vezin was willing to express the view that the livestock farmers of the department, who had in his view hitherto shown little interest in mutualist solidarity in general and in insurance societies in particular, had at last come to understand that mutual insurance was the only way of obtaining a guarantee against risks at minimum cost.[22] As an enthusiast for the cause of livestock insurance, Vezin's assessment was perhaps unduly pessimistic, reflecting his own disappointment: it appears to have ignored the extent to which insurance societies had developed spontaneously during the nineteenth century and to have undervalued the massive expansion in their number during the early years of the twentieth century, in some measure no doubt in response to the stimulus of his own efforts: between 1902 and 1912 the number of new societies established was 100. Perhaps Vezin's disappointment stemmed from the fact in his terms the farmers of the Loir-et-Cher had shown little interest in the idea of a federation of insurance societies: by 1913 there were in the department only six cantonal societies belonging to the Union Fédérale des Associations Cantonales et Communales de France and it was not until 1913 that he was able to persuade a large meeting of farmers that a federation of local societies should be established in the department.[23]

The distinctive characteristics of societies c. 1900–1914

Those societies founded after *c.* 1900 can be viewed as a new generation: while they no doubt to some extent represented a continuation of the nineteenth-century movement, they also reflected the changed context in which they developed, promoted by authorities within the department and subsidised by grants from the State. The intention here is not to examine the basic characteristics of these societies, as was done in the case of those created in the nineteenth century, but instead to focus upon the novel and more importantly the significantly distinctive new wave of societies. The statutes of the post-1900 societies tended to be much more detailed and comprehensive than had previously been the case, sometimes running to a dozen or so closely printed pages (as at Billy, est. 1908) or to almost 100 articles (Brévainville 1905).

Insurance cover

Of the 128 livestock insurance societies clearly established between 1898 and 1913, almost equal proportions catered either for cattle or for horses (29% and 30% respectively); a few (7%) were for cattle and donkeys; but most of these new societies offered insurance for both cattle and horses (34%). Some societies made it clear that they were intended for agricultural livestock and excluded others such as riding horses, racehorses and commercial haulage horses (Avaray, 1907; Billy, 1908). Some associations explicitly excluded livestock traders (Coulommiers, 1901). Some explicitly excluded hacks (hired horses) from being admitted into cover and made it clear that any member who hired out his horse after joining the society would only be covered in that case for an indemnity of 30 per cent (Blois, 1910; Cellettes, 1909).

Some societies specified the age ranges of animals accepted for insurance purposes (Avaray, 1907; Blois, 1900); some specified the maximum value for which particular kinds of animal could be insured (Avaray, 1907; Averdon, 1912; Chaumont-sur-Loire, 1910). Societies generally reserved the right not to provide cover or indemnities for the death of or injuries to an animal resulting from a lack of care on the part of its owner. Some societies excluded from their cover the death of an animal as a result of *force majeure*, such as war, riot, flood, lightning, fire, the collapse of buildings and transportation on the railway (Avaray, 1907; Blois, 1910; Dhuizon, 1909). In some cases death resulting from an animal's castration was only covered if it had been carried out by a state-qualified vet (Avaray, 1907); in others it was partially covered even if the owner undertook the task, provided that he had alerted his local *expert* in advance (Blois, 1900). During an epidemic or while a contagious disease was officially recognised in a society's commune, no new animal could be accepted for insurance cover. If a horse died outside the society's own area of operation, an indemnity would only be paid if the owner produced a death certificate signed by a vet, stating the cause of death, and/or by the mayor of the commune in which the animal had died (Blois, 1900; Chailles, 1910; Chouzy-sur-Cisse, 1908).

Geographical coverage and membership

A few societies were still being established during the early 1900s covering only sections of a commune: for example, a society for *bétail* appears to have been founded for owners in the hamlet of Herbilly in the commune of Mer in 1904; the three hamlets of Granges, Villejoint and Villiersfin in the commune of Blois established a horse insurance society in 1910, and a society covering horses, donkeys and mules was created for the *section* of Les Noëls in the commune of Vineuil in 1911. But these were exceptions. Most of the societies created between 1898 and 1913 were single-commune based, and there was also a detectable tendency towards an enlargement of the geographical area covered by an individual society.

A few societies were willing from their foundation to admit local livestock owners from outside of their commune of operation: that established at Cellettes in 1909 was for owners living in that commune and for owners living within a radius of five kilometres from its *bourg*, and quite a few societies created for a particular commune were also from their beginnings open to owners in unspecified neighbouring communes or hamlets (Chailles, 1910; Huisseau-en-Beauce, 1903; Les Montils, 1902 and 1904; Nouan-sur-Loire, 1903; Talcy, 1907). Some societies initially established in relation to a single commune subsequently expanded their field of activity to include owners from surrounding communes. For example, that established at Chambon-sur-Cisse in 1900 included members from four neighbouring communes by 1907.

Some societies from the outset were groupings of owners in two or three named communes: such was the society created in 1904 for the inhabitants of Concriers, Talcy and Roches. (Also Contres, 1910; Contres, 1912; Faye, 1909; Gy, 1906; Monteaux, 1911; Montrichard, 1911; Le Plessis-Dorin, 1909; Thoury, 1899). In a few cases the grouping involved four communes (Selommes, 1910 and 1912). But the existence of a strong preference for a very locally based society, with its own activities and the actions of its members being closely monitored, was exemplified in that established in 1910 for livestock owners in the two communes of Chaumont-sur-Loire and Rilly, where from the outset the society was run with a single set of statutes but, because of the size of the territory it covered, by two committees, one for each of the communes, an arrangement which unsurprisingly soon led (in 1911) to conflict between the two sets of members (over the extent to which the society should rely upon vets to check animals being presented for insurance cover).

Some societies were based upon cantons. The earliest was that for the canton of Droué, for cattle and horses, established in 1899. Open not only to farmers but also to merchants throughout the twelve communes of the canton (although livestock merchants could only insure their own animals, not those they were trading), its membership grew steadily from 58 at its foundation to 315 by 1911, from which it then fell back to 296 by 1913. With an average of only 26 members per commune at its peak, La Mutuelle de Droué was never a large society despite the fact that it had no local competition, for none of the canton's twelve communes had its own livestock insurance society. But it functioned satisfactorily, raising subscriptions from members when a claim required it to do so and consistently paying an indemnity of 80 per cent.

The other cantonal societies which developed in Loir-et-Cher were each established as a Caisse de Prévoyance Mutuelle of the Union Fédérale des Associations Cantonales et Communales de France, which had been founded in 1898. The first was that set up at the very end of 1900 for the canton of Blois and which in 1902 extended its area of operations to include the neighbouring canton of Herbault. Similar associations were formed for other cantons: Contres (1902), Mer (1902), Marchenoir (1904), Bracieux (1907), and jointly for Montoire and Savigny-sur-Braye (1909). Each *caisse* insured both cattle and horses aged four months to fifteen years against death, unless caused by neglect or ill-treatment, by castration (unless carried out by a state-qualified vet), or by enemy invasion, civil war, fire, lightning or flooding. It never insured cattle over the age of fifteen but it would exceptionally insure to the age of twenty those animals of the *espèce chevaline* which were in the first of three categories for insurance purposes. Membership of each *caisse* was open to farmers living in the named canton and to those in the neighbouring cantons, provided that any such canton did not already have its own *caisse* and that it belonged to the same zone for premium purposes, as designated by the Union Fédérale. Premiums of fixed but different percentages were paid in relation to animals in five categories, two for cattle (*espèce bovine*) and three for donkeys, mules and horses (*espèce chevaline*) (in 1905 the rates were 1.5 per cent for bullocks and calves (*boeufs* and *veaux*), 1.7 per cent for cows and heifers (*vaches* and *génisses*), and 2.5 per cent, 2.75 per cent and 4.5 per cent for the first, second and third category *espèce chevaline*). Most of each premium went to the *caisse* but part (equivalent to a premium of 0.2 per cent) went to the Union, essentially to contribute to a reinsurance fund. When an insured animal fell ill or had an accident, its owner was required to inform the treasurer of the *caisse* and two of its local *commissaires*, and at his own expense summon a vet. These federated cantonal associations relied upon the local policing of their affairs just as did the communal societies. They did, however, have significantly higher administrative costs than the latter and so new members had to pay an entry fee equivalent to 1 per cent of the value of their livestock. They paid out indemnities of 80 per cent. Losses incurred by an association were expected to be met normally from the premiums paid to its cantonal *caisse* but exceptional losses would be indemnified from the reinsurance fund of the Union. Such an association thus spread its insured risk first among its members throughout the canton and then secondly among the members of the federation of cantons throughout France. It cannot be said, however, that they were spectacularly successful. The membership of the association for Blois and Herbault grew from 9 in 1901 to 258 in 1913; that of Contres from 72 in 1902 to 446; that of Mer from 21 in 1902 to 194; that of Marchenoir from 13 in 1904 to 95; and that of Bracieux from 11 in 1907 to 49, while the association for Montoire and Savigny-sur-Braye began with only 2 members in 1909 and still had only 8 by 1911. The seven cantonal associations in 1913 accounted for slightly more than one-fifth of the total membership of Loir-et-Cher's livestock insurance associations, while the ten cantons which they collectively covered represented slightly more than two-fifths of the department's cantons.

It is, however, noticeable that with only two exceptions (those of Droué and of Montoire and Savigny) all of the cantonal associations straddled the Val de Loire and its immediate environs, stretching from Marchenoir in the north through Mer and Blois cantons on the right bank, and through Bracieux and Contres on the left bank. The location of these cantonal associations was a reinforcement of the overall distribution of communal societies. It is also noteworthy that the cantonal associations were (with two exceptions) located in the *arrondissement* of Blois.

Attempts to promote unions of communal societies at the level of their *arrondissements* were made by the professors of agriculture for Vendôme and Romorantin, and at the level of the department by its principal professor of agriculture at Blois, but none of them appears to have made much progress. Five communal societies established in 1904 (Houssay, Lancé, Morée and Saint-Rimay) belonged to the Union des Sociétés d'Assurance de l'Arrondissement de Vendôme and a few others were added later (Brévainville, 1905; Vendôme, 1907). Local societies within this Union pooled their funds, so that from a financial point of view they constituted a singe insurance society. An attempt to encourage the more explicit form of reinsurance by federating local societies was made with the creation in 1907 of the Fédération des Mutuelles-Bétail de l'Arrondissement de Romorantin. Each local society was autonomous within the federation but at the end of each financial year those societies which had incurred losses of less than 3 per cent of the value of their insured livestock were expected to contribute to the federation sums of money (calculated in proportion to the total value of their insured live-stock, but with a ceiling of 3 per cent embracing both indemnities paid to their own members and the contribution to the federation) which would be distrib-uted proportionately to those societies which had suffered losses in excess of 3 per cent of their insured livestock. At least four communal societies (Vernou, Billy, Lanthenay, Romorantin) joined that Fédération in 1907–8 and one more (Soings) in 1913, but there is no evidence to suggest that it became hugely popular among the farmers of the *arrondissement* as a whole. Similarly, the professor of agriculture at Blois was keen to promote a Fédération Départementale of livestock insurance societies and he included references to such a federation in the model statutes which he prepared to encourage the formation of local societies. But almost all of the local societies which used those statutes when being established in fact deleted the references to the Fédération Départementale, and those few which did not found themselves with articles in their statutes which were inoperable, because such a federation had not actually been established in Loir-et-Cher.

Some societies based upon single communes nonetheless recognised in their statutes that there might be advantages in joining a union of societies in due course but did not in fact do so (Courmemin, 1906). Occasionally, a society's officer when reporting on its business to the prefect claimed that although it was functioning satisfactorily there might be some advantage in joining a federation of societies (Faye, 1909; Membrolles, 1908; Saint-Romain, 1910).

But the fact remains that most livestock insurance societies established between 1898 and 1914 were based upon single communes and that the idea of a more broadly-based federation seems to have had little appeal for the farmers of Loir-et-Cher before 1914. Their conception of fraternity, of mutuality, was community-based, grounded in a very local geography. In consequence, they remained quite small societies. Taking the 81 societies for which it is possible to determine membership numbers in 1913, the average (mean) size of a society was then 87 – remarkably close to the figure of 80 in 1898. If we exclude from the calculation the societies based on cantons or *arrondissements* and consider just those based on one commune or a small group of communes, then the average size of such a society in 1913 falls to 73. The largest was the cattle and horse insurance society established at Saint-Georges-sur-Cher in 1909, with 201 members by 1913; the smallest was the cattle society at Epuisay, with only 8 members in 1913 (which was only three fewer than it had at its foundation in 1903: astonishingly, this society functioned satisfactorily – as its president reported in 1906 – with a membership which was never more than 15 and with indemnities of 75 per cent).

Given the nature of the sources, it is impossible to quantify with accuracy the proportion of Loir-et-Cher's farmers who were members of livestock insurance societies at particular dates. But some 'best estimates' may nonetheless be calculated. The number of societies in existence can be determined for some dates with reasonable accuracy, and multiplying that number by the mean membership of those societies whose sizes can be determined provides an estimate of the total membership. That figure can then be compared with the total number of heads of farming households recorded in some, but by no means all, of the censuses of population. Such a procedure leads to the suggestion that around 1870 about 5 per cent of farmers were members of livestock insurance societies; by around 1890 the proportion had increased to about 10 per cent, and by 1914 to approximately 40 per cent.

Functioning and financing

Although most societies drew up their own statutes, quite a few based them to some degree – and some to a substantial degree – upon model statutes provided by one of the department's three professors of agriculture. The Syndicat des Agriculteurs de Loir-et-Cher also promoted the formation of livestock insurance societies, initially doing so by encouraging its existing members to establish them in their own communes and offering them a system of reinsurance with the Syndicat. One such was the cattle insurance society created at Sargé-sur-Braye in 1900, but on checking its statutes the Ministry of Agriculture pointed out that such societies could only take advantage of state subsidies if they were completely independent of any institution in terms of their organisation and administration. The society at Sargé-sur-Braye accordingly modified its statutes to be independent from the Syndicat and to make it clear that it was open to all livestock owners (as it fact it had been since its foundation). Such a restriction on the role of the Syndicat meant that it was in future limited to providing its members with encouragement and advice about livestock insurance, including supplying model statutes, but it could not itself be more closely involved.

The administrative costs of the societies founded in this period were kept low, largely because those who ran them gave their services without charge (although small honoraria were paid to one or two officers for their work). They were voluntary associations which demanded considerable effort and trust from their members: for example, given that the value of an animal could change during the course of a year, with age, condition and market prices, their valuation was crucial to the proper functioning of a society and it was normal for it to be undertaken regularly once or more times each year by a society's *experts*. The costs of veterinary fees and medicines were met by the owner, not by the society. But these societies came increasingly to be cash-based, deriving the bulk of their incomes from premiums paid by their members and from state subsidies, with entry fees and fines for indiscipline providing additional resources.

On being admitted to a society a new member was usually required to pay an entry fee of the order of 1 per cent of the value of the animals he wished to insure. These admission fees, together with grants awarded by the State, the department and the commune, and legacies and donations from private individuals, constituted a society's reserve. The department's professor of agriculture, M. Vexin, had made it known throughout his territory that these societies could be seen as philanthropic institutions and that they were legally entitled to receive charitable gifts. There is, however, no evidence that such gifts ever constituted a significant part of their revenues.

The most distinctive characteristic of societies founded from 1898 onwards was their move away from demanding from their members uncertain payments at irregular intervals to requiring virtually fixed annual premiums. There was a very clear and very marked shift towards the elimination of uncertainty, to the minimisation of risk, through a system of regular money payments. While the more primitive methods of paying indemnities continued for a while after 1898, they were very soon replaced by more sophisticated methods.

Of the approximately 110 societies founded from 1898 onwards, only 5 (Blois, 1898; Chambon-sur-Cisse, 1900; Les Montils, 1902; Nouan-sur-Loire, 1903; Chouzy-sur-Cisse, 1906) raised the indemnity by requiring their members to purchase portions of meat from the animal's carcass or to pay equivalent sums if the meat were not fit for consumption. The cow insurance society created in 1898 in the hamlet of Bas-Rivière in the commune of Blois had a nuanced system for distributing the meat among its membership, which grew from 42 at its foundation to 58 by 1913. The price of meat per kilo was agreed by the society's committee on four occasions through the year, according to the market; the meat itself would be put into three categories by quality, with meat coming from an animal being within two months from calving or having calved within the previous two months being placed in the first class, and the rest of the meat being placed in the second and third classes at the discretion of the committee. The dead animal's owner was duty-bound to assist the butcher in preparing the meat for distribution, an event which had to take place at the premises of the owner. Each member of the society had to be present (absence incurred a fine of 50 centimes, and failure to present himself within twenty-four hours resulted in the member's being struck off the membership list altogether). The portions of meat were distributed by lot in the customary fashion, with each carcass being cut into as many portions as there were cows insured with the society and each member being required to draw in the lottery, and to pay for, as many portions as he himself had cows insured. From its foundation until 1906, the society at Bas-Rivière paid indemnities of 100 per cent but, following a ministerial requirement for a change of statute, that figure was reduced to 80 per cent thereafter, to enable the association to continue to be eligible for state subsidies. A similarly organised cow insurance association established at Chambon in 1900 with 151 members had by 1907 extended its operations into three adjoining communes and had increased its membership by 1909 to 216 (reduced to 170 by 1912), paying indemnities each year of a little over 70 per cent but managing nonetheless to attract a state subsidy in almost every year. The State did not object to the money for indemnities being raised through such distribution of meat but it did to the payment of indemnities other than at the 80-per-cent rate. The cow insurance society created at Nouan-sur-Loire in 1903 paid indemnities raised in that way but at the full rate of 100 per cent until 1912, and received a state subsidy in almost every year, despite ministerial instructions that indemnities should not exceed 80 per cent. The cow insurance society founded for the two communes of Les Montils and Monthou-sur-Bièvre in 1902 saw its membership grow from 79 to 140 by 1906, when it was also operating in two other communes, but it had fallen back to 112 by 1913. The last meat-distributing society to be founded was that for cows at Chouzy-sur-Cisse, established as late as 1906: its membership grew from 83 to 124 by 1912 (reduced to 117 in 1913). Its statutes exhibited a transitional characteristic, no doubt influenced by an awareness of practices elsewhere and perhaps of ministerial policies, because they stipulated that the cash payments made by members for their shares of meat were to be viewed as annual subscriptions and that an additional subscription would be levied if the sums paid for the meat amounted to less than 1 per cent of the capital value of animals insured with the society.

The transition to insurance based on regular and more-or-less fixed cash premiums is identifiable in two kinds of society. The first in fact represented continuity of practice because, as was the case before 1898, some societies founded after that date required their members to pay no subscription at all (or at most a nominal subscription) until an insurance claim had to be met, but they were not required to purchase portions of meat. The animal's carcass, if fit for human consumption, was instead sold on the open market, with the proceeds generally going into the society's funds and sometimes being credited to the animal's owner and so deducted from the indemnity paid to him. Only if the sum of money thus raised were less than the indemnity due to the owner were members of the society required to make cash payments to make up the shortfall. Whereas about one-third of societies founded before 1898 functioned in this way, raising sums from members in effect only when there was a claim to be indemnified, between 1898 and 1912 the proportion was less than one-tenth and, moreover, most of these societies were in existence by 1904. This system was adopted by societies, for example, for horses at Blois (1900), for horses and cattle at Billy (1908), and for cattle at Concriers (1904). Although such a system of insurance shared risk equitably among a society's members, it did not spread it evenly through time: there remained the uncertainty of being required to make substantial cash payments at irregular, unpredictable intervals. Some societies limited the uncertainty by imposing an upper limit on the subscription which could be raised, of 3 per cent of the member's capital value insured: this was, for example, the rule employed by those societies in the federation for the *arrondissement* of Romorantin.

The second kind of transitional society reduced that uncertainty somewhat by requiring its members to pay both a small, fixed annual premium and a supplementary premium in those years when the losses incurred by members necessitated it. The annual premium was low, between one-fifth and one-half of 1 per cent of the insured capital value (i.e. only 20 to 50 centimes per 100 francs insured), but this had indeed to be supplemented frequently, because the normal level of livestock losses averaged over wide areas and over long time periods was between 1 and 2 per cent of their value. Approaching a dozen such societies seem to have been established in Loir-et-Cher between 1898 and 1912 (for example, at Coulommiers, 1901; Ménars, 1901; Lunay, 1902; Les Montils, 1904; Montlivault, 1907; Talcy, 1907; Fresnes, 1908; and Saint-Georges-sur-Cher, 1909).They seem to have been able to function for the most part satisfactorily, as even did the horse insurance society at Les Montils which had to cope in its first year with an exceptional six claims from among its 74 members, requiring them in practice to pay a premium which amounted to 5 per cent of the insured value of their animals. The desirability of reducing the element of uncertainty about the amount which each member would in practice have to pay presumably lay behind the decision of the society established at Fresnes in 1908 on the basis of an annual premium of 0.2 per cent (i.e. 20 centimes per 100 francs insured) to change its statutes in 1909 and to require an annual premium of 1 per cent.

The really distinctive feature, however, of those societies established between 1898 and 1914 was that most – probably about three-quarters – of them were based on the payment by members of annual cash premiums which varied within an agreed and known range. Each member was expected to pay a minimum premium at the beginning of the financial year and then at the end of the year to pay a second premium, which would vary according to the level of losses sustained during the year and to the level of the society's reserve fund but which would not exceed the agreed maximum percentage. The range of premiums to be paid varied from one society to another and was usually different for cattle and horses, but it was typically between 0.6 and 1.6 per cent for cattle and between 0.7 and 2.4 per cent for horses. Many societies collected an initial premium of 1 per cent and then made a later adjustment, as necessitated by the level of losses.

From 1902 onwards this system of insurance, based upon annual cash premiums which varied around 1 per cent within a defined range, became the norm for societies being newly founded and a model towards which many already established societies gradually moved by modifying their statutes, often urged or even required to do so by local and central authorities. The progressive views of the department's professors of agriculture gradually became better known and more and more societies came to base their statutes on models which they had provided. The professor for the *arrondissement* of Vendôme was, it seems, particularly active and influential during 1903, when five new societies were created on the basis of his model statutes. But the professors were most influential from 1910 onwards, when model statutes drawn up by the department's principal professor of agriculture, based at Blois, for use throughout the department came to be widely accepted at least as an initial framework around which to construct an association. At least two-dozen societies were founded from 1909 to 1913 on that basis.

The Ministry of Agriculture also came to have an increasing impact upon the character of societies. In principle, the award of subsidies by the State was frequently made conditional upon a society's having statutes of which the ministry approved and associations were often instructed to change their statutes to bring them into line with what the Ministry regarded as necessary practice. It came to insist, for example, that in order to be eligible for grants a society had to charge an annual premium of at least 1 per cent of the capital value insured, and to pay indemnities of not more than 80 per cent. In addition, the Ministry wanted the statutes of a society to state that, in the event of its dissolution, the balance of funds remaining in its account after all expenses had been paid would be allocated to 'une oeuvre agricole voisine' (a related agricultural venture) decided upon at a general meeting of the society and then approved by the Ministry. In practice, communication between the Ministry and a society was conducted through the prefect and it was often protracted; moreover, societies were not always willing to change their rules as soon as they were asked by the prefect, on behalf of the Ministry, to do so and the Ministry, for its part, not infrequently awarded a subsidy to a society which had not brought its rules into line with necessary practice despite being asked to do so. Practice was much more untidy than policy intended.

Nonetheless, viewing the period between 1898 and 1914 as a whole, the impact of the Ministry and of the local professors of agriculture was gradually to reduce the diversity of practices among societies. At the beginning of that period, for example, indemnities being paid varied considerably from society to society, ranging from 50 per cent (Cheverny, 1904) to 100 per cent (Blois, 1898), although most societies were paying indemnities of between 70 and 80 per cent (although from that sum was usually deducted any sum raised from selling the animal's skin or meat, if those moneys went to the owner and not to the society). From 1907 onwards most of the newly founded societies adopted the 80-per-cent rate and from 1908 they all did so. Similarly, during these years and then especially from 1910, when the Ministry tightened its control of subsidies, many of the older societies, those established before 1898, also modified their statutes to conform to the 80-per-cent rate.

Subsidies awarded to societies which requested them were calculated not only on the basis of their rules and practices but also in relation to their financial needs and resources. Requests were, for example, sometimes refused on the grounds that a society was not making a sufficient charge upon its own members or on the grounds that the losses it had experienced in the year concerned were not exceptionally heavy. Unsurprisingly, many societies included in their applications for grants complaints about their financial difficulties but, equally unsurprisingly, the Ministry seems to have treated such pleas with equanimity. The overall level of grants made to societies in Loir-et-Cher was not inconsiderable: between April 1909 and April 1914 the almost 200 grants awarded to livestock insurance societies in Loir-et-Cher by the Ministry amounted to almost 900,000 fr., but the sums allocated to individual societies were usually quite small, at 50 fr. on average. State subsidies no doubt did encourage the livestock insurance movement as a whole, but given (from the point of view of an individual society) their unpredictable, often delayed, receipt by the societies and the relatively small sums involved it is difficult to argue that they played a key role in the creation and functioning of very many individual societies.

How successful were these insurance societies perceived to be by contemporaries? Of course, those committed to the project sought perfection and wanted to see all owners of livestock insuring all of their cattle and horses: and as that utopian target was never reached they did from time to time express their frustration at the lack of progress. Thus the president of the cattle insurance society established at Morée in 1904 with 30 members and a capital insured of 24,000 fr. complained to the prefect in June 1907 that that sum represented only about one quarter of the value of livestock in the commune: he whined that such a low proportion showed clearly how few farmers yet understood their own true interests and the difficulties encountered in attempts to overcome their doubts. In fact, by 1907 membership had risen to 55 and the capital value assured to 34,260 fr.; and by 1912 there were 72 members insuring cattle valued at 44,750 fr., so that by then probably about one-half of the farmers and their livestock were within the society's embrace.

But there are also reports of even higher levels of interest and, more importantly, of participation. Thus the cow insurance society established at Ménars in 1901 had 21 members by 1903 when it reported to the prefect that that number included all except two of the cow owners in the commune. At Fresnes in 1909, the mayor reported to the prefect that almost all of the commune's farmers were members of the horse and cattle insurance society which had been created the previous year: its membership in fact grew from 83 in 1908 to 103 in 1901, falling back to 88 by 1913. In 1908, the mayor's letter to the prefect in support of a request for a subsidy for the association pointed out that almost all of its members were small tenant farmers or owner-occupiers who could be heavily hit by the least *sinistre* (calamity).

From the insurance of peasants' livestock it is now appropriate to turn to the question of their own, personal insurance. If the livelihood of a peasant household depended considerably upon the welfare of its livestock, it was also highly dependent upon the welfare of its head. The incapacity of a peasant farmer, whether through illness or accident, was another risk which could be managed to some extent by the formation of another kind of voluntary association, the mutual aid society. It was clearly the case that farmers came increasingly to insure their livestock against serious injury and death, and they did so by forming and joining voluntary associations explicitly and exclusively for that purpose. To some extent, the successful development of insurance societies in relation to one particular risk could be expected to have encouraged the growth of similar associations in relation to other risks.[24] Farmers did also become increasingly aware of the need to protect their own persons, to counter the uncertainty of illness or injury which would prevent them from working their farms. For this, they developed another set of voluntary associations, the mutual aid societies.

5

Mutual aid societies

'L'Union, l'Amitié, la Paix, l'Égalité et la Fraternité'

Slogan of the Société des Amis Réunis de Contres (1848)

Context

The historiographical context

The history and geography of mutual aid societies (*sociétés de secours mutuels*) in France remain largely ignored and unknown. It has recently been argued by Allan Mitchell, an historian, that 'one of the obvious lacunae in the historiography of modern France is the absence of a thorough study of mutual aid societies, which played a crucial part in public health throughout the nineteenth century and for much of the twentieth'.[1] Although it is recognised that in France by 1911 almost 20,000 mutual aid societies embraced almost 3,500,000 members,[2] quite recent studies of the mutualist movement still admit that its evolution remains a 'puzzle' and its social impact 'largely unexplored',[3] that its long and complex history remains poorly known,[4] that its history is only now coming to be accessible,[5] and that in particular few studies have been devoted to the creation and development of mutual aid societies in rural, agricultural contexts.[6]

Perhaps the best recent general account of the history and geography of mutual aid societies is that provided by André Gueslin as an integral part of his wider survey of social welfare in France during the nineteenth century.[7] But the diversity of the origins, forms, functions and evolution of mutual aid societies which he emphasised served itself to highlight the need for more local studies to put flesh on to the bones of the national skeleton. Such studies are gradually being produced and in due course will almost certainly necessitate a revision of the current orthodoxy about the history and geography of mutual aid societies.[8] This present study of mutual aid societies in rural Loir-et-Cher, for example, takes as its point of departure that orthodoxy but will have seriously questioned at least one component of it by the time it reaches its destination.

The fundamental aim of a mutual aid society, it is generally agreed, was to protect its members against the consequences of not being able to work either because of catastrophic, unexpected risks (such as illness or invalidity through accident or injury) or because of life-cycle, expected hazards (such as maternity, old age and death). A mutual aid society achieved its aims by providing its members with medical diagnosis and medicines (free or at a reduced cost), with a cash indemnity during the recognised period of illness, incapacity or retirement, and with a contribution towards the funeral expenses of a member. In addition to providing these material benefits, a society offered opportunities for sociability, for both public demonstrations of fraternity (for example, by attendance at a member's funeral by other members of the society and by attendance at the Mass to celebrate the festival of a society's patron saint) and for private affirmations of solidarity (for example, through visiting the sick and at occasions arranged mainly for social drinking and eating). Furthermore, many promoters of mutual aid societies saw them as a source of social integration, as a way to spread moral values about thrift, order and responsibility, and as means of encouraging 'good' behaviour by excluding from membership those who engaged in 'bad' conduct such as brawling, thieving and drinking. Finally, mutual aid societies are seen as contributing to the development of working-class consciousness and as providing one limited way in which workers could defend themselves and offer some resistance to economic and social exploitation in a period when other workers' organisations were closely monitored or even repressed.

Although the orthodox view of mutual aid societies has come to recognise that they had roots in the *confréries* (confraternities) and *compagnonnages* (trade guilds) of the eighteenth and early nineteenth centuries, it continues to emphasise their modernity as fundamentally urban, industrial organisations which played a role in the formation and development of class consciousness among the better-paid craftsmen and male workers. It has been argued, for example, that the functional flexibility and diversity of mutual aid societies enabled them to play an important transitional role between the traditional *compagnonnages* and the modern *syndicats professionnels* (trades unions) which were to develop in the late nineteenth century.[9] Similarly, although many mutual aid societies initially drew inspiration from religious ideas and practices, they became increasingly secularised during the nineteenth century and increasingly expressive of the growing discourse on socialism and solidarity.[10] But they were also, from the mid-nineteenth century onwards, encouraged by the State, which saw mutual aid societies as promoters of social order, as potentially effective intermediaries between the individual and the State. That alliance between the State and the mutualist movement, it is argued, was reflected in the exceptional growth in the number of societies in the years immediately following new, significant legislation favouring their formation (particularly laws passed in 1852 and 1898, which will be considered in detail shortly).

Thus the orthodox view is that mutual aid societies were an integral part of the urbanisation and industrialisation of France during the nineteenth century, and of the development of urban working-class organisation and consciousness. Until 1830, such societies were only to be found in Paris and in the largest towns like Lille, Marseille and Grenoble. At mid-century – by which time there were about 2,500 societies and about 400,000 members – the mutualist movement was concentrated in urban and industrial France, that is to say, in Paris, the Nord region, the Lyon region, and in major urban centres like Marseille. Although by 1910 the geographical pattern of mutual aid societies had changed in detail, the principle underlying it had not: looking at France as a whole, most mutual aid societies were still small and in urban and industrial locations, while there were few of any size in rural France. By 1910, about one-third of societies still had fewer than 50 members, another third fewer than 100, and fewer than five-hundred societies had more than 1,000 members. Approximately three-quarters of the communes of France did not then have a mutual aid society and they were largely absent from the rural regions of the country.[11]

The conventional wisdom is that, despite their diversity and dynamic character during the nineteenth century, mutual aid societies in France were fundamentally 'voluntary associations of like-minded wage earners' in urban centres of commerce and manufacturing.[12] By implication, the relative paucity of concentrations of wage earners in rural, pre-modern France can thus be called to account for the scarcity of mutual aid societies in its smaller towns, villages and hamlets. But a more profound interpretation has also been offered, emphasising the fundamental role of peasant individualism. Both contemporaries and historians have argued for a long time that it was the inertia and egoism of individualistically inclined peasants that checked the spread of mutual aid societies in rural France.[13]

That view is now being challenged. André Gueslin, for example, argues that peasant individualism itself is not an adequate explanation of the virtual absence of mutual aid societies from rural France, except perhaps in the short term, because it ignores the fact that by the end of the nineteenth century other rural welfare societies of various kinds were flourishing. He argues instead that the tradition of professional *confréries* which mutual aid societies extended was essentially urban, and that spontaneous expressions of solidarity could have long-term, productive consequences in urban settings while in rural contexts they checked the development of organised societies. The success of mutual aid societies in towns stemmed, according to Gueslin, from their multiple functions, which were perhaps less useful in rural contexts. Moreover the kinds of formal assistance offered to individuals by mutual aid societies were traditionally provided informally in rural areas by their relatives and neighbours. Additionally, he suggests that while the *notables* of rural France were much engaged not only in agricultural innovation but also in promoting societies related to the development of agriculture, they were much less inclined to become involved with an institution which was not directly part of the agrarian system.[14] Similarly, Solange Goldman has argued that while ideas about mutuality and insurance diffused readily in the towns of nineteenth-century France, they encountered many obstacles to their development in the countryside: the dispersed nature of rural settlements; the small population sizes of rural communes; their remoteness from doctors; the high costs of medicines; the limited use of cash in the economy; and especially, the depopulation of the countryside. The first attempts to promote them in rural areas were, according to Goldman, taken by the departments' professors of agriculture and by the *syndicats agricoles* (agricultural associations) themselves established in the wake of the law of 21 March 1884.[15]

All of which leads one to expect to find few mutual aid societies in rural France during most of the nineteenth century and it is indeed the case that the recent, major discussions of the French peasantry during the nineteenth century make no reference at all to the role of mutual aid societies.[16] But before searching for any in Loir-et-Cher, it is first necessary to become acquainted with the legal framework and to recall the political discourse within which such societies would have developed.

The legal context

Given the post-revolutionary dominance of the principle of the liberty of the individual and the continuing impact of the Enlightenment's opposition to social institutions intermediate between the State and the individual, it is hardly surprising that for a considerable part of the nineteenth century the State's attitude towards mutual aid societies was ambivalent, repressing them when they were seen to present political threats and to be societies of resistance, and encouraging them when they were recognised as promoters of social order and welfare. During the first half of the nineteenth century, however, the widely held view was that the State ought not to intervene directly in the field of social welfare because to do so would be to infringe upon the liberty of the individual. For most of that period, therefore, such societies were tolerated rather than being legally recognised and encouraged; moreover, they were policed when necessary within the clearly defined legal framework provided by articles 291–4 of the Penal Code of 1810.

The 1848 Revolution brought temporarily to the surface a developing discourse about the role of popular associations, and – with the repeal of article 291 of the Penal Code of 1810 – the freedom to establish mutual aid societies without preliminary formalities. But the return to a conservative liberalism after June 1848 left unanswered questions about the extent to which, and the ways in which, the State should intervene in the field of social welfare, as well as questions about the roles of popular associations in that arena. The *coup d'état* of 1851 put a brake on the development of fraternal associations in general, on the grounds that many of them had been hearths of socialism or at least of republicanism, and a decree of 25 March 1852 re-established the repressive and restrictive legislation of article 291 of the Penal Code of 1810. Even so, in relation to the question of social welfare, the debate continued between liberal thought committed to the principle of the liberty of the individual and opposed to the notion of state intervention to protect individuals, and socialist thought which – in its various guises – wrestled with the relation between the State and society and often concluded that the former had some responsibility for the condition of the individuals who constituted the latter.

Gradually during the second half of the nineteenth century new legislation reflected a perceptible shift in general attitudes towards the problem of social welfare, away from the supremacy of the principle of the freedom of the individual in such matters towards that of limited state intervention through the provision of financial aid for private initiatives, and then progressively towards the creation of a system of social welfare based upon legal obligations. In effect, until the middle of the nineteenth century mutual aid societies evolved somewhat pragmatically and empirically, rather than being fashioned ideologically and theoretically through legislation. Then during the second half of the century their position and standing were rationalised, their role and functions clarified and regularised.

That process really began during the Second Republic. Although in so many ways he was to prove himself to be an authoritarian and a military dictator, Louis Napoléon Bonaparte – even before he became Napoléon III at the head of the Second Empire – had a genuine interest in problems and issues of social welfare. In 1844 he had published a book on the extinction of pauperism and he became an influential advocate and supporter of mutual aid societies, seeing them both as a means whereby differences among social classes might be reconciled and as a way of encouraging prudence, thrift and good conduct by individuals. In the wake of the 1848 Revolution, and in the midst of the dislocations associated with the developing industrialisation of the French economy, there was wide interest in ways of promoting social stability and of tackling questions of social inequality which would enable both state and private initiatives to play a part. Gradually, a framework of new legislation was constructed.

Key legislation was passed in the early 1850s. The law of 15 July 1850 envisaged three kinds of mutual aid society: first, *sociétés libres* (free societies), which could be established simply by lodging a set of their statutes with the authorities and which would not be monitored by the State, although of course such societies had to operate in conformity with the general law; secondly, *sociétés approuvées* (approved societies), whose statutes had to have ministerial approval, were monitored by the authorities and were eligible to receive grants from the State and from the department; and thirdly, *sociétés reconnues* (recognised societies), those acknowledged as being *d'utilité publique* (in the public interest) on the grounds that they provided temporary assistance to their members when they were unable to work because of illness, injury or infirmity. That legal status entitled recognised societies to receive bequests and to benefit from the protection of the State. Commune councils were required to provide free to such societies a meeting place and the books and registers necessary for their functioning, while the State itself exempted them from stamp and registration duties. In return for such support and concessions, a society had to accept certain restrictions on its activities: the commune's mayor had the right not only to attend but also to chair its meetings; the statutes of a society had to be approved by the government in Paris; the provision of assistance to unemployed members, or of pensions to retired members, was not permitted; and the operational rules of a society had to be based on tables of morbidity and mortality. Such societies were to have a minimum of 100 members (although the figure could be lower in rural communes) and a maximum of 2,000 members. Given that the law imposed such restrictions while providing no practical incentive to the formation of mutual aid societies, it is hardly surprising that it had almost no impact, especially when one takes into account the relative complexity of the statistical calculations upon which such societies were meant to be founded.[17]

During the early years of the Second Empire, debate continued about how best to tackle the problem of social welfare and insurance. Napoléon III considered that mutuality should be compulsory. Social Catholics, aware that many workers could not afford to pay the necessary premiums, argued in favour of relying upon the patronage and subscriptions of honorary members, and upon gifts and legacies, which it was thought would have the added, moral advantage of creating links among the classes and so of undermining socialism. A compromise solution was eventually expressed in the decree of 26 March 1852, which modified earlier legislation relating to mutual aid societies. While allowing for the continued existence both of 'free societies' as straightforward associations with no legal rights beyond that of being permitted to keep their funds in savings banks and of societies recognised as 'being in the public interest', the 1852 decree favoured the creation of 'approved societies' which would be entitled to receive subsidies from the State as well as legacies from benefactors and subscriptions from both honorary and ordinary members. It was for the mayor and priest to promote the formation of such a society in their commune and to submit a proposal to the prefect of the department for his – and ultimately the government's – approval. The president of a society was to be appointed by the Head of State. The rules and conditions for running such a society were effectively those set out in the law of 1850, although the limitation on the number of members a society could have was removed. The societies were not allowed to own property and they could only have as objectives the provision of temporary assistance to members who were sick, injured or disabled and the payment of the funeral expenses of a deceased member. A decree of 16 April 1856 permitted approved societies to extend their role by establishing pension funds.

The legislation of the 1850s established the official but voluntaristic framework within which mutual aid societies developed and spread in France during the second half of the nineteenth century. There were some modifications (for example, a decree of 27 October 1870 provided for the election of presidents of societies by their own members) but that general framework remained in place almost until the end of the century. But from the early 1880s there developed both an increasing pressure for more independence to be given to the societies and a growing body of other legislation which reinforced their role in offering some degree of social protection. The outcome was the law of 1 April 1898, which released societies from close administrative control and extended their field of activities: societies could be formed without having to seek prior approval (they merely had to deposit a copy of their statutes with the prefect) and they could provide life insurance, pay benefits even to their unemployed members, and also set up work training schemes and professional courses. Moreover, mutuality could henceforth embrace every family member and not just the head of a household. The 1898 law applied not only to societies based upon communes but also to those on factories and even to those created within an agricultural syndicate itself established under the law of 1884. Although the 1898 law relaxed the surveillance of societies by local and central authorities, it did impose greater control over their financial operations in order to ensure that the benefits promised by a society could be appropriately funded, so that this restriction was imposed in order to give greater protection to a society's members.

Other laws of the 1890s also need to be noted, both because they had a direct impact upon the fields of activity which had been covered by mutual aid societies and because they constituted significant steps away from reliance on the provision of social welfare upon private initiatives encouraged by state aid and towards the creation of legal obligations. For example, the law of 15 July 1893 gave to the poor a right to free medical aid, each commune being required to establish a *bureau de bienfaisance* (welfare bureau) to maintain lists of those judged entitled to aid and to arrange for the payment of doctors and nurses; and the law of 9 April 1898 on compensation for accidents at work established the employers' liability for accidents at work. The latter law was initially applied only to industrial workers but was extended in 1899 to agricultural workers, in 1906 to all workers in commercial enterprises, and in 1914 to those in forestry. Employers reacted, of course, by taking out private insurance against the consequences of accidents suffered by their employees in the course of their work.[18]

So, the legal framework which developed in relation to mutual aid societies during the nineteenth century was to some extent a response to discourses about the freedom of the individual and the security and social responsibility of the State, and about individualism, socialism and solidarism. It was also to some extent a response to the spontaneous and pragmatic development of diverse mutual aid societies operating to some degree outside of, or even unaware of, the legal position. Some legislation was simply catching up with the reality of social welfare provision. But the legal framework for mutual aid societies was also intended to be, and is usually assumed to have been, a stimulus to their further but controlled development. That national legal framework is, for all of those reasons, also an appropriate context within which initially to consider the historical geography of mutual aid societies regionally and even locally in Loir-et-Cher.

The development and spread of mutual aid societies

Data

There exist massive unpublished sources and also a limited amount of published information from which to reconstruct an historical geography of mutual aid societies in Loir-et-Cher during the nineteenth century, but the task is daunting not only because of the voluminous nature of the unpublished materials but also because most of them have not yet been catalogued. Although there are numerous bundles of documents stored at the Archives Départementales at Blois which include the description 'Sociétés de secours mutuels' on their labels, there does not exist an inventory of those bundles and *a fortiori* there does not exist a catalogue of the contents of those bundles – and use of them soon revealed that their contents were not simply uncatalogued but also largely unsorted. This has protracted the task of collecting data about the societies and, more importantly, it has meant that the work has been conducted without being able to be confident that all of the relevant materials for the project have been consulted. To a greater extent than is normally the case with any research in historical geography, therefore, the research findings presented here about the mutual aid societies of Loir-et-Cher during the nineteenth century cannot be seen as being definitive. Nonetheless, the sheer volume of the material consulted does permit the claim to be made that the general outlines of that historical geography can be determined with a reasonable degree of confidence, even if not all of its details can be similarly viewed.[19]

From the middle of the nineteenth century onwards, the central authorities took an increasing interest in the development of mutual aid societies. This resulted in voluminous correspondence between mayors and prefects, and between prefects and ministers in Paris. Each society was required when it was being established, and from time to time thereafter, to provide a copy of the statutes and rules according to which it functioned; lists of ordinary and honorary members; sets of financial accounts, and information about the nature and extent of the benefits provided by the society. Each change of statute or regulation was subject to prefectoral and ministerial approval, about which there was often protracted correspondence as there was also over claims by societies for state subsidies. Quite a number of societies existed for a period and then ceased to function, for a variety of reasons, and on each such occasion the dissolution of society had to follow certain legal procedures which themselves had to be properly documented. The prefect therefore maintained a dossier on each society – but of course the surviving records, however voluminous, are not necessarily complete in relation to any one society. In addition to these dossiers on societies, the prefect reported from time to time to the Conseil Général of Loir-et-Cher about the overall situation with regard to mutual aid societies in the department, and he also submitted from time to time to the Minister in Paris lists of, and some statistical information about, the department's mutual aid societies. It is from this copious documentation – of running files on individual societies and of summary surveys of all of the department's societies at particular dates – that the present account has been critically reconstructed.

Historical development

While it is clear that the earliest mutual aid societies in Loir-et-Cher date from the 1840s and that by the 1910s there were about 200 of them, it is difficult to be sure just how many societies were functioning at different times in the intervening period. During the early part of that period some societies were being approved and registered by the prefect although they had, in one form or another, already been in existence for some months or even, in some cases, years. We cannot always be sure, therefore, that the date recorded as being that when a society was approved was also the date when it was established. Some societies existed informally for some years, beyond the ken of the prefect, before being formally recognised; and in many other cases, a society was established, began to function and sought official approval but did not receive prefectoral approval until after a considerable delay. In addition, it was some while before the distinction between free or authorised societies and approved societies came to be consistently appreciated and the number of each clearly and separately recorded by the department's administrators. The task of monitoring the societies and of recording their number accurately was also made difficult by the fact that some of them had relatively short lives, so that it was not simply a matter of recording the births of new societies but also of noting the deaths of existing ones.

For these reasons, all of the sources have to be used with circumspection and no single source can be viewed as being entirely reliable on its own. Indications about the number of mutual aid societies existing at particular times are to be found in some of the published reports of the prefect to the Conseil Général from 1854 onwards; in the prefect's summary tables submitted to the minister in Paris, listing the number of societies registered by the department's administrators in 1868, 1902 and 1907; and in an unpublished report by the secretary of the Société des Sciences et Lettres de Loir-et-Cher, listing those societies existing in 1860. The date for the establishment and authorisation or approval of an individual society can also sometimes be determined from documents in its own dossier, while the lists already mentioned for 1860, 1868 and 1902 also indicate the date of foundation and/or authorisation or approval for each society existing at the given date.

From careful analysis of this body of information has been compiled a 'best estimate' of the number of mutual aid societies existing in Loir-et-Cher at nine dates from 1848 to 1907 (figure 5.1). The overall picture of the growth in the number of societies provided by that 'best estimate', based on documents from throughout the entire period, differs only marginally from the picture derived retrospectively from the list for 1902, with the number for 1907 added from the list for that year. Although no figure for any particular year should be regarded as being entirely accurate, the general pattern of growth thus depicted may be taken as being a good representation of reality. There were, it seems, slightly more societies in existence at any time than were actually known to the authorities, but the gap was never very wide and was hardly significant from the mid-1860s onwards. The mutual aid societies of Loir-et-Cher, then, had their origin phase during the late 1840s and there were 7 in operation by 1851. Their number doubled to 14 by 1855, after which it tripled to 44 by 1868. Thereafter there was steady growth in the number of functioning societies, to 80 by 1883, to almost 150 by 1902, and to almost 200 by 1907. While it is clear that the two decades of the 1850s and 1860s – the period of the Second Empire – saw rapid growth in the number of mutual aid societies, it is also equally clear that the mutualist movement in Loir-et-Cher had significantly taken off before the decree of 26 March 1852. Furthermore, the statistical picture also suggests that the slower but steady rate of growth in the number of functioning societies from 1870 onwards was continued through to at least 1907, and that the rate of growth was not significantly affected by the law of 1 April 1898. The clear implication is that national legislation had remarkably little impact upon the rate of growth of mutual aid societies in Loir-et-Cher. This is an issue which will need to be examined more closely in due course, given the conventional wisdom that such legislation – and thus the role of the State – was crucial in the promotion of mutual aid societies.[20]

The growth of mutual aid societies was, of course, also reflected in the increasing total size of their membership. The total number of ordinary

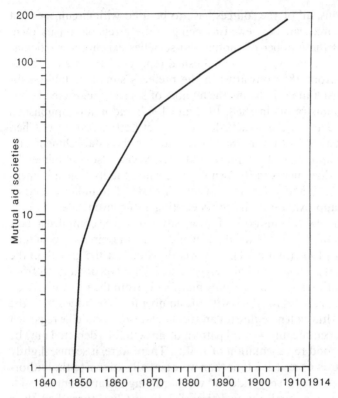

Fig. 5.1 Mutual aid societies in Loir-et-Cher, 1840–1914
Sources: AD Série X (uncatalogued files, listed in bibliography)

members in the societies of Loir-et-Cher in 1863 was reported by the prefect to be 2,305; by 1868 it had increased to 4,694 and by 1907 to 18,504. Those numbers represented respectively 0.85 per cent, 1.70 per cent and 6.70 per cent of the total population of the department in each of those three years, almost an eight-fold increase between 1863 and 1907 which compares well with the six-fold increase noted nationally between 1856 and 1911.[21]

Geographical distribution

There is evidence that nine mutual aid societies had been established in Loir-et-Cher before the 1852 law. They exhibited a distinctive geography. There were four in the department's three principal towns: one in Blois (with a population of 17,749 in 1851), two in Vendôme (population 9,235) and one in Romorantin (population 7,962). Of the other five, there were two in the valley of the Cher, in the cantonal centres of Mennetou-sur-Cher (population 947) and Saint-Aignan (population 3,434); and three in the Petite Sologne, at the cantonal centre of Contres (population 2,575), at Pontlevoy (population 2,580) and in the two adjacent communes of Fougères-sur-Bièvre and Feings (combined population 1,104).

The earliest to be established was the society at Romorantin, founded in January 1845 by twenty-nine master craftsmen (*maîtres ouvriers ou chefs d'ateliers*) of the town, in direct imitation of societies established in various other towns of France. The Société Générale de secours mutuels de la ville of Blois was given ministerial approval in January 1853 but it had its origins in a *caisse de secours mutuels* founded in November 1848 by the Second Company of the National Guard: initially for those on the role of the Company, it underwent a transformation when the National Guard was dissolved and from 1853 it was open to all *membres des classes ouvrières*. At Vendôme, the Société Vendômoise d'association mutuelle received prefectoral approval in October 1853 but it had been established in April 1849, as a result of the philanthropic inspirations of former pupils of the Collège de Vendôme: it was open to men of any occupation. Also at Vendôme, a Société d'assistance pour les femmes had been established by the mayor's wife in September 1850, to come to the aid of women when they were unable to work because of illness, pregnancy or involuntary unemployment, although it was not formally approved by the prefect until 1854 (and then, of course, only in relation to illness and pregnancy, not unemployment).

In the valley of the Cher, an *association mutuelliste* was established at Mennetou-sur-Cher in 1845 by its new *curé,* in the name of St Vincent, the patron saint of *vignerons.* In 1857 when it was seeking, and in 1858 when it received, official approval it was known as the Société de secours mutuels des vignerons of Mennetou-sur-Cher. In 1860 its founder, Monseigneur l'Abbé Lecaniat, was awarded by the Emperor a medal for his exemplary Christian and philanthropic work. Further down the valley, a mutual aid society was established in 1850 for the artisanal, craft-working population of the cantonal centre of Saint-Aignan. A very similar society had been established in March 1848 in a cantonal centre of the Petite Sologne, as the Société de secours mutuels des Amis Réunis of Contres. That society drew its inspiration explicitly from the 1848 Revolution and from the enthusiasm of its nineteen founding members. Nearby and just over a year later, in May 1849, a mutual aid society was founded by the mayor of Fougères-sur-Bièvre and by the *curés* of Fougères-sur-Bièvre and Feings for the populations of those two adjacent communes. Then in September 1849 was founded the Sociétés des Amis Réunis for the working population of Pontlevoy, with M. Monier, the professor of mathematics at the Ecole de Pontlevoy, as its first president.

The earliest mutual aid societies in Loir-et-Cher thus had not only a distinctive geography but also diverse origins (a characteristic which will be considered again in due course).

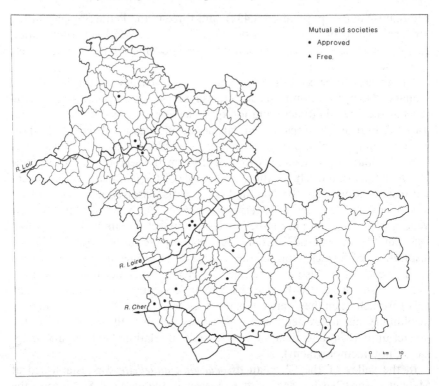

Fig. 5.2 Mutual aid societies in Loir-et-Cher in 1860
Source: AD 11J 31

By 1860 the number of functioning societies in the department had increased to twenty but the earlier geographical distribution persisted (figure 5.2). Two further societies were operating in Blois: one had been established in 1853 in its *faubourg* on the left bank of the Loire; under the patronage of St Joseph and initiated by the vicar (*vicaire*) of the church of St Saturnin in Vienne, it was open to all workers in the parish and included many gardeners and market-gardeners among its members; the second was a mutual aid society for teachers in the *arrondissements* of Blois and Romorantin, established during the winter of 1857–8 by the inspector of primary schools at the instigation of the prefect, and in direct response to the creation in 1856 of a mutual aid society for teachers of the *arrondissement* of Vendôme as a result of the efforts and enthusiasm of its sub-prefect, bringing to three the number of societies in that town in 1860. No new society was established at Romorantin between 1850 and 1860.

Three new mutual aid societies were established in the valley of the Cher: one in 1854 for the workers of the cantonal centre of Selles-sur-Cher (1851 population: 4,544) and one in 1855 for those of the cantonal centre of Montrichard (1851 population: 2,854), but one had also been created earlier, in 1853, under the patronage of St Vincent, for the *vignerons* of Chissay-en-Touraine (1851 population: 1,056), a commune adjacent to Montrichard. Elsewhere, there were few but significant additions to the overall picture. In the Petite Sologne, a society was established in 1860 for the craftsmen of Cour-Cheverny (1861 population: 2,328); in the Grande Sologne societies were established in 1855 in La Ferté-Imbault (1861 population: 877) and in adjacent Selles-Saint-Denis (1861 population: 1,113), in both instances promoted by their *curés;* in the Val de Loire, a society under the patronage of St Joseph was created in 1854 at Chouzy-sur-Cisse (1856 population: 1,392) with the encouragement of the mayor and the *curé;* and finally, the first society to be established in Perche was founded in 1855 at Beauchêne (1856 population: 411), at the instigation of its mayor.

Thus by 1860, twelve of Loir-et-Cher's mutual aid societies were located in the department's three principal towns and in five of its other twenty cantonal centres, four of them in the valley of the Cher and one in Petite Sologne. There was undoubtedly an 'urban' bias to the origin phase of the diffusion of mutual aid societies in the department, but it was by no means uniquely 'urban': many cantonal centres still had not established any such society whereas some rural communes with only small populations had done so. Moreover, two of the twenty societies functioning in 1860 had been established explicitly for the *vignerons* of their communes.

By 1868 there were almost fifty societies functioning and their geographical distribution had both intensified, with many more societies being established in localities where some already existed in 1860, and extended, with some societies being founded in localities where none, or only one, existed in 1860 (figure 5.3). The societies continued, to some extent, to develop and diffuse within the department's 'urban' framework. Of the twenty-seven new societies established from 1861 to 1868, eight were located in cantonal centres so that by the latter date sixteen of the department's twenty-three cantonal centres had acquired at least one mutual aid society. Furthermore, most of the communes which acquired societies during the 1860s each had populations in excess of 1,000 and it is clear that some such 'critical mass' was generally a necessary, but certainly not a sufficient, condition for the creation of a mutual aid society in Loir-et-Cher at that time. New societies were also being founded during this period in rural communes with much smaller populations, especially in the Val de Loire (in Maslives and Mesland, with populations of 508 and 688 respectively in 1866) and in Petite Sologne (in Ouchamps, Sambin and Chitenay, with populations of 769, 833 and 883 respectively in 1866). These societies were generally open to men of any occupation, which meant in these rural communes that they could have included not only artisans and craftsmen but also farmers and *vignerons*, and one society – that established at Mesland in

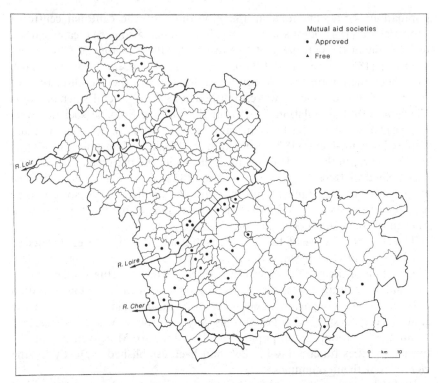

Fig. 5.3 Mutual aid societies in Loir-et-Cher in 1868
Source: AD Série X Sociétés de secours mutuels: affaires diverses de 1853 à 1870

1867 under the patronage of St Vincent – was explicitly and exclusively for those *vignerons* of the commune who owned at least one hectare of vines.

Most of the mutual aid societies newly founded between 1861 and 1868 were located in the Val de Loire and the Petite Sologne, so that by the end of that period there was a clear and developing concentration of societies in those two localities. In the former, new societies were established upstream of Blois on the right bank at Mer and Suèvres and on the left bank in the adjacent communes of Saint-Dyé-sur-Loire, Maslives, Montlivault and Saint-Claude-de-Diray; downstream, new societies were founded on the right bank at Onzain and Mesland. On the edge of the Petite Sologne, a new society was created at Bracieux and within it new ones in the adjacent communes of Cellettes, Chitenay, Ouchamps, Les Montils and Sambin. Those two clusters of new societies in rural communes – on the left bank of the Loire and in the Petite Sologne – indicate that a process of contagious diffusion was operating in the 1860s simultaneous with that of a partial hierarchical diffusion through the settlement system. Elsewhere in the department a few new societies were

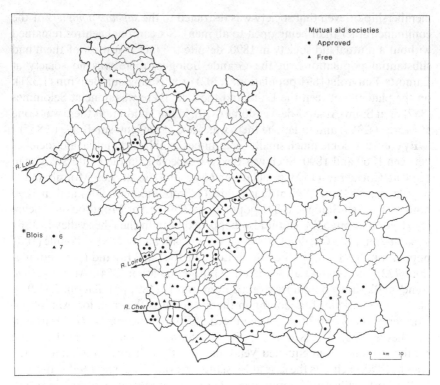

Fig. 5.4 Mutual aid societies in Loir-et-Cher in 1890
Source: AD Série X Sociétés de secours mutuels: états récapitulatifs 1905–1907–1908

added during the 1860s to those already existing in the valley of the Cher, in the Vendômois and in Perche, but by 1868 they were still largely absent from Beauce and the Gâtine tourangelle (in effect, from the plateau country between the Val de Loire and the valley of the Loir), and also from most of the Grande Sologne.

By 1890 the geographical concentration of mutual aid societies in the towns of Blois and Vendôme and in the Val de Loire, the Petite Sologne and the valley of the Cher had become very marked indeed, as had also their continuing scarcity in Perche, Beauce, the Gâtine tourangelle and much of the Grande Sologne (figure 5.4). The general picture of mutual aid societies in Loir-et-Cher in 1890 was undoubtedly dominated by Blois, which then had thirteen such societies, but in the other principal towns there were far fewer: three in Vendôme and only one in Romorantin. The declining significance of the 'urban' framework for the development and spread of societies during the two preceding decades is also suggested by the fact that only one more cantonal centre had established a society by 1890 (at Herbault in 1872, but membership of even that society was restricted to the *sapeurs-pompiers* of the commune rather than being open to all men). Six cantonal centres remained without a mutual aid society in 1890, despite the fact that most of them had substantial populations: in the Grande Sologne, there was no society at Lamotte-Beuvron (1891 population: 2,202) or at Neung-sur-Beuvron (1,321); on the plateau between the Loire and the Loir there was none at Selommes (862) or at Saint-Amand-de-Vendôme (810); in the Vendômois there was none at Morée (1,354) and in Perche there was none at Savigny-sur-Braye (2,855).

By contrast, some much smaller communes established mutual aid societies between 1869 and 1890. Such was the case especially in the Petite Sologne in 1874 at Cormeray (1873 population: 654), in 1886 at Les Montils (1886 population: 874), in 1887 at Ouchamps (1886 population: 906) and in 1890 both at Tour-en-Sologne (1891 population: 726) and at Monthou-sur-Bièvre (1891 population: 652). But it also happened both in the Cher valley in 1869 at Saint-Julian-de-Chedon (1866 population: 548) and in 1881 at Pouillé (1881 population: 832), and in the Val de Loire in 1881 at Mesland (1881 population: 722) and in 1883 at Monteaux (1881 population: 754). Also, by 1890 some small, rural communes rivalled Romorantin (1891 population: 7,812) by having two societies functioning in them: such was the case, for example, at Cormeray (666), Mont-près-Chambord (599), Monteaux (787), Monthou-sur-Bièvre (652), Les Montils (955), and Saint-Dyé-sur-Loire (841). Some rural communes even equalled Vendôme (9,538) with three mutual aid societies in 1890: such was the case at Mesland (801) and Suèvres (1,995), and also in the essentially rural communes of Saint-Georges-sur-Cher (2,233) and Onzain (2,476).

Thus the geographical distribution of mutual aid societies in Loir-et-Cher in 1890 retained the general outline of that of the 1860s and earlier. But by the end of the century that pattern had been nuanced in two significant ways: firstly, the earlier urban dominance, or at least shaping, of the spatial patterning of societies had been considerably diluted by their spread into many rural communes; secondly, although societies were distributed throughout the department, there had developed a marked concentration in the Val de Loire, the Petite Sologne and the Cher valley.

The concentrations of mutual aid societies in those three localities remained clearly detectable in 1907 (all three had by then seen significantly more societies established within them) but the very considerable growth in the absolute number of functioning societies (from just over 100 in 1890 to approaching 200 in 1907) was accompanied also by a wider spread of the mutualist movement (figure 5.5), notably into much of the Grande Sologne and on to the plateau between the Loire and the Loir (although here more societies were founded in the Gâtine tourangelle than in Beauce). Indeed, by 1890 it was really only the Perche district which persisted in showing scant interest in the movement. In 1907 there were mutual aid societies functioning in at least 114

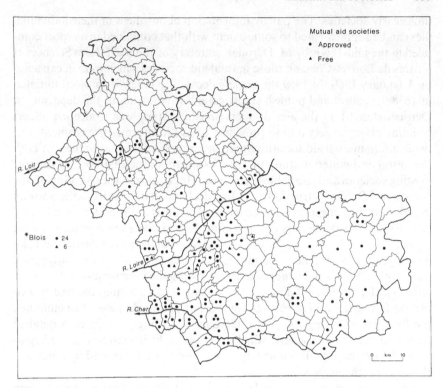

Fig. 5.5 Mutual aid societies in Loir-et-Cher in 1907
Source: AD Série X Sociétés de secours mutuels: états récapitulatifs 1905–1907–1908

communes of the department. About 40 per cent of the communes of Loir-et-Cher thus had at least one mutual aid society. There was, unsurprisingly, a very considerable concentration of societies – in fact, 30 out of the department's total of 182 – in the town of Blois, but the overall distribution of societies in 1907 was not a reflection of the department's urban or settlement geography. On the contrary, there was both a high density of societies in particular localities – in the Val de Loire, the Petite Sologne and the Cher valley – and a generally wide spread of societies throughout most of the department. That distinctive geographical distribution, together with the salient features of its historical development, now needs to be considered carefully in terms of the fundamental characteristics of the societies themselves.

Characteristics of the earliest mutual aid societies

Some mutual aid societies existed before the law of 15 July 1850 and it is worth examining them closely, to consider the origins, structures and functions of those early societies. The patchy information about them in their individual files can be supplemented to some extent with that contained in a report compiled in the mid-1860s by M. Derouet, secretary of the Société des Sciences et Lettres de Loir-et-Cher, on those mutual aid societies which were in existence on 1 January 1860. In 1859 the Société decided to renew its project, initiated in 1838, to collect and publish statistical information about the department. Derouet stated that the aim of such statistical study was to deepen understanding about society and to provide a basis for informed government: statistics about mutual aid societies would, he argued, enhance awareness of both the moral and material utility of these associations, needed both to support existing societies and to establish new ones. He concurred with M. Guizot, the Minister of Public Instruction, who had declared that 'il est très utile, sous un grand nombre de rapports, de regulariser les règles qui président à la formation des sociétés de secours mutuels et de vulgariser les renseignements à cet regard' ('It would be very useful, on many grounds, to regularise the rules which govern the founding of mutual aid societies and to popularise such information'). Guizot had advocated comparing the functioning of such societies in different parts of France, as a means of determining the best principles and practices and 'à éclairer l'avenir par l'étude du passé' ('to enlighten the future by studying the past'). Derouet accordingly produced a detailed report on the twenty societies which he found to be in existence at the beginning of 1860, basing his report largely on information provided by officers of the societies themselves.[22]

The two earliest societies established in Loir-et-Cher would seem to have been those founded in Romorantin and in Mennetou-sur-Cher in 1845. A meeting in the *hôtel de ville* of Romorantin on 19 January 1845, attended by twenty-nine craftsmen (*maîtres ouvriers ou chefs d'atelier*) and presided over by the town's mayor, discussed the creation of a mutual aid society in explicitly stated imitation of those already founded in several towns of France. The society's statutes were approved by its members on 1 February 1845, but it was not until 19 June 1847 that they were officially authorised by the Minister of the Interior. Aiming to provide monetary assistance to its members when they were ill or physically disabled, and also in their old age, the society admitted men of any occupation between the ages of 21 and 45 provided they were in good health at the time of their admission and that they were, and continued to be, of good moral character. On admission, there was payable an entry fee graduated according to age and thereafter a monthly subscription. Individuals who wished to support the society financially for philanthropic reasons could be admitted as honorary members: they could not participate in the running of the society or vote at its meeting, but they were permitted to offer advice. By the end of 1853, the Romorantin society had 193 active members and 12 honorary members.

Romorantin was at that time the third largest town in the department, with a population of 7,962 in 1851. Mennetou-sur-Cher, location of the second and very different society established in 1845, was much smaller, with a population of 1,413 in 1851. According to Derouet, in 1845 a newly installed *curé* at Mennetou-sur-Cher promoted among its *vignerons* the creation of an *association mutuelliste* under the patronage of St Vincent, with its members paying an annual subscription of only 50 centimes to cover administrative costs and accepting an obligation to undertake work collectively for any member who fell ill. Derouet reported that the *curé* had been awarded a medal by the Emperor in 1860 for his promotion of mutual aid societies in the department. In fact, this society at Mennetou-sur-Cher seems to have had even more remote origins. In 1857, when the commune's municipal council was asked by the Confrérie des Vignerons dite Saint-Vincent for permission to regularise itself under the law as a mutual aid society, the council noted that this *confrérie* had converted itself spontaneously in 1845 into a mutualist association which would provide labouring coverage for its members who were sick and attendance at the funerals of those who died: the society in existence in 1860 had evolved from an earlier association which in turn had mutated from a *confrérie*. The society at Mennetou-sur-Cher had deeper historical roots than, and was very different in character from, that at Romorantin.

Different again were the origins of the early societies established in Blois and Vendôme. In late 1848 a mutual aid society was founded at Blois as a *caisse de secours* for members of the Second Company of the town's National Guard. The moving spirit behind this society, according to Derouet, was an architect, Jules de la Morandière, who was keen to promote the idea of mutuality in Blois at a time when its inhabitants were experiencing economic stresses and social strains. Any member who was ill for more than three days would receive free medical attention and medicines, and a daily benefit of 1 fr. 25 c., and the society would pay the funeral costs of a member who died. When the National Guard itself was dissolved, the society had also to be dissolved but it was reconstituted in 1852 as the Société Générale de secours mutuels de Blois as a society for the *classes ouvrières* of the town.[23] At Vendôme in the spring of 1849 a group of former pupils of the town's college gathered for their annual reunion dinner: Derouet suggests that they might well have been moved both by the fraternal spirit of the occasion and by the revolutionary spirit of the time when deciding to establish a mutual aid society which was to be open to all of the town's citizens, whatever their occupations. When officially approved by the authorities in 1853, it took the name of the Société Vendômoise d'association mutuelle.

A number of other mutual aid societies had been founded during the early years of the Second Republic, established by the Revolution of February 1848. At Contres in March 1848 a doctor and officer of health, M. Jauze, was the moving spirit behind the creation of the Société des Amis Réunis de Contres. Its initial active membership of 20 had increased to 30 by the end of 1848. Its 49 active members at the end of 1849 embraced twenty-five different occupations, reflecting the wide range of artisanal and service activities to be found in such a cantonal centre of 2,500–3,000 people.[24] By the end of 1849 the society had acquired 18 honorary members, who were mainly landowners (some of them titled) but who also included the *curés* of Contres and of Sassay (an adjacent commune) and a pharmacist. In the mid-1860s Derouet suggested that the creation of this society might have been influenced by the revolutionary climate of the late 1840s but acknowledged that it was animated by the best of fraternal intentions to provide assistance to its members unable to work because of illness, injury or old age. The society issued each member with a magnificent printed certificate which proclaimed that the society was for the glory of God and the benefit of humanity, and that the member being admitted had provided proof of his fitness, morality and good conduct.

When writing to the Minister of Agriculture and Commerce on 1 May 1852, M. Jauze claimed that after the creation in 1848 of the society at Contres (of which he was president) similar ones had been established on his advice at Fougères and at Pontlevoy, as well as at Saint-Aignan and Valencay, all of them having adopted in part the statutes of the society at Contres: the latter two need not concern us here, because the society at Saint-Aignan was established after the law of 15 July 1850 and that at Valancay lay outside of Loir-et-Cher, in a neighbouring department. According to Derouet, a society for the two communes of Fougères and Feings was founded in May 1849 by M. Bailloux, a spinner who was mayor of Fougères. At Pontlevoy a Société des Amis Réunis was established in the autumn of 1849, specifically acknowledging it was the same sort of society as those created at Contres and Romorantin to provide assistance to members during periods of illness, infirmity or old age. Its rules stated boldly that 'l'homme laborieux et prévoyant trouvera dans cette association une ressource utile aux époques les plus pénibles de la vie, et l'homme riche un moyen de secourir les semblables sans sacrificier à la vanté et à l'ostentation ce qui appartient à la fraternité bien entendue' ('the hard-working and far-sighted man will find this association to be a helpful support during the most difficult periods of his life, and the rich man a way of helping fellow-men without sacrificing to boasting or ostentation the essence of fraternity'). A member falling sick was to receive 75 c. per day for up to ninety days, and then 50 c. for up to a further ninety days: if the indisposition continued beyond six months, there was no benefit for the next three months after which it would be resumed at a level fixed in relation to the society's resources. There was an entry fee of either 5 fr. (for those aged 21–35 years) or 8 fr. (for those aged 36–45 years), and an annual subscription of 8 francs. With an initial membership of only 12, by 1852 the Pontlevoy society had 37 ordinary members, including 6 stone-cutters, 3 metal-workers, 2 joiners, roofers, rope-makers, cloth merchants and wig-makers, and 1 person from each of a wide range of other artisanal occupations, as well as 1 doctor, 1 post office employee and 1 farmer: members were drawn from throughout Pontlevoy's population, which was 2,580 in 1851. In addition, the society had 17 honorary members, including 9 landowners, the vice-principal and 3 teachers (of music, of physics and of mathematics) from the College of Pontlevoy, the teacher from the commune's own school, a doctor, and the director of the local farm school at La Charmoise.[25]

The seven mutual aid societies known to have existed in Loir-et-Cher before mid-1850 were diverse in character. There was one in each of the principal towns of the three *arrondissements,* a fourth was in a cantonal centre, and a fifth (that in Pontlevoy) was in a local centre as large and economically active and socially vibrant as many a cantonal centre. The largely artisanal membership of all of these five societies reflected their 'urban' contexts and men from the liberal professions were instrumental in establishing them. But the two other societies – those at Fougères and at Mennetou – were located within much smaller settlements of less than 1,000 people; and that at Mennetou, situated within a community of *vignerons*, had the deepest historical roots.

These early societies were often explicitly called Sociétés des Amis Réunis: promoted by the Church and supported by notables and members of the liberal professions as a means of encouraging social order and even of countering republican idealism, these societies have been seen as endeavouring to offer a sort of replica of republican idealism. Muller has argued that they flattered themselves that they created a solid fraternity among their members, who included not only manual workers but also professional people and notables.[26] There was, clearly, a degree of paternalism about some of these fraternal associations.

Characteristics of mutual aid societies 1850–1914

There was, as has already been shown, a remarkable development and diffusion of mutual aid societies in Loir-et-Cher during the second half of the nineteenth century: by 1907 there were 182 societies recorded by the authorities as being in operation. It is clearly inappropriate to consider each of them individually; instead, an attempt will be made to identify their principal characteristics in terms of their structures, functions and evolution. This endeavour will be based primarily on the detailed, more or less standardised, surveys of such societies by the local authorities in 1868 and in 1907, supplemented with non-standardised evidence from the files of particular societies.[27]

The structures of mutual aid societies

Membership
Societies varied considerably in size, as measured by their memberships. They were comprised of both honorary and ordinary members, with the former providing management skills, moral guidance and monetary support and the latter deriving benefits from the resources of the society.

The 1868 survey recorded a total of forty-six societies in the department at the end of that year, forty-four of them 'approved' and just two of them 'free'. The approved societies collectively had 1,126 honorary and 4,259 ordinary members. The average approved society thus had 26 honorary and 97 ordinary members, the former comprising about one-fifth of the total membership. Of course, the societies varied considerably in the number of their ordinary members, ranging from the Société Générale de secours mutuels at Blois, with 753, to the society at Chissay, which had just 17 members. The median number of ordinary members in an approved society in 1868 was 67. The two 'free' societies – those at Saint-Aignan and at Romorantin – had respectively 56 and 15 honorary members and 241 and 194 ordinary members.

The 1907 survey recorded a total of 182 societies in the department at the end of that year, 149 of them 'approved' and 33 of them 'free'. The approved societies had a collective membership of 2,263 honorary and 16,320 ordinary members, while the free societies totalled 643 honorary and 2,184 ordinary members. The average 'approved' society thus had 15 honorary and 110 ordinary members, while the figures for the average 'free' society were respectively 19 and 66. Taking both groups together, the average mutual aid society in 1907 had 16 honorary and 102 ordinary members. As in the earlier period, the variation in the size of societies was considerable. The smallest 'approved' society, for carpenters at Blois, had only 7 ordinary members and four societies (those for *vignerons* at Mesland and – probably for *vignerons* – at Pouillé, and those for *sapeurs-pompiers* at Couture and at Couffi) had only 14 each. The largest societies were based in Blois, where the society for bakers had 864 members and the *Union Départementale des Sociétés de Secours Mutuels* (in effect, a reinsurance society) had 3,663 members. The median number of ordinary members in an 'approved' society in 1907 was 55. The smallest 'free' society, for *vignerons* at Ouchamps, had only 12 ordinary members; the largest, La Prévoyante du Faubourg de Vienne, based in the left-bank suburb of Blois, provided death benefits for its 391 members and the next largest provided sickness benefits for its 235 members in the commune of Theillay. The median number of ordinary members in a 'free' society in 1907 was 48.

While the spectacular growth in the membership of some mutual aid societies between 1868 and 1907 is obvious, less striking but just as important was the growth in the number of societies with between, say, 40 and 70 ordinary members. The increase in the number of such middling-sized societies in the rural localities of Loir-et-Cher was just as important as the dramatic growth in the size as well as number of societies within its urban districts.

Geographical and occupational identities

Most societies were small, with fewer than 100 members, grouping together working men (and sometimes women) who shared a geographical and, in some cases, an occupational affinity. The need to monitor carefully the claims of their members for assistance and to administer the system of benefits equitably and efficiently underpinned the local, geographical, basis of almost all societies. But some societies associated members whose trust in each other was based more upon their occupational identity than upon their geographical propinquity and personal familiarity. While the former group of 'open' societies built upon a sense of community, the latter 'closed' societies promoted a consciousness of class while still being grounded within a particular locality. These two groups existed in almost equal numbers in Loir-et-Cher between 1850 and 1914, with 'open' societies being more important in the earlier part of that period and 'closed' societies becoming more significant in the latter part.

Both sets of societies relied upon trust among members living in the same locality and, for the most part, known to each other. Almost all societies, 'open' and 'closed', operated within the framework of a single commune, a geographical scale at which a society's activities – especially its careful monitoring of members' claims – could be undertaken economically and equitably, with relative simplicity and considerable transparency. By law a mutual aid society was restricted to operating within its own commune but exceptions to that rule could be and were authorised by the central authorities. Such a geographical limitation might have had an underlying political motivation – central control over peripheral activity – but it also made sense pragmatically. The administration of a society was more readily and effectively accomplished within the territory embraced by a single commune than in any larger area. Almost invariably, a mutual aid society had a commune as its operational, geographical, basis. Within such a spatial frame, a claimant could be visited relatively easily and quickly, and the course of his (or her) illness or injury, treatment and recovery could be monitored by a society's representative.

The significance of this geographical constraint upon mutual aid societies is highlighted in those communes where it was argued that more than one society was needed because of the problems of ensuring acceptable surveillance within a commune comprising a number of settlement units rather than a single village. For example, in March 1875 the mayor of Saint-Georges-sur-Cher informed the prefect that he was having difficulty in promoting a society for the whole commune because of its great extent and that the population of each *quartier* (locality) within it would wish to run its affairs in its own way. At Onzain, where a society had been established in 1867, a second society was founded in 1878 because of the distance which separated the principal hamlets of the commune from its central village. Mutual aid societies were founded upon trust, so that their geographical reach could not extend beyond the area in which such trust could be fostered. In March 1855 the mayor and the priest at La Ferté-Imbault told the prefect that it had become necessary to establish two societies rather than one because of long-standing rivalries between the inhabitants of the commune's two principal settlements.

In most cases, a mutual aid society in Loir-et-Cher was based upon a single commune but in a few cases it had a smaller geographical base and in a few others its base was larger. In Blois, with a population of almost 20,000 in 1872, most of its mutual aid societies embraced the town as a whole geographically but in November of that year there was established one society for the workers of just one locality within it, the parish of La Chapelle des Grouëts. By contrast, some rural commune societies sought to increase their small size by admitting members from adjacent communes. The society at Chitenay, with 36 ordinary members at its foundation in 1865 but only 20 in 1868, received authorisation in 1888 to extend its operations into three adjacent communes and in 1902 it recorded 78 members. The society created in 1860 for workers at Mondoubleau from the outset also admitted those at Courmenon, receiving official approval to do so on the grounds that the two communes were in effect a single settlement, focused upon tanneries which employed men from them both indiscriminately. Earlier, the Société des Amis Réunis created at Contres in 1848 and officially approved in 1852 had been established for workers (*la classe ouvrière*) not just of the commune itself but of the canton of Contres. Wishing to encourage the formation of societies, the ministry in Paris had given approval despite the legal restriction on communes joining together to form a single society unless the participating communes each had a population less than 1,000: the 1851 census recorded a total population of 13,615 for the canton's seventeen communes, four of which each had populations in excess of 1,000. When this society's statutes were revised in 1882, the ministry noted that since its foundation a separate society had been established in 1861 at Cour-Cheverny, admitting members from Cheverny as well as from Cour-Cheverny, and ruled that accordingly membership of the society based at Contres would have to be restricted to workers from the other fifteen communes of the canton. In practice, however, membership of the society had not extended throughout the canton and the society's general meeting on 12 November 1882 itself decided to restrict membership to workers from just four communes (Contres, Fresnes, Sassay and Oisly). Even that restriction was then tightened further, because the following month the ministry ruled that, as the society had not had any members from Oisly for a long time, the society would only henceforth be permitted to cover the other three communes. The ambitious, idealistic, endeavour to create a mutual aid society for the workers of an entire canton had to be pruned back to a more realistic practice. The Contres society was the only one in Loir-et-Cher set up at the level of the canton. The next administrative tier above the canton was the *arrondissement* and it was only in relation to a specialist professional group, that of teachers, that mutual aid societies were established at that level: for the *arrondissement* of Vendôme in 1856 and for those of Blois and Romorantin together in 1864. Those two societies merged in 1864 into a single society for teachers from throughout the department.

But such a society – for a specific group of professionals throughout the department – was highly exceptional. Most were much more open in terms of their memberships and much more restricted in their geographical coverage. About half of the societies founded between 1850 and 1914 were 'open' to workers of any occupation – such a society admitted working people (but usually only working men) residing in the commune in which it operated. These societies generally stated in their statutes that they were being established for the benefit of *citoyens de toute profession* or for the *ouvriers* (or members of the *classe ouvrière*) and others (*autres*) in the commune, or for its *ouvriers et cultivateurs*, or for its *travailleurs, cultivateurs et vignerons*. The membership of such societies thus reflected the general occupational structures of their communes, with the societies in the more populous communes and cantonal centres having many more artisans and craftsmen among their members than had the societies in the less populous and dominantly agricultural communes. Some societies recognised that their memberships would, because of local circumstances, be dominated by men of a particular occupation but explicitly stated that workers of all occupations could join: thus the society established in 1853 at Vienne, a suburb of Blois, was principally but not exclusively for its market-gardeners; that at Mondoubleau (1862) for its tanners and other working men; and that at Fréteval (1867) for those employed in its blast furnaces, foundries and enamel works, but also for other working men.

Of the 'closed' societies, the single largest, most important, group by far comprised those created for *vignerons*. Some societies were for *vignerons* and others: the society at Montlivault (1863) had two categories of members, those who were *vignerons cultivateurs* and those who were *ouvriers de différents métiers et les industriels*; the society at Châtillon-sur-Cher (1870) was technically 'open' but of its sixty-four initially participating ordinary members all but three were *vignerons;*[28] and the society approved in 1910 at Romorantin and Lanthenay was for both *vignerons* and *cultivateurs*. But at least a dozen societies newly founded during the second half of the nineteenth century were explicitly for the *vignerons* of their communes.[29] The statutes of the mutual aid society for the *vignerons* of Onzain approved in 1867 stated clearly that, in order to be admitted to membership, a candidate must have experience of looking after at least one hectare of vines: such knowledge and experience was vital for membership of such a society, because the assistance provided to sick members was not in cash but in kind, with the vines of an indisposed member being cared for by other members of the society. The 'closure' of such societies was grounded in a practical requirement.

A second group of 'closed' societies were those whose memberships were restricted to men who were volunteers in a commune's fire brigade. These societies provided benefits to their members not only when they were ill but also when they were unable to work because of injuries sustained while carrying out their fire-fighting duties. Half a dozen or so such societies were founded in Loir-et-Cher, mainly between the early 1860s and the early 1870s.[30] In addition, an unusual association was the Société des Membres Honoraires du Corps de Sapeurs-Pompiers of Salbris: founded in 1909 and open to both sexes, this society provided assistance to members of the corps who were unable to work because of illness or injury, whether or not the illness or injury was related to fire-fighting, and the society also subsidised the costs of the corps' uniforms, equipment, training and participation in competitions.[31]

The remaining 'closed' mutual aid societies were more of a miscellany. Two – those at Soings (1887) and Chitenay (1888) – were exclusively for members of the agricultural syndicates operating in those communes. Two were for schoolteachers: one for those in the *arrondissement* of Vendôme (1856) and one for those in the *arrondissements* of Blois and Romorantin (1864). A few other societies were for specific groups of workers in particular places: such were the societies for the building craftsmen of Cour-Cheverny (1861); for the clog-makers of Mennetou-sur-Cher (1861), and for the workers in the shoe factory of the Rousset brothers at Romorantin (1884).[32] The town of Blois was the seat of a number of such 'closed' societies: it had a society for shoe-makers (1869), a society for both men and women workers in the shoe factory of M. Guéritte (1871), and a society for those employed at the town's stud (1907). Two other societies at Blois were established to provide pensions, one for old members of a school in the town (1908) and one for both owners and employees in the baking industry throughout the *arrondissement* of Blois (1908).

The generally larger size and greater degree of membership exclusivity of the mutual aid societies in the towns of Blois, Vendôme and Romorantin, and the significant numbers of artisans and craftsmen in some societies in smaller (but, in rural terms, still relatively large) centres of population, can too readily be allowed to disguise the fact that mutual aid societies were founded in many small, very agricultural, communes during the second half of the nineteenth century. In 1868, 36 per cent of the approved societies were located in communes each with a population of less than 1,000; in 1902 the comparable figure was 32 per cent. There were some societies in communes with populations of less than 500, as was the case at Beauchêne in 1868 and at La Chapelle-Vendômoise, Coulanges, Cour-sur-Loire, Huisseau-en-Beauce and Villetrun in 1902.[33] The mutual aid movement in Loir-et-Cher was very far from being exclusively, or even principally, an urban phenomenon: it was also very active in the countryside, especially so in viticultural localities. But in both town and country, it affected men much more than women.

Gendered membership

Ordinary membership of mutual aid societies was strikingly gendered. In 1860, of the 20 societies identified by Derouet 11 were only for men, 8 admitted men and women, and 1 was only for women. In 1868, of the 43 approved societies for which the information is available, 30 restricted their ordinary membership to men, 12 were open to men and women, and 1 was restricted to women only. Women comprised only 8.1 per cent of all ordinary members in the 43 societies, and only 19.5 per cent of the ordinary membership of those societies open to both women and men. Some mixed societies catered for men and women of all occupations, some were specifically for workers and labourers, and one was for the schoolteachers of the department – but this last society had only 3 women members alongside 160 men. The society exclusively for women was the Société d'assistance mutuelle des femmes de la ville de Vendôme. Launched in 1850 by Mme Tremault, wife of the town's mayor, the society was officially approved by the prefect in 1854: its aim as set out in its statutes was to come to the aid of women unable to work because of illness or pregnancy, providing for them the care of a doctor or a midwife, and similarly for any of their daughters under the age of sixteen. Unfortunately, the survey of 1907 does not provide details about membership by gender but an earlier one, of 1902, does do so.[34] In 1902, there were 105 approved societies and 42 free societies registered with the local authorities. Of the total of 147, only males were admitted as ordinary members to 116, males and females to 29 and only females to 2. Females comprised 8.4 per cent of all ordinary members in the 147 societies, and 18.7 per cent of the ordinary membership of those societies open to both males and females.[35] The two societies whose memberships were restricted to females were the society in Blois called Notre Dames des Aydes, established in late 1868 for working women (*femmes de la classe ouvrière*) as a branch of the already-established Société Générale de secours mutuels, and the Société d'assistance mutuelle de femmes at Vendôme, previously noted. In 1902 of the 29 mixed societies, 11 admitted not only men and women but also boys and girls; of the restricted societies, 1 admitted only boys (La Prévoyante, for schoolboys at Saint-Aignan) but there was no comparable society for girls. The total number of children who were society members was small (337, constituting 2.7 per cent of the total of ordinary members).

In the societies which had mixed memberships, women did not have an equal standing with men. For example, the printed statutes of the mutual aid society of Bracieux founded in 1864 made it clear women could be members not in their own right but only as the wives, widows or daughters (if they were running the household as *filles maîtresses*) of male members. Furthermore, while men in the Bracieux society benefited both from a *per diem* sickness allowance and by the society's payment of doctors' fees and the costs of medicines, women were covered only by the latter and were not provided with any direct sickness payment (and even medical costs were not covered in relation to problems associated with childbirth). The different level of benefit provided by the Bracieux society was reflected in the regular subscription, with that for a woman being only half that for a man. The society established at Theillay in

1888 admitted not only men but also women and children (but not more than four children for any one family): there was a single subscription rate for all members and all members had equal rights to assistance, but this covered only medical and funeral costs and did not include any direct sickness benefit. Although they were members, women (and children) were not permitted to take part in any of the society's discussions or administration, this despite the fact that certainly by the early 1900s women out-numbered men in the society.[36] The running of mutual aid societies was a male domain. Similarly, men were clearly the principal promoters of these societies.

Paternalistic and democratic underpinnings
In terms of the principles underlying the formation of mutual aid societies, two basic kinds of society can be identified: one group rested upon the principle of paternalism, another upon that of fraternity. Let us consider each group in turn, recognising that there were also variations within each group.

Many mutual aid societies were grounded in patronage by the social power brokers of their localities: they reflected, therefore, the growing struggle for power between the Church and the State, as well as the desire on the part of some wealthy individuals to exercise beneficial influence within their own communities. The traditional concept of patronage came to be combined with more modern philanthropic impulses to provide a rationale for the creation of these welfare societies. Practical expression of such underlying motives was provided by the honorary members, who contributed to the societies but who did not receive any direct benefit from being members: they all provided financial support (in the form of annual subscriptions and occasional donations), some also provided professional experience and/or social prestige, but none derived any of the benefits provided by the society to its ordinary members. Honorary members were significant numerically and financially: in 1868 they comprised 26 per cent of the total membership of mutual aid societies in Loir-et-Cher and they contributed 22 per cent of the societies' subscription income (and they also very probably contributed substantially to the donations made to the societies that year, which were equal to a further 5 per cent of their subscription income). Thus each honorary member could be said to be providing support for every three or four ordinary members.

While ordinary members, as we have seen, were drawn generally from a wide range of craftsmen, cultivators and commercial men, honorary members came from a narrower and more distinctive social group. When a mutual aid society was established at Saint-Aignan in 1850 with 51 ordinary members, its 25 honorary members included 5 landowners, 5 merchants, 3 notaries, 3 law students and 1 medical student, as well as 1 justice of the peace, 1 doctor, and the *curé* and the curate of Saint-Aignan.[37] At Pontlevoy in 1852 the society's 37 ordinary members were supported by 17 honorary members, among whom were 9 landowners, 1 vice-principal and 3 teachers from a private college, the commune's schoolteacher, as well as the director of the local farm school, 1 doctor and 1 grocery merchant. The society created at Mont in 1855 had at its foundation 24 ordinary members and 15 honorary members, including in this latter category the Viscount de Bizemont (owner of the local château) and 2 other aristocrats, 5 landowners, 1 retired notary, 1 flour merchant and the *curé*. The society established at Villefranche-sur-Cher in 1868 with 56 ordinary members had among its 18 honorary members 5 landowners, the mayor of the commune and his deputy, the parish priest, 1 notary, 1 justice of the peace, the owner of a tile-works and 1 hotelier. The mutual aid society established in 1872 for the volunteer firemen of Herbault had among its 24 honorary members 3 landowners, 3 law officers, 3 tax collectors, 2 doctors, the mayor and his deputy, the *curé*, 1 estate manager and 1 schoolteacher. Honorary members of mutual aid societies thus came principally from the landowning class, the liberal professions and the (presumably wealthier) tradesmen. When a society was being founded at Selles-Saint-Denis in 1855 the mayor reported to the prefect that it rested on the enthusiasm, enlightened attitude, generosity and devotion of some of the commune's *honorables notables* and public servants. There can be little doubt that a similar combination of factors was also at work to varying degrees in other communes.

Such patronage by a social elite and by philanthropically minded individuals was often orchestrated, at first by the Church but increasingly also by the State. The Church had, of course, a long tradition of providing charity to the poor and needy in its parishes, and the promotion of mutual aid societies came to be seen as one way of modernising such charitable activity while at the same time engaging the support of local communities for the church itself. As has been seen already, *curés* were often honorary members of the societies founded in their localities and there is additional evidence of their involvement in the creation of such societies. For example, the statutes of the society founded at Ouchamps in 1867 stated explicitly that it was being established in order both to provide mutual Christian aid for its members and to contribute to the maintenance and improvement of public morality in the commune. Quite a few societies (at least 27, more than one in ten of all societies) operated explicitly under the patronage of a named saint. Although such saintly patronage might be indicative of a direct promotional role by the Church, it might alternatively have simply reflected the (unthinking) continuance of a tradition because no less than thirteen such societies were in the name of St Vincent, the long-standing patron saint of *vignerons*, while only three were in the name of St Joseph, with whom Christian charity was traditionally associated, and the remaining societies operated mainly in the names of the patron saints of their communes. That the religious foundation of some societies might have been more apparent than real is suggested in the case of the society of *vignerons* at Saint-Romain-sur-Cher. In 1862 the *curé* had been instrumental in obtaining official approval for the society but only two years later he informed the prefect that he was resigning as its president and that, because of the persistence of the spirit of 1848 in the commune, it would be better for the society to have as its president the mayor, because he was both devoted to the imperial regime and the local symbol of its authority.

There are some cases of explicit co-operation between priests and mayors in the creation of mutual aid societies. For example, at Chouzy-sur-Cisse in 1854 the *curé* and the mayor worked together to set up a society in the name of St Joseph, convinced that in doing so they would both be acting in the charitable spirit of the emperor himself and be moralising a social group within the commune in need of such instruction. At La Ferté-Imbault in 1855 the *curé* and the mayor joined forces to create a mutual aid society in the name of St Thaurin. In their letter of 7 March to the prefect, they pointed out that while the benefits of such societies were recognised in towns, it was difficult to establish them in rural districts accustomed to their isolation and characterised by individualism and by populations resigned to their fate. The statutes of this society were headed by a proclamation explicitly enjoining God and mutual aid and combining four moral instructions into a lesson for life: 'Amour des hommes, Union des coeurs, Ordre dans la conduite, Economie dans sa famille – Faites cela et vous vivrez.'

Although such collaboration did happen, there is much more evidence of mayors acting independently of priests. Sometimes they were acting directly on their own initiative – or at least with respect to some local proposal – but they were often responding to pressures imposed upon them by the prefect. Following the law of 15 July 1850 and decree of 26 March 1852, the Ministry of the Interior in Paris instructed prefects throughout France to encourage the creation of mutual aid societies, but also to monitor them carefully to ensure that they did not become political organisations. By law, commune councils were obliged to provide approved mutual aid societies with the books and registers needed for accounting purposes, and the Ministry provided prefects with a standardised list of the materials to be supplied to each society. In addition, it provided a set of model statutes to be furnished to those considering founding a society. Accordingly, from the early 1850s the prefect of Loir-et-Cher transmitted this information to the mayors in the *arrondissement* of Blois and to the sub-prefects of the *arrondissements* of Romorantin and Vendôme (who in turn distributed the information to the mayors of communes under their respective jurisdictions). A similar, renewed, official endeavour to promote mutual aid societies was made in the early 1860s. There were, undoubtedly, considerable official attempts in the 1850s and 1860s to encourage the creation of mutual aid societies. They met, however, with a mixed response. While information about such societies was, because of the efficiency of the administrative system, uniformly available throughout the department, the use to which it was put varied from place to place and from time to time.

When reporting to the minister on 2 July 1859, the prefect claimed that mutual aid societies existed in Loir-et-Cher in those localities where there was an orderly and charitable spirit among the population. But official exhortations to establish such societies were resisted in many localities of the department, especially those in which the population was considered to be too small to be able to benefit from such a society, those where the working population already had informal mutual arrangements, and those where there were too few potential honorary members (a group perceived by the prefect to be indispensable to the formation of a society). Given the inconsistent nature of the surviving records of societies, it is impossible to assess with certainty the efficacy of official efforts. That they were instrumental in some cases is indisputable. For example, on 9 January 1855 the mayor of Selles-Saint-Denis informed the prefect that in response to the latter's circular a society was being established: the mayor had contacted some key workers in the commune, selected because of their good character and sound economic standing, to form the initial core of ordinary members, and he had also acquired support for the society as honorary members from some of the commune's leading individuals. In some cases the prefect's exhortations might have tipped the balance in an existing debate about whether or not to establish a society. Such appears to have been the case at Selles-sur-Cher in April 1853, because the mayor then told the prefect that the commune's council had considered the official circular and had decided to establish a society, but he added that the idea had been under discussion in the commune even before the decree of March 1852. In fact, the debate continued: the society's draft statutes were not ready until September 1853 and were not formally approved by the prefect until October 1854. In other communes the prefect's exhortations met negative responses. For example, in January 1855 the mayor of Mont-près-Chambord returned to the prefect the official copy of the model statutes for a mutual aid society, because despite all of his efforts he had not been able to arouse enough interest in the idea. In some communes, official efforts to promote mutual aid societies were neutralised by the long time which it often took for the official procedures for establishing a society to be completed. For example, on 26 May 1864 the mayor of Bracieux complained to the prefect that delays in considering proposals to create a society in his commune were being very badly received: such delays, he said, disappointed even the devoted and charitable supporters of such a society who nonetheless did not submit themselves willingly to bureaucratic controls, while they allowed egoists in the commune to persist in their denigration of such an institution.

By the 1860s some communes were, however, responding more positively to official attempts to foster societies, but they were not always doing so simply because they had come to accept their advocacy by the central authorities. At Montoire-sur-le-Loir there had been an unsuccessful attempt to establish a society in 1852, in the wake of the prefect's circular. But a similar circular in September 1863, coupled with visits to the canton by the sub-prefect of

Vendôme during which he had explained the merits of mutual aid societies, resulted in Montoire's council deciding to create a society. But it did so in part, but quite explicitly, on the grounds that other communes considered to be of lesser importance than Montoire itself had founded such societies: local pride tipped the balance in favour of creating a society. The prefect's circular of 8 October 1864 certainly seems to have been the trigger for the creation of societies at Cellettes, at Droué and at Les Montils.

But many localities remained sceptical about the advice emanating from the central authorities. In the early 1860s successive mayors at Ouzouer-le-Marché reported to the prefect that it was difficult to set up a society there, because the commune had an agricultural rather than a 'working' population: a society was eventually approved in January 1863. It was not unusual for the debate about, and the procedures for, establishing a society to take years rather than months. The mayor of Saint-Secondin (Molineuf) told the prefect in February 1865 that he and the *curé* were trying to found a society in the commune but that there were difficulties and obstacles to overcome. That was no understatement: on 28 April 1868 the mayor again wrote to the prefect, informing him that the commune had been discussing his circular of 4 February 1865 on the utility of mutual aid societies and that at last it had been agreed to establish one in the commune. That the considerable efforts of the central authorities over a long period of time to promote mutual aid societies were by no means consistently successful is made abundantly clear in a report of 22 February 1866 by the sub-prefect of Vendôme. He recorded that he had sent the prefect's circular of 11 January on mutual aid societies to the mayors of twenty-eight selected communes in the *arrondissement* of Vendôme: he had received only twelve replies and only two of those (for Le Plessis-Dorin and Fréteval) expressed a positive interest in the idea of setting up a society (and only Fréteval was in due course to do so). The commune council of Lunay acknowledged the utility of mutual aid societies but declined to establish one for the same (unspecified) reasons that it had decided against doing so seventeen months previously. The replies from the councils of the other nine communes are more illuminating, providing a range of reasons for their negative responses. Five councils argued that there were too few 'workers' in their communes to justify such societies;[38] two stated that there was not enough enthusiasm among potential ordinary members and especially not enough support from potential honorary members;[39] two claimed that their communes mainly comprised well-to-do farmers who had no need of mutual aid and another explicitly stated that its comfortably off farmers could afford to pay for exceptional help on their properties if the need arose;[40] one council said that there was no-one in the commune with sufficient influence who would be capable of leading such a society;[41] another argued that the most numerous group in the commune was the *vignerons* and that they already had an informal arrangement for helping each other in practical ways when necessary;[42] and two councils stated that they had insufficient public funds to support such societies.[43] These instances clearly indicate that the efforts of the central authorities could only be productive where local circumstances permitted them to be. They suggest that there had to be a locally authentic sense of fraternity and that while mutuality might be encouraged from the centre it could not be imposed.

While some mutual aid societies in Loir-et-Cher were part of a patronal tradition, many were spontaneous, even democratically inspired, associations – and this is clear despite the fact that the former, 'top-down', movement has left more records than the latter, 'bottom-up', movement has done. It is suggested, for example, at Saint-Georges-sur-Cher where the mayor wrote to the prefect on 8 June 1865 telling him that the *vignerons* of the commune wanted to form a mutual aid society and asking him (despite having been inundated with prefectoral circulars on the subject) how to proceed.

It is also suggested by those cases where a society which had existed prior to the legislation of the early 1850s came to seek official approval, in order to regularise its position. Such was the case with the society of *vignerons* at Chissay which was formally approved by the prefect in July 1853 but which had by then been in existence for some years. It was also the case at Mennetou-sur-Cher, where a *confrérie des vignerons* had been established spontaneously in 1845, as a voluntary association providing assistance in the form of replacement labour for those of its members unable to work. In May 1857 the *confrérie* sought, and in March 1858 received, official approval as a mutual aid society. A similar *confrérie* of *vignerons* at Mesland had its manuscript statutes drawn up in January 1861 but it appears (from a note added to the statutes) to have been founded earlier, in July 1850. In March 1867 this *confrérie* sought and received prefectoral approval as a mutual aid society. These pre-existing *confréries* were not confined to *vignerons*, although they do seem to have been more significant among that occupational group than in any other. There had been, for example, at Cour-Cheverny a Confrérie de Saint-Joseph for craftsmen and workers in the building industry; in March 1860 it sought, and in September 1861 it received, prefectoral approval to convert itself into a mutual aid society.

Additional evidence of the spirit of grass-roots fraternalism which underpinned some mutual aid societies is provided by the titles and 'mission statements' which they adopted. For example, the rules of the Société des Amis Réunis et de secours mutuels of Pontlevoy drawn up in early 1850 referred explicitly to the concept of fraternity upon which the society was based, as has already been noted.[44] Similarly, the 1850 statutes of the society at Saint-Aignan stated that its aim was to unite all of its members 'par des liens de fraternité et de bienveillance réciproque' ('by bonds of fraternity and reciprocal welfare').[45] At Cormeray in November 1873 the commune council supported a proposal for a mutual aid society for the commune's *sapeurs-pompiers* on the grounds that it would draw tighter the 'bons sentiments de fraternité' ('good fraternal feelings') which motivated them. In January 1895 the *sapeurs-pompiers* and the farmers of Saint-Dyé-sur-Loire created a mutual aid society with the explicitly stated single aim of 'union et fraternité réciproque' ('union and mutual fraternity').[46] The society created among the *vignerons* at Ouchamps in the spring of 1899 took as its title La Fraternelle:[47] an indication that the traditional *confrérie* had become the modern fraternal association.

The functions of mutual aid societies

General perspectives on the 1860s and the early 1900s

A general picture of the aims and objectives, as well as of the ways and means, of the mutual aid societies in Loir-et-Cher can be reconstructed from the broad surveys undertaken in the 1860s and in the early 1900s.

In his report on the situation in 1860, M. Derouet identified the existence of 20 societies: 18 aimed to provide assistance to members when they were unable to work because of illness or accidental injury and to contribute to the funeral costs of members, but the other 2 also gave financial help to their members when they were unemployed.[48] The duration and level of sickness and injury benefit provided varied considerably from society to society. Retirement pensions were provided in principle by 13 societies, subject to their having available sufficient funds, while the other 7 accepted the principle but had no explicit rules about its practice. Each society was administered by a committee whose president was appointed by the Emperor in the case of approved societies and by the society's members in the case of authorised societies. Derouet provides a detailed statement of the financial position of each and all of the societies. Overall, the entry fees and annual subscriptions of ordinary members comprised about 60 per cent of the societies' income, while donations and subscriptions of honorary members represented about 25 per cent of the total, and grants from the authorities and interest from funds on deposit each contributed about 6 per cent. The remaining balance of about 5 per cent came from special subscriptions collected to assist orphans and widows, and to meet funeral costs of members, as well as from fines and miscellaneous donations. The expenses of societies were, of course, mainly payments of sickness benefits directly to members (about 52 per cent of the total) and medical costs (with the fees of doctors and midwives and the costs of medicines amounting to about 32 per cent of the total). Benefits to widows and orphans, together with contributions towards funeral costs, comprised about 6 per cent of expenditure, while unemployment benefit was an insignificant amount (only 0.02 per cent). Administrative costs represented about 9 per cent of expenditure. All of the societies had incomes which exceeded expenditures: overall, expenditure amounted to 70 per cent of the societies' income. The number of societies claiming to have established a reserve fund (*caisse de securité*) was 15, while 4 others explicitly ruled out such a fund.[49]

The 1868 survey of the 44 approved societies is as informative about their functioning as about their membership, and it furnishes some new insights into their activities. The survey provides for each society a statement of the number of its members, men and/or women, who were paid sickness or injury benefit during the year ending 31 December 1868, as well as the total number of days for which such payments were made. These figures reveal that the societies were generally providing many of their members with assistance but that they did so only for a relatively short period of time. Almost one in three (32.5 per cent) of the 3,913 men and just over two in five (42.1 per cent) of the 346 women members were in receipt of benefits: overall, one in three members (33.3 per cent) were assisted during the year. But if the benefits were widely shared, they were also somewhat thinly spread. The men receiving assistance did so for a total during the year of, on average, 15.4 days, and the women for an average of 13.5 days: overall, the average was 15.2 days. In addition, in 1868 societies provided assistance to the relatives of those of its members who had died: the death toll of 43 during the year represented 1 per cent of their total membership. The 1868 survey also makes it clear that the payment of retirement pensions was still not an important function of these societies: only 3 of the 44 societies paid such pensions, to sixteen individuals who constituted only 1.5 per cent of the membership of those societies and only 0.4 per cent of the membership of all of societies then surveyed.[50]

The 1868 survey also sets out the financial position of the societies. Overall, the annual subscriptions of ordinary members comprised about 67 per cent of the societies' income, while those of honorary members represented about 19 per cent of the total. Grants, gifts and legacies together with entry fees produced about 9 per cent of total income, while interest from funds on deposit contributed about 3 per cent. The remaining balance of about 2 per cent came from fines and miscellaneous other sources. The expenses of societies remained, of course, significant payments of sickness benefits directly to members (about 38 per cent of the total), but the indirect medical costs were greater (with the fees of doctors and midwives and the costs of medicines amounting to about 43 per cent of the total). Benefits to widows and orphans (presumably including contributions to funeral costs) comprised about 5 per cent of expenditure. No unemployment benefit was paid by any of the societies and only insignificant amounts (representing 0.4 per cent of the total) were paid out to members deemed to have a incurable medical problem or permanent infirmity.

Administrative costs represented about 5 per cent of expenditure. Most of the societies had incomes in 1868 which exceeded expenditures during that year: for all of the societies together, expenditure amounted to 96 per cent of their income. However, 15 of the societies had expenses which exceeded their incomes that year, so that they had to call upon their reserves to meet the deficits. Most were able to do so, however, without much difficulty. Overall, at the end of 1868, the reserves of the societies collectively were slightly in excess of their total expenditure in that year. But 3 societies (7 per cent of the total), very different in the size of their memberships, had expenditures which exceeded their incomes in 1868 by amounts which were in excess of the balances remaining in their reserve funds at the end of the year (after taking into account the deficits for that year) and so were clearly vulnerable rather than financially secure.[51]

The precise benefits provided by these societies for their members varied. For example, the medical costs were not borne invariably by the societies, for in about two in five of them such costs had to be borne by the individual members. The daily rate of sickness benefit provided itself varied, for men from 0 fr. 50 c. to 2 fr. (but it was normally 1 fr.) and for women from 0 fr .50 c. to 1 fr. 50 c. (but it was normally 0 fr. 50 c.). Not all societies provided benefits exclusively as money payments: some, perhaps as many as 7 (16%) of the 44 societies covered by the 1868 survey, furnished not a sickness benefit in cash but instead provided replacement labour for specified numbers of days.[52] The 1868 survey records old-age pensions as being payable by 21 (48%) of the 44 societies, with pensions normally being payable when a member reached the age of sixty-five, normally on condition that s/he had been a member for at least fifteen years.[53]

By the early twentieth century the picture of mutual aid societies in Loir-et-Cher had become much more complex, as the survey of them in 1907 revealed. The authorities then recorded 180 societies in operation, of which 147 were approved and 33 were free. Of the approved societies, 65 provided just sickness benefits; 72 provided sickness benefits and retirement pensions and 5 provided only retirement pensions (these 77 societies represented 52 per cent of the total); and 5 provided various other benefits, like reinsurance and death benefits. Of the free societies, 21 provided just sickness benefits; 1 provided sickness benefits and retirement pensions; 1 provided only retirement pensions; and 10 furnished various other benefits.

The first two categories of societies – those providing just sickness benefits or sickness and pension benefits – numbered 159 with 13,663 members, representing 88 per cent of the total of all societies and embracing 75 per cent of their membership in 1907. It is, therefore, worth looking at their functioning most closely. During 1907, just over one in four (28%) of members of the sickness societies and just over one in three (35%) of members of the sickness and pension societies were provided with sickness benefit of some kind. Cash payments were made to one in three (32%) of the beneficiaries of the sickness societies, and to two in five (42%) of those of the sickness and pension societies. Presumably others received aid in the form of labour replacement. Such cash benefits were paid out by half (50%) of the sickness societies and by two-thirds (69%) of the sickness and benefit societies. Those receiving such assistance in 1907 did so each on average for a total of twenty-two days in the former societies and for twenty days in the latter. Doctors' fees were met directly by two in five (41%) of the former societies and by three in four (77%) of the latter societies, while the costs of medicines were met directly by just over one in three (37%) of the sickness societies and by three in four (73%) of the sickness and retirement societies. Very clearly, the benefits provided by the sickness and retirement societies were not only more comprehensive but also more generous than were those provided by the sickness societies. A further illustration of this point is the fact that the sickness societies met the funeral costs of only just over half (57%) of those of their 87 members who died during the year, whereas the sickness and pension societies met the costs of almost all (96%) of those of their 135 members who died that year.[54]

The 1907 survey also records 5 approved societies and 1 free society providing just pensions. Together the 6 societies embraced only 413 members: the smallest societies – one for the *sapeurs-pompiers* of Authon and another for members of the agricultural syndicate at Coulommiers – each had only 17 members, while the largest – that for veterans of the *Armées de terre et de mer 1870–71* at Montoire – had 216. Such specialist societies clearly had a local, but not a general, significance.

The financial situation of the societies was detailed in the 1907 survey. Once again, a comparison is possible between the two main sets of societies. In both, the annual subscriptions of ordinary members was the largest single source of income, representing 56 per cent of the income of sickness societies and as much as 62 per cent of that of sickness and pension societies. The annual subscriptions of honorary members were of virtually the same proportion in the two sets of societies (being 10% and 9% respectively). Grants, gifts and legacies together with entry fees produced about 14 per cent of the income of sickness societies and about 10 per cent of that of sickness and pension societies, while interest on invested capital contributed about 13 per cent of income in both cases. Overall, the combined sickness and benefit societies were more self-reliant financially than were the sickness societies. The expenditures of the two sets of societies also showed some significant differences, as well as some similarities. In both cases, the indirect medical costs (doctors' fees and medicines) were the largest single item, amounting to about 41 per cent in the sickness societies but to about 57 per cent in the sickness and pension societies (presumably reflecting the different age profiles of the two sets of societies). The direct costs of cash payments to members comprised 32 per cent of expenditure in the former societies, but only about 23 per cent of that of the latter. Meeting funeral costs represented respectively about 4 per cent and about 5 per cent of expenditure, and administrative costs about 3 per cent and about 4 per cent. Most of the societies had incomes in 1907 which exceeded their expenditures during that year: for all of the societies together, expenditure amounted to 78 per cent of their income. Overall, at the end of 1907 the reserves of these societies amounted to more than five times their total expenditure in that year. Those societies providing just sickness benefits had expenditures which amounted to about 78 per cent of their income, whereas the figure was about 85 per cent for those societies providing both sickness and pension benefits. At the end of the year, the former societies had reserves which amounted to more than five times their collective expenditure in 1907, while the latter's reserves amounted to more than four times their expenditure. Three of the free societies and 22 of the approved sickness and sickness and benefit societies – 14 per cent of the total – had expenditures in 1907 which exceeded their incomes for that year, so that they had to call upon their reserves to meet those deficits. All were able to do so, most without difficulty, and all 25 societies had accumulated balances remaining at the end of the year which were larger than the working deficit for the year itself. But 10 societies (6 per cent of the total) had outgoings in 1907 which exceeded their annual receipts by amounts which were larger than the balances remaining in their reserves at the end of the year (after taking into account the deficits for that year): they could not be considered as being financially sound.[55]

These general perspectives on the functions of mutual aid societies between 1850 and 1914 now need to be brought into sharper focus. Constraints of space do not allow detailed consideration of specific societies at work; instead we will turn to a consideration of the social roles of such societies in general.

The social roles of societies
The initial organisation and subsequent management of mutual aid societies allowed well-established individuals within particular localities to use their social standing, their financial resources and their leadership qualities for the benefit of a wider community. But running such societies also provided other individuals with opportunities to develop their managerial and personal skills, to make them more aware of the extent to which they could both control and change their worlds. Mutual aid societies also had importance not only for their individual members but also for the communities within which they developed.

The primary role of mutual aid societies was self-evidently to provide assistance to their members when they were unable, for good reason (usually ill-health or injury) to work. Such societies in effect modernised, rationalised and institutionalised the provision of charity and of informal mutual help systems within a community or among a self-identifying social group. They provided aid either in the form of substitute labour or in cash. In addition, they met or contributed towards the costs of an attending doctor and/or of prescribed medicines: in this regard the societies were instrumental in aiding the diffusion of scientific forms of treating ill-health and injuries, and the gradual displacement of traditional and even superstitious practices. They sometimes made some provision for permanent disability and/or old-age pensions for members and often aided their widows and orphans. In doing so they encouraged the development of an awareness of the social benefits of organised, rather than informal, systems of personal insurance.

The provision of substitute labour rather than cash benefits was particularly characteristic of the societies of *vignerons*: the practice of labour substitution was, of course, only practical in a society whose members practised the same occupation. Such labour substitution was a simple method of providing mutual aid, building upon a tradition of co-operation within agricultural localities and involving opportunity costs rather than what might otherwise have been perceived as 'real' costs to members. In viticultural districts it also meant that such formal mutual aid would only be required from members during the period when vines had to be carefully attended. At Chaumont-sur-Loire, for example, the society's statutes (1872) specified that cover would be provided between 22 January and 10 July.[56] At Mesland (1861) the relevant period was specified as being from 12 January until 12 July; at Onzain (1864) from 22 January until 22 July; at Les Montils (1886) from 1 November until 8 July; at Ouchamps (1899) from 1 November until 12 July.

A secondary role of mutual aid societies was the moralisation of communities. When the municipal council of Les Montils expressed the view in November 1864 that a mutual aid society would bring both material and moral advantages to its members, it was echoing a widely held view. Often intentionally, and almost always, such societies acted as a source of social integration: they brought together to their mutual benefit members who identified themselves as a group either geographically or occupationally, and at the same time they brought together the well to do and the less well off respectively as honorary and ordinary members. At the same time, societies established standards of socially acceptable behaviour and excluded from membership individuals who were considered not to meet those standards. Illnesses and injuries resulting from the consumption of alcohol were sometimes explicitly excluded from the cover provided by societies.[57] Membership imposed social discipline upon individuals, for there were rules which had to be obeyed and obligations which had to be fulfilled. In the earlier period of their history some societies endeavoured to spread religious values, as previously noted, but that role gradually declined. Instead, mutual aid societies fundamentally promoted the bourgeois values of the family, and of good personal conduct and thrift. In many cases, as has been mentioned already, they were grafting such values on to the fundamental principal of mutual aid which had long been put into practice in agricultural, and especially in viticultural, localities.

A third role performed by mutual aid societies was that of providing a community with opportunities for sociability. Some were essentially of a private kind, involving visits to the sick. Others were more public demonstrations of fraternity, such as attendance at the funerals of members (as required by some societies), attendance at the general meetings of a society (as required by most

93 — **Salbris** (L.-et-C.) - Fête de la Société de Secours Mutuels

Photograph 5.1 The *fête* of the mutual aid society of Salbris, in the Grande Sologne
Source: 3 Fi 6087 photo, Archives Départementales de Loir-et-Cher

societies), and on certain other occasions (such as the fête of a society's patron saint) to celebrate Mass (photograph 5.1) or for social drinking. It was, of course, mainly adult males who were able to participate in such social activities but they were, no doubt, at least observed from time to time by women and children as local spectacles. While almost all societies required members to contribute towards the costs of funerals, some obliged them to attend those ceremonies. Thus members of the society of *vignerons* at Mennetou-sur-Cher (of whom there were 76 ordinary members and 10 honorary members in 1868) were required by their founding statutes of 1858 to attend funerals. Members of the societies of *vignerons* at Onzain (established 1864), at Chaumont-sur-Loire (established in 1872), and at Mesland (established 1867) were required to attend the funerals of their respective members and also to celebrate 'with dignity' the fête of St Vincent on 22 January each year.[58] An apparently more extravagant celebration of the same fête was made by members of the mutual aid society of the canton of Contres in 1852. The society had been founded in 1848 by M. Jauze, a doctor, who was to be its president for almost thirty years. He was totally committed to the society, producing for it an elaborate, decorated, printed membership certificate and writing for it a song of nine verses to be sung exclusively at its meetings. He arranged for the society's patronal fête on 19 July 1852 to be presided over by the prefect; the society met at the

Hôtel de Ville of Contres at 9.30 a.m., then processed to the church for a celebratory mass at 10.00 a.m. Money raised at the collection during the service
was then used to pay for a distribution of bread to the poor, in front of the
mairie, by the society's committee members. Then at 2.00 p.m. began a
banquet for members of the society – at which no doubt M. Jauze led the
singing.

The evolution of mutual aid societies

During the second half of the nineteenth century mutual aid societies in Loir-
et-Cher became both more numerous and more widely spread throughout the
department. But they also changed in terms of their structures and functions,
while at the same time retaining some persistent characteristics: in effect, they
evolved, adapting to changing circumstances.

The more than trebling in the number of societies between 1850 and 1914
was not accompanied by a similar increase in the size of societies. As has
already been stated, the societies in 1868 had an average of 97 members and
in 1907 they had an average of 103 members. The median-sized approved
society in 1868 had 67 ordinary members, while in 1907 it had 55 members and
the median-sized free society then had 48. The overall growth in the number
of societies during the second half of the nineteenth century was most marked
in those with below-average numbers of members and the average size of societies decreased. The increase in the number of relatively small societies is one
of the striking features of the evolution of mutual aid societies in Loir-et-Cher
during this period. Many of these new societies were in rural areas, some in
communes with relatively small populations. In 1868, for example, there were
16 approved societies with fewer than 50 or fewer ordinary members each and
7 of them were located in communes whose populations were less than 1,000
in 1866. In 1907 there were 50 such societies and 30 of them were located in
communes with populations of less than 1,000 in 1906.[59]

Another striking feature is the declining numerical significance of honorary
members, from an average of 26 per cent of the total membership of a society
in 1868 to only 16 in 1907. From being about one in four of the total membership of all of the department's societies in 1868, the proportion of honorary
members fell to about one in six by 1907. Such numerical decline no doubt
reflected the diminishing social role which honorary members played in societies: they came to rely less on patronage of the *notables* and more on the
enterprise of ordinary members. The contributions of honorary members to
the financial health of societies declined during the period, representing about
one quarter of the income of societies in the 1860s but only one-tenth in the
early 1900s. Correspondingly, the societies came to rely increasingly on the
financial contributions of their ordinary members. They became, accordingly,
even less patronal and more fraternal than they had been.

But the overall growth of mutual aid societies between 1850 and 1915 was not matched by a growing participation in them by women. On the contrary, the proportion of societies admitting women as members declined from just over two in five in 1860 to only one in five by 1902. Women comprised 8 per cent of the total membership of societies in 1868 and the proportion had only increased marginally to 9 per cent by 1902. There is no indication that as these societies evolved they did so by extending the concept of fraternity to embrace that of sorority.

Two major changes in the finances of societies between the 1860s and early 1900s were the considerable increase in the significance of medical costs, from about one-third to about one half of the societies' expenditure, and the enormous increase in the reserves of societies from being about the same size as the annual expenditure of societies to being between four and five times that amount. Societies appear to have become more cautious in the distribution of their benefits. In 1868, approximately four out of every five societies had incurred expenditure in meeting doctors' fees, in contributing to the costs of prescribed medicines, and in making cash payments to members; but in 1907 the proportion was only one in two. While there was a general development and spread of mutual aid societies in Loir-et-Cher, they seem to have come to focus their benefits more narrowly.

While the striking growth of mutual aid societies in Loir-et-Cher between 1850 and 1914 may be taken as an indication of their overall success, the fact that some struggled to survive and others actually failed needs also to be considered. Such struggles and failures can be made to shed new light on the nature of these associations themselves.

Struggles for survival

Some societies ran into difficulties because of what some contemporaries considered to be the indifference of their members, although in practice it was probably related to their inability – because of their small size – to deliver the promised benefits. Five years after its foundation in 1853, the society of *vignerons* at Chissay was reported by the mayor not to be functioning properly. Most of its members had resigned, because – according both to the mayor and to the society's president – they were not willing to meet their obligation to provide replacement labour. In 1858–9 the society was reformed, with revised statutes and a new president, but he resigned in September 1861 on the grounds that it was impossible to run a society of only a dozen or so members, most of whom refused to obey orders given to them by its officers. The only way for the society to survive, in the view of the resigning president, was for it to change to levying subscriptions rather than to continue to rely on replacement labour. For some years thereafter the prefect endeavoured to keep the society going. He did so in 1863 by trying to engage the support of the *curé* of Chissay and of the local *juge de paix*, but both declined on the grounds that the society was in effect dead. In September 1863 the prefect told the mayor that the society would have to be dissolved formally, but in November 1864 the mayor told the prefect of renewed interest among many of Chissay's residents in reorganising the society. The following month a general meeting approved new statutes, incorporating money subscriptions and benefits. Even so, the society then had only 25 ordinary members and its future was not assured. In January 1866 the president reported that the society had been cruelly tested by the sickness of the majority of its members: one member had contracted smallpox and passed it to eight others, and the cost of meeting that crisis had almost entirely used up the resources of the society, just when it was beginning to achieve a new vigour. The sick members had alleviated the society's difficulty, by accepting only half of the benefits due to them. The society's request for a grant of 500 fr. was in practice met by an award of 200 fr. from central government. In 1868 the society was recorded as having no honorary members and only 17 ordinary members. Despite its difficulties, the society survived: in 1907, still without any honorary members, it had 22 ordinary members.

The incidence of an exceptional number of claims for sickness benefit did at times put pressure upon the resources of societies. For example, at Chitenay in July 1891 the mayor reported that the financial position of the commune's society had been undermined by a large number of serious and prolonged illnesses, and that during the first half of 1891 expenditures had substantially exceeded the society's income from subscriptions. The society's bid for a subsidy from the central authorities was unsuccessful; its problem proved to be only temporary (and might have been exaggerated in order to bolster the claim for a subsidy), because it was still operating in 1907, when it had 84 ordinary members.

Some societies ran into difficulties because their members were only willing to provide replacement labour, not to pay monthly subscriptions From 1865 onwards the mayor of Saint-Georges-sur-Cher tried to establish for the commune's *vignerons* a society which would both provide replacement labour and meet the medical costs of an illness. When he was at last successful, in 1875, he could only be so by setting up a society which limited its benefits to replacement labour: potential members were unwilling to pay the cash subscription necessary to underwrite the payment of medical costs. At Saint-Julien-de-Chedon the society for *vignerons* established in 1869 allowed its members to opt either to undertake replacement labour themselves or to make a payment in lieu at the rate of one franc for each one-third of a day's work.

Replacement labour was itself not always forthcoming to the extent that it was needed – an exceptionally high demand for replacement labour (because of an exceptionally high number of claims for benefit) could place a strain upon the resources and loyalty of a society's fit members. That was the case at Mennetou-sur-Cher in mid-1857, where unusually high demands for replacement labour and an alleged inability to afford a change to a system of cash subscriptions led the *vignerons* to seek official approval for their society in the hope that it would enable it to obtain subsidies from the authorities.

The central authorities certainly preferred societies to be established on the basis of subscriptions by members which would be sufficient for all benefits to be paid in cash, eliminating the need for replacement labour. When a society was being created for farmers in Montlivault during the summer of 1862, the prefect tried unsuccessfully to persuade its founders to rely upon subscriptions and not upon replacement labour. In 1863 the society was expanded and divided into two sections, one for *vignerons* and farmers (based upon replacement labour) and a second for other workers (based upon cash subscriptions and benefits). There was clearly support in farming localities, and especially among *vignerons*, for the creation of societies relying substantially upon replacement labour rather than upon cash subscriptions. New societies based upon the former continued to be founded throughout the second half of the nineteenth century:[60] for example, such a society called La Fraternelle was created by the *vignerons* of Ouchamps in 1899 (and continued to function in that manner until its dissolution in 1920).[61]

Some societies encountered problems – or were warned by the central authorities that they might do so – because they levied subscriptions at too low a rate to meet their obligations. For example, when the society at Mer was seeking approval in 1865 the Minister of the Interior pointed out the proposed monthly subscription of 50 c. per member was unlikely to be sufficient to meet the proposed benefits (doctor's fees, costs of medicines, daily sickness allowance of 50 c., funeral costs). The society itself demurred from that view, arguing that any deficit could be met from the contributions of the large number of honorary members. It was certainly the case that in 1868 the subscriptions of 62 honorary members represented almost one-third of the society's income, significantly underpinning the subscriptions of 102 ordinary members and allowing the society's accounts to show a small surplus for that year. In 1907 the society's 52 honorary members contributed 49 per cent of its income, while its 204 ordinary members contributed a mere 7 per cent (the other main source of income was interest on investments, amounting to 29 per cent of the total).

By contrast, other societies encountered problems because their subscriptions were considered by members to be too high. Thus the annual general meeting of society at Saint-Dyé-sur-Loire in February 1867 decided that the monthly subscription, set when the society was founded in 1863 at one franc, was too high: many members considered it to be excessive, some had resigned because it was too high and it was argued that many people had not joined the society for the same reason. Accordingly, the rate was reduced to 75 c. per month with marked effect: during 1866 the society's membership was declining and at the end of the year it had only 28 ordinary members, but by the end of 1868 it had increased to 83 (and in 1907 it had 104).

Some problems stemmed from the competitive existence of two societies within a single commune, with each potentially limiting the membership and threatening the viability of the other. Two societies were established at Mont in 1873. The Société Générale was recognised by the commune council at the outset as having the larger number of members and wider social reach, while the Union – with more honorary members, including the Viscount de Bizemont and the *curé* of Mont-près-Chambord – was seen as being and as likely to remain a smaller society with members drawn from a narrower social base. Proclaiming its belief in the freedom of association, the commune council declared that it was 'not opposed' to the formation of two societies – although it did not expressly accept the need for both. In 1907 the nine honorary members of the Union contributed 13 per cent of the society's income, compared to the 46 per cent coming from ordinary members and 27 per cent from interest on investments; by contrast, the one honorary member of the Société Générale contributed less than 1 per cent of that society's income, while its 115 ordinary members contributed 54 per cent and interest 40 per cent. At Onzain in 1877, when a second society was being proposed, the prefect pointed out that while it was perfectly legal for there to be two societies within a commune it was not entirely sensible, because each would weaken the other and it would be more advantageous to have just one society. Similarly, the Ministry of the Interior argued that a single society would have more resources and be better able to act than would two separate societies. Nonetheless, a second society was established in early 1878 for the *vignerons* of two hamlets in Onzain, eventually rationalised by the prefect to the minister because of the geographical distance separating the hamlets from the main *bourg* of Onzain. This second, free, society used the system of replacement labour. It never had many members: at the end of 1902 it had a total of 16 honorary and ordinary members and at the end of 1904 the remaining members of the society agreed to its dissolution.[62]

Sometimes a society malfunctioned if it lost, for whatever reason, the services of a key individual in its organisation. Thus after the death of its active president in 1875, the society at Salbris – founded in 1868 – virtually ceased to function: meetings were no longer held regularly, its income was not properly collected, and payments to the doctor for his services were not made. On 5 October 1875 Dr Jourdain drew the matter to the prefect's attention, pointing out that he was already owed his fees for the last nine months. He was still attending to members of the society but asked the prefect whether he should still do so or whether he should require them to pay personally. He recognised that, if he were to do the latter, the society might collapse completely, but he did not want that to happen because the society had provided a good service to the poor labourers (*pauvres journaliers*) of the commune. The problem proved to be temporary: the prefect arranged, through his sub-prefect at Romorantin, for the mayor to use his influence to revive the society and to have a new president nominated. He must have been successful: by 1907 the society had 206 ordinary members.

Collectively, these cases indicate that mutual aid societies could run into a variety of problems. Often the difficulties were overcome, sometimes they were not.

Admissions of failure
Societies had to be formally established; if they failed, then they had also to be formally dissolved in accordance with their statutes. Records relating to these extreme cases throw further light on the difficulties some societies faced – difficulties which in some cases meant that a society had ceased to function properly or indeed at all for a year or even more before it was formally dissolved, either by its own members or by the prefect.

Some societies were said to have failed because they had insufficient resources. This was the prefect's explanation for the dissolution of the societies at Azé (1865–74), Beauchêne (1855–75) and at Châtillon-sur-Cher (1869–74).[63] The society at Azé was listed in the 1868 survey but no information was provided about it and it might be that the society had already by then ceased functioning. The society at Beauchêne had only 20 members in 1868 and a deficit at the end of that year of 81 fr. 5 c., which left it with reserves of only 6 fr. 73 c. It seems to have been too small to be viable as a mutual aid society. The medical costs which societies had to bear certainly increased significantly during late nineteenth and early twentieth centuries, imposing a growing burden on their resources. The president of the society at Theillay (1888–1909) blamed its dissolution principally on the increasing number of surgical operations, but it is arguable not only that the society's expenses were increasing (which they undoubtedly were) but also that the society levied unrealistically low charges upon its members. The society initially required ordinary members to pay an entry fee of one franc and a monthly subscription of only 25 c., while honorary members had to pay 40 fr. annually. In order to cope with increasing costs, the monthly subscription of ordinary members was raised in 1894 to 50 centimes. The society in 1907 had only 6 honorary but 235 ordinary members. Its actual income that year from the former amounted to 200 fr., while 1,454 fr. 50 c was the amount from the latter. The society's deficit on the year, 681 fr. 15 c., had to be charged to its reserves which amounted to 727 fr. 40 c at the end of the year. The society's committee agreed in August 1907 that the high annual fee for honorary members might be deterring some individuals from joining and agreed to create instead two categories, with founder and former members paying 40 francs and new honorary members paying 20 francs. No suggestion was made for changing the level of subscriptions collected from ordinary members.[64]

Some societies became non-viable as their membership numbers fell, for whatever reason. That was the case with the society for the *vignerons* of two hamlets in Onzain (1878–1906), which had only 86 fr. 65 c. in its reserves when it was dissolved. The society founded in 1886 for the *vignerons* of Rilly-sur-Loire dissolved itself in 1900, when it had only 13 members. Its president reported that the society had operated on the system of replacement labour between 1 January and 14 July each year and that the new way of tending vineyards with ploughs (*à la charrue*) had almost entirely suppressed the need for manual labour.[65]

Unsurprisingly, some of the difficulties encountered by societies resulted from personal clashes. The society created at Maslives in the spring of 1864 immediately had problems. According to the commune's mayor, the society's president was a troublemaker who was leading opposition within the commune to the council's decision to purchase a new bell for the church. For two years the mayor unsuccessfully pressed the prefect (and, through him, the minister) to demand the president's resignation. At the end of 1868 the society was recorded as having 28 ordinary members but there were no financial transactions recorded for it during that year, and it must be assumed to have been inactive. In October 1873 the prefect informed the minister that he had dissolved the society, because it had really only existed in name since its foundation.[66] The prefect had taken more drastic action in the summer of 1854 in dissolving the Sociétés des Amis Réunis at Pontlevoy, because of a major dispute between its honorary members and its ordinary members about whether or not the society's celebratory church service should be followed by a banquet. To the prefect, the dispute reflected the hatred of the working class for the rich and was animated by the democratic ideas of the commune's stone quarry workers. He dissolved the society, informing the minister that he would endeavour to reconstitute it with better people (*avec meilleurs éléments*).[67] A new society, with the same name, was indeed approved in June 1855. Political differences also underpinned the prefect's dissolution of the society at Saint-Aignan in November 1891. In February of that year, according to the mayor, twelve honourable citizens had been refused admission to the society's annual meeting on political grounds, because they were republicans. The mayor reported that the society was in contravention of its statutes, that it had become a coterie, and that it was a nest of reaction (*foyer de réaction*). Having made further enquiries, the prefect dissolved the society under articles 291 and 292 of the law of 10 April 1834.[68]

There were other miscellaneous reasons for the failure of mutual aid societies. The society founded at Soings with about 50 members in 1887 dissolved itself only four years later, when it had only 32 members. It did so in part because of the difficulty of obtaining medical care in a district with so few doctors and in part because it was thought that its members were unwilling to increase the subscriptions to the level necessary to meet the society's costs. This latter explanation seems to have been inappropriate, because when the society's assets were liquidated it was found that over the four years of its existence the society had an income of 1,919 fr. and an expenditure of 1,013 fr. 77 c., resulting in an overall balance of 905 fr. 23 c. The crisis at Soings seems to have been one not of cash but of confidence. The mutual aid society established in 1863 for the volunteer firemen of Marchenoir ceased to function during the Franco-Prussian War of 1870–1 and attempts to reorganise it afterwards were unsuccessful, so the society was dissolved in May 1874. The society created in 1900 for the volunteer firemen of Couture was dissolved in 1908. Most of its honorary members had stopped paying their subscriptions because the commune's council had decided to insure its firemen against accidents.

While it has been appropriate to consider in detail some of the mutual aid societies which failed and were dissolved before 1914, as well as some of the struggles encountered by those which survived them, it needs to be emphasised that most mutual aid societies were successful.

Between 1815 and 1914 mutual aid societies in Loir-et-Cher came to be both numerous and significant. During that period one or more societies were established in approximately one in two of the communes of the department. Although the first formally approved societies were founded in the towns and larger settlements, many societies were created in rural communes and some of them were building upon pre-existing informal systems of mutual aid, especially among *vignerons*. While it is clear that there were serious attempts by the authorities – of both Church and State – to promote mutual aid societies, as well as by some *notables* and by some members of the liberal professions, it was also the case that in some instances the creation of societies reflected a broader democratic spirit and sense of fraternity (but not of sorority). Many mutual aid societies of Loir-et-Cher had patrons, but few of them – as has sometimes been argued more generally – were heavily dependent upon state subsidies: in 1868 subsidies and donations together accounted for less than 4 per cent of the societies' income, and in 1907 subsidies and gifts each accounted for less than 5 per cent, together less than 10 per cent, of the societies' income. In Loir-et-Cher it was not the case, as has been argued generally,[69] that mutual aid societies were underpinned by massive state subsidies and that without state support they would scarcely have been able to make ends meet. Their financial position was not as precarious as it has sometimes been claimed.

To be successful, a mutual aid society had to have a critical mass of honorary and especially of ordinary members: as institutions intended to manage risk by distributing and sharing it, societies could not be expected to function properly unless the risks were spread sufficiently widely throughout a population or locality. It could be argued, in the case of the smallest societies, that the fraternal sentiments of their members outran their economic logic. Even so, total failures were few.

Finally, while some of the mutual aid societies of Loir-et-Cher had memberships which were closed to particular occupational groups, many of them were open to all (male) workers within a community. While all such societies were defensive and some were exclusive in their posture, it would – in this case – be wrong to claim (as has been argued elsewhere)[70] that they were fundamentally elitist, intended to preserve the status of the few and not to serve the needs of many. The concept of mutual aid was put into practice in Loir-et-Cher by social groups who identified themselves as either an occupational or a geographical community; they did so in both rural and urban settings; and they often did so as a direct expression of the spirit of fraternity.

6

Fire-fighting corps

'Le règlement de discipline est accepté par les sapeurs-pompiers désirant mainte-
nir la bonne harmonie et la vrai fraternité'

Rules of the Compagnie de sapeurs-pompiers of Mulsans (1863)

Next to diseases and accidents affecting livestock and to those affecting
people, fire was probably the greatest single threat to social welfare in early
nineteenth-century rural France (as in other pre-industrial rural societies). It
was a perpetual environmental hazard, whether caused deliberately or accidentally. The risk of fire was continuous and widespread: from fire torches and
candles used for lighting; from unswept chimneys and badly maintained
ovens; from burning coals carried from room to room or neighbour to neighbour to light a domestic hearth; from furnaces and forges; from bonfires (both
festival and utilitarian); sometimes from lightning and from arson.
Furthermore, many of the organically derived materials of daily life were
highly combustible: buildings were substantially or partially of timber, many
roofs were of thatch, straw provided household bedding and litter for animals,
hay was stored in lofts for fodder and for insulation as well as in stacks in fields.
Fires could spread catastrophically (especially in towns and larger villages,
where buildings were cheek-by-jowl), causing not only physical destruction to
buildings and their contents but also social destitution for their owners and
occupiers (photograph 6.1).[1]

During the morning of 13 November 1803 at Pezou, for example, a fire
caused by a spark from a chimney setting light to the thatched roof of a neighbouring house, burned down four houses, four barns, three stables and a pigsty
as well as killing a horse and a pig and destroying stores of hay, straw and
wheat; at Vineuil just before midnight on 31 May 1818 a fire, caused by the
overheating of millstones not being fed with grain, totally destroyed a wind-
mill and its store of 150 bushels of wheat in less than an hour; in May 1860 at
Dhuizon the commune's inhabitants, lacking fire-fighting equipment, had passively to watch a fire which destroyed buildings along a 200 metres' stretch in

EN SOLOGNE. — *Coin de Ferme*   *Collections ND Phot*

Photograph 6.1 Farm buildings in the Sologne
Source: 3 Fi 6375 photo, Archives Départementales de Loir-et-Cher

the *bourg*, as well as stores of wheat and animal fodder and a considerable amount of agricultural equipment.[2] However started – whether by carelessness, by arson or by lightning – fires caused considerable damage to buildings and their contents, the cost of which often amounted to hundreds and sometimes to thousands of francs at each incident in the 1860s. Fire damage to buildings in 1907 varied between 10 fr. and 16,000 fr., but was usually 1,500–3,000 fr., while that to harvested crops and fodder stored in the buildings ranged between 30 fr. and 13,000 fr., but was usually 500–1,500 fr.[3]

As in most of France outside Paris, fire-fighting techniques in Loir-et-Cher were at best primitive and at worst useless at the beginning of the nineteenth century, so that the scene of a fire could easily degenerate into panic. In the absence of fire-insurance companies and of organised fire brigades, the only possible response to a fire was for friends and neighbours to form a chain of buckets (sometimes leather buckets intended for the purpose, but often just any available container) from a source of water such as a pond, stream or well to the site of a fire, constituting a pragmatic expression of *fraternité,* expressing a basic form of mutual aid. Individuals responded to an outbreak of fire by forming themselves into a spontaneously created fire-fighting group whose existence was not intended to outlast the incident which occasioned it. When hand-operated pumps became available, then organised and equipped associations of firemen – of volunteer fire-fighting associations (*corps de sapeurs-pompiers*) gradually came on to the scene in significant numbers.[4] In 1800 there

Photograph 6.2 An officer reviewing a *corps de sapeurs-pompiers*, and its pump
Source: 3 Fi 7988 photo, Archives Départementales de Loir-et-Cher

was only one such fire brigade in Loir-et-Cher, but between then and 1914 a further 160 were created (photograph 6.2).

The development and spread of fire brigades

Conscious of the damage which fires could inflict not only upon individuals but also upon communities, and aware of the need for a militaristic discipline in the training and operating of fire-fighting *corps*, central authorities in France were anxious both to promote and to control the activities of local fire brigades. This concern resulted in considerable correspondence (between and among officers of the brigades, mayors of communes, sub-prefects, prefects and ministers), as well as in numerous circulars and enquiries, rules and accounts. From this mass of unpublished – and mainly uncatalogued – records it has been possible to reconstruct the development and spread of fire brigades in Loir-et-Cher during the nineteenth century.[5] They have hitherto received no analytical attention: they were, for example, entirely neglected by Roger Dion in his classic study of the Val de Loire, by George Dupeux in his scholarly study of the social history of Loir-et Cher between 1848 and 1914, and – perhaps more surprisingly – by Christian Poitou in his recent survey of peasant life in the Sologne during the nineteenth century.[6] A useful but essentially factual survey of the history of the *corps de sapeurs-pompiers* in Loir-et-Cher has been compiled by a local historian, André Prudhomme: his descriptive account provides a back-cloth for a more interpretative and contextual analysis and synthesis of the historical geography of Loir-et-Cher's fire brigades during the nineteenth century.[7]

Historical development

In France, the first fire brigades – *corps* of men trained to use manually operated water pumps equipped with leather hoses – were established in Paris in the early eighteenth century. In Loir-et-Cher, only a very tentative step in such a direction was taken in that century when, on 5 June 1773, the town council of Blois decided to purchase two small pumps and to put one *pompier* in charge of each of them. The first fire brigades – with pumps operated by teams of firemen wearing uniforms or insignia which set them apart from the crowd – were established during the summer of 1805 in the three principal towns of Loir-et-Cher: Blois acquired two pumps and a *corps* of 40 men; Romorantin one pump and 14 men; Vendôme one pump and 16 men. During the early 1800s the formation of brigades was being actively encouraged by Loir-et-Cher's central administration: in 1806, the *Annuaire* of the department reported that the prefect wanted to see the benefits of such fire-fighting teams not restricted to its most populous communes but extended to all communes by the formation of *corps* which would serve groups of neighbouring communes whose populations could not each provide an effective complement of men; and in 1810 the prefect appointed one of the officers of the brigade at Blois as Inspecteur des compagnies de pompiers du département de Loir-et-Cher, charged with the responsibility of checking the proper functioning of the pumps and training of the *pompiers* in the existing brigades. By the end of 1810 there were ten brigades in the department: to those at Blois, Romorantin and Vendôme had been added *corps* at Savigny-sur-Braye (1807), at Contres and Mondoubleau (1809), and at Mer, Montoire, Morée and Saint-Aignan (1810). All brigades existing in 1810 had been established in *chefs-lieux* of cantons (but not all such communes had set up brigades by then).[8]

The efforts of the prefect to promote the formation of brigades during the opening years of the nineteenth century had only limited success and can hardly be said to have led to the sustained 'take-off' of fire-fighting services in Loir-et-Cher. During the 1810s and 1820s only one new brigade was formed (at Cheverny, where the owner of the château purchased a pump in 1811 and placed it at the disposal of the commune and its neighbouring commune, Cour-Cheverny, although a brigade of *pompiers* was not formally constituted until 1817). Local interest in the formation of brigades lay dormant for two decades, awakening in the early 1830s when brigades were established at Oucques and Saint-Claude-de-Diray (1832) and at Chaumont-sur-Tharonne and Selles-sur-Cher (1834). By 1835 there were fourteen brigades in existence: most (to be precise, eleven) of them had been set up in cantonal *chefs-lieux*

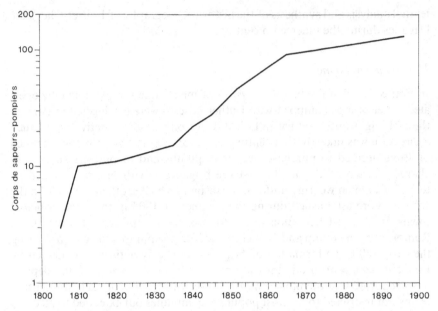

Fig. 6.1 *Corps de sapeurs-pompiers* in Loir-et-Cher, 1800–1900
Sources: AD Série R Sapeurs-pompiers (uncatalogued files, listed in bibliography)

(although it was still the case that slightly fewer than half of such centres had established brigades by 1835).

What might be described as the origin phase of the development of *corps de sapeurs-pompiers* in Loir-et-Cher before 1914 extended from the late eighteenth century through until the mid-1830s, when the founding of brigades quickened and continued apace until the mid-1870s (figure 6.1). One-half of the brigades founded in the department by 1914 were created between 1850 and 1875: more than one-quarter were formed during the 1860s alone. In sum, 161 brigades were recorded as having been founded in Loir-et-Cher before 1914: more than half (53 per cent) of its communes saw fire-fighting associations established. Created unevenly in time, they were also distributed unevenly in space.

Geographical distribution

Until the mid-1830s brigades were almost entirely restricted to cantonal *chefs-lieux* located in the more populous and better connected communes, and they were distributed fairly widely throughout the department. Of the fourteen brigades operational in 1835, two were located in Perche, two in the Loir valley, two in Beauce, three in the Val de Loire, two in the Cher valley and three

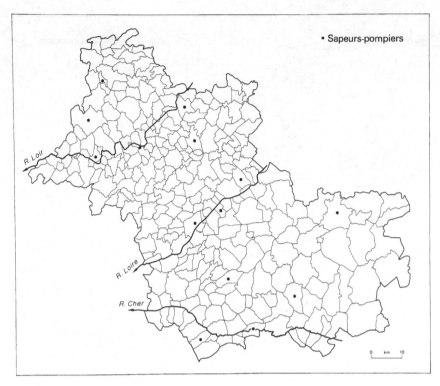

Fig. 6.2 *Corps de sapeurs-pompiers* in Loir-et-Cher in 1835
Source: AD Série R Sapeurs-pompiers: organisation de services d'incendie 1807–42

in the Sologne (figure 6.2). From the mid-1830s, the diffusion of brigades was decreasingly structured by the settlement system (between 1835 and 1842, for example, eight new brigades were founded but only one was established in a cantonal centre); moreover, from the mid-1830s one can begin to detect an increasing localisation of brigades within and proximate to the Loire Valley (figure 6.3). By 1852 a concentration of fire brigades in communes of the Val de Loire, in communes of the Petite Sologne closest to Blois, and to a lesser extent in communes of Beauce, was apparent (figure 6.4). Thereafter, during the 1850s, 1860s and 1870s the formation of a further eighty brigades significantly altered the density but only marginally changed the distribution of the early 1850s (figures 6.5 and 6.6). By the end of the nineteenth century (figure 6.7) the concentration of brigades in Beauce, in the Val de Loire, and in the Petite Sologne was most marked, although by then the 'lagging' *pays* of the Grande Sologne and of Perche had done some catching up.

The uneven development of fire brigades in time and over space thus identified needs to be explained and interpreted. It is relatively easy to reach

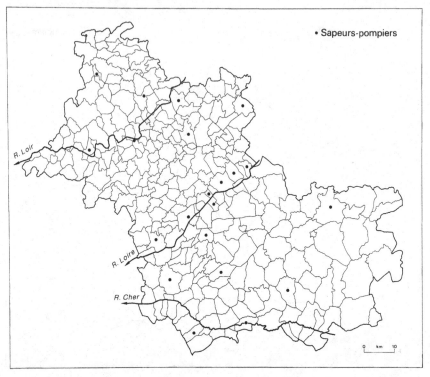

Fig. 6.3 *Corps de sapeurs-pompiers* in Loir-et-Cher in 1842
Source: AD Série R Sapeurs-pompiers: organisation de services d'incendie 1807–42

three negative conclusions. Firstly, the growth of fire brigades was not related to a growing incidence of fires: although a few communes did found brigades immediately following serious fires on their territories, so that a few brigades could be interpreted as direct responses to particularly disastrous events, there is no indication or suggestion that the overall historical geography of fire brigades was related to any patterned incidence of fires. Outbreaks of fire in the department varied in number and in location very considerably from one year to the next.[9] *Corps de sapeurs-pompiers* were indeed formed immediately after fires in the communes of Courbouzon (1842), Cour-Cheverny (1847), Dhuizon (1860), Coulanges (1865) and Saint-Amand (1869), but attempts to establish a brigade at Mondoubleau after a fire there in 1804 were unsuccessful and in most communes the creation of a brigade was not an immediate response to a local fire. Secondly, the formation of brigades was only indirectly a response to technical improvements in fire-fighting: information about manually operated water pumps and leather hose technology was available centrally and had been diffused locally by the end of the eighteenth century, and

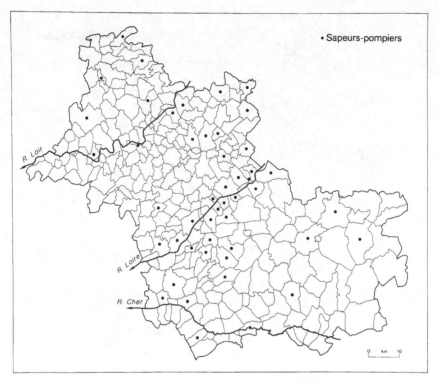

Fig. 6.4 *Corps de sapeurs-pompiers* in Loir-et-Cher in 1852
Source: AD Recueil des actes administratifs du département de Loir-et-Cher 1852

technical innovations of the nineteenth century (such as the rubber hose and the steam-powered pump from the mid-1880s) did not give rise to a wave of brigade formation in Loir-et-Cher. Fires and technical innovations were both necessary but not sufficient conditions for the creation of *corps de sapeurs-pompiers*. Thirdly, the organisation of local fire brigades cannot in general be attributed to the evident enthusiasm on the part of the central administration for the formation of effective fire-fighting corps: such enthusiasm persisted throughout the nineteenth century and its peaks – in the years immediately following legislation in 1831 and 1851 promoting and controlling *corps de sapeurs-pompiers* – do not correlate with those periods when brigades were most actively being organised. Most significantly, although the early spread of brigade formation shows some of the characteristics of an hierarchical diffusion (in effect, from Paris to Blois and onward to cantonal centres within the settlement system of Loir-et-Cher), its geographical extension during the main phase of expansion exhibited an uneven, locally specific, distribution.

The timing and spacing of *corps de sapeurs-pompiers* in Loir-et-Cher during

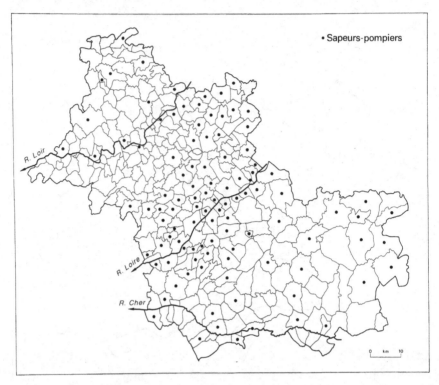

Fig. 6.5 *Corps de sapeurs-pompiers* in Loir-et-Cher in 1865
Source: AD Série R Sapeurs-pompiers: caisses communales de secours 1864–69

the nineteenth century appears not to have been closely correlated with the incidence of fires, with the promotional activities of the authorities, or with the pattern of settlement and its degree of nucleation. Consequently, what is needed is a more finely grained, contextual, interpretation of the historical geography of *corps de sapeurs-pompiers* than has so far been essayed.

The promotion of fire brigades

Significant agents

A handful (fewer than 10 per cent) of brigades owed their existence very directly to the benevolence of particular individuals, some of whom no doubt acted disinterestedly while others did so in order to further their own standing within a particular locality. Such generosity on the part of the *notables* was not always immediately appreciated by the commune councils or the *peuple* whom they represented. For example, the owner of the Château de Cheverny

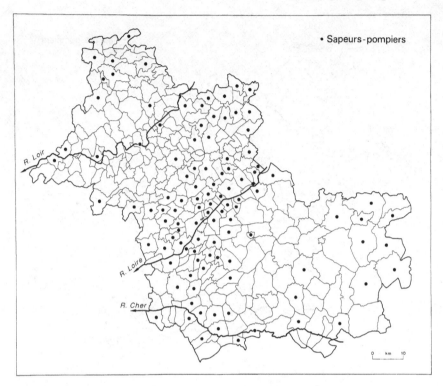

Fig. 6.6 *Corps de sapeurs-pompiers* in Loir-et-Cher in 1887
Source: AD Série R Sapeurs-pompiers: instructions, enquêtes, effectifs, matériel,
accidents, subventions 1876–97

put his pump at the disposal of the two communes of Cheverny and Cour-
Cheverny in 1811 but not until 1817 was a brigade formally established (and
it lasted only for thirteen years, being dissolved in 1830). Later, in November
1847, the gift by the Marquis de Vibraye of another pump to the commune of
Cheverny was only hesitantly followed by the formation of a brigade, eventually founded in August 1849. Gifts – either of pumps or of the cash needed to
purchase them – came from a variety of local and remote patrons: the King
and the Duc d'Orléans at Cellettes (1837); the Prince de Chimay at Ménars
(1841); the Duc d'Avaray at Avaray (1842); the Comte de Gomnegnier at
Salbris (1854); the Emperor Napoléon III at Vouzon (1860); Monsieur Tassin
de Beaumont at Villermain (1861); and an anonymous rich benefactor at
Coulanges (1865). Such benefactors came generally from the 'traditional' élite
and the *corps* which their gifts in effect created can hardly be interpreted as
being 'modern', spontaneous and local expressions of fraternity on the part
of those who became their first *sapeurs-pompiers*.

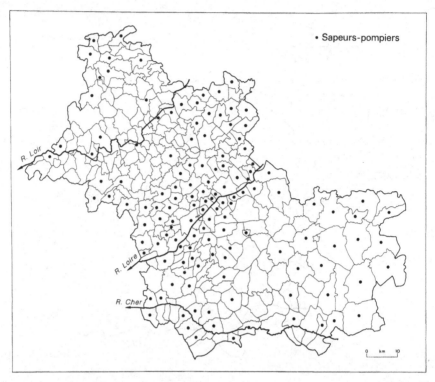

Fig. 6.7 *Corps de sapeurs-pompiers* in Loir-et-Cher in 1896
Source: AD Série R Sapeurs-pompiers: instructions, enquêtes, effectifs, matériel, accidents, subventions 1876–97

Central encouragement

That the central authorities actively promoted the formation of brigades is evident: the acknowledged public utility of brigades coupled with the need to control their activities, given the quasi-military nature of the *corps*, created a considerable, if inconsistent and incomplete, archive relating to them. Spurred by both encouragement and instruction from the Ministry of the Interior in Paris, prefects in Loir-et-Cher during the nineteenth century were constantly – but to varying degrees and with varying results – trying to stimulate the organisation and to ensure the regulation of *corps de sapeurs-pompiers* throughout the department.

In the 1830s, for example, the prefect and sub-prefects were keen to promote fire brigades through the department but especially in its more populous communes, and most especially in the *chefs-lieux* of cantons, but they were also well aware that many communes had limited financial resources. In the summer of 1835, therefore, they persuaded the Conseil Général to make some funds available to a cantonal centre or other populous commune deciding to acquire a pump. On 10 November 1835, the prefect wrote to the mayors of selected communes in the *arrondissement* of Blois, informing them that the department would provide them with some financial assistance if they were to establish fire brigades. Of the ten communes thus directly approached, only two had organised brigades within two years of receiving the prefect's letter and it was to be thirty years before all of the ten had established brigades. It is worth examining the reactions and experiences of those ten communes more closely.

A quick and positive response to the prefect's letter was seen at Onzain, whose council decided on 15 November 1835 to establish a brigade, raising through voluntary donations and a loan from an insurance company some of the 375 fr. needed for purchasing a pump, and receiving in 1836, when the brigade became operational, a grant of 200 fr. from the department. By 1842 the pump was manned by a brigade of twenty-five men, and served not only Onzain but also six other communes, claiming to reach four of them within one hour of being alerted and the other two within one and a half hours.[10] The second positive response came from Suèvres, but not until the autumn of 1837 when the mayor reported that a group of more than forty public-spirited men wanted to form a brigade, that they were each willing to donate 10 fr. for the purpose and also to clothe and equip themselves (except for helmets) at their own expense. The mayor told the prefect that he was planning a public appeal for funds to pay for the pump, but he could not count on a majority of the members of the commune's council agreeing to the purchase of helmets from commune funds and he feared that some members would use the issue as a tactic to delay the formation of the brigade. So the mayor asked the prefect – referring to his letter of 10 November 1835 – if the department would contribute financially to the project. On 25 November the mayor, reporting to the prefect that the public appeal launched on 29 October and contributed to by 279 people had raised 1,609 fr., asked how to progress the matter. On 19 November the prefect wrote to congratulate the mayor on the remarkable success of the appeal and said that he had submitted to the Ministry of the Interior a request for helmets. When the mayor on 5 December sent to the prefect a list of the names of forty-three men willing to serve as *pompiers*, the prefect pointed out that the brigade had to be formally established by the commune council. Divisions within the council about the project meant that it had to meet three times before, on 17 January 1838, finally approving the creation of a brigade. On 9 March the prefect told the mayor that he had still not been able to obtain helmets; on 8 May the mayor, informing the prefect that a pump had been acquired, asked directly for money as he had himself authorised the purchase of helmets so that the brigade could begin to operate, telling the prefect that they would both be embarrassed if the prefect did not keep his promise to provide help in creating the brigade; already embarrassed, the prefect replied immediately on 9 May that he had unsuccessfully tried to obtain helmets from the Ministry of the Interior and from the Ministry of War and that he would now, without any guarantee of being successful, seek a grant for the brigade from the department's Conseil Général; furthermore, on 1 June the prefect sent the mayor a personal contribution of 100 fr. as a token of his own appreciation of the efforts being made to establish a brigade at Suèvres. In August the Conseil Général approved a grant of 300 fr. Nonetheless, the full cost of the pump was not paid for some time: the mayor was still fund-raising during the winter of 1839–40, but by March 1840 there was still 160 fr. in payment outstanding. Two pleas for financial aid for the brigade by the prefect to the Ministry of the Interior were rejected, but the prefect eventually managed to persuade the Conseil Général to make a further grant of 160 fr., to enable the debt to be settled. That was in November 1841, so the prefect's 'success' at Suèvres had taken four years to achieve if looked at optimistically, six years if looked at pessimistically. But at least the brigade there was able to tackle that year a fire started by children playing with matches.[11]

In the other eight communes approached directly by the prefect in 1835, the outcome was even less positive. A brigade was not established at Ouzouer-le-Marché until 1840 and at Montrichard until 1843, each receiving a grant from the department of 300 francs. When a brigade with two pumps, funded by a supplementary communal tax of 2,000 fr., was established in 1847 at Vineuil, the mayor wrote in February to the prefect to seek a grant; in September, no grant having been received, he wrote again to say that he had read in *Le Journal de Loir-et-Cher*, the local twice-weekly newspaper, that some communes had been awarded grants towards the purchase of pumps. Not until 1848 did Vineuil receive a grant from the department of 200 fr., alongside a similar sum from the Compagnie d'Assurance Mutuelle.[12] At Bracieux, a brigade was created in 1852, financed from a public appeal (550 fr.) and grants of 200 fr. each from the department and the Société d'Assurances Mutuelles.[13] At Herbault the prefect's letter of 10 November 1835 was debated by its commune council on 19 May 1839. It was agreed that the commune itself could not afford to establish a fire brigade but that it should do so by seeking to meet the cost of a pump and its accessories by charging one-third to the commune and asking the department on the one hand and a group of neighbouring communes on the other hand each similarly to contribute one-third of the cost. When the prefect on 17 June 1839 informed the mayor that any grant from the department would be far less than one-third of the expected costs and that he had no authority to act in relation to the other communes concerned, the council of Herbault dropped the idea and it was not until many years later that a brigade was established in that cantonal centre, in 1853 when it had a population of just under 800.[14]

At Marchenoir, the commune council in August 1842 agreed in principle to establish a fire brigade: the necessary funds were to be raised by seeking a grant from the department, by public appeal, and – if insufficient funds came from those two sources – by an hypothecated communal tax. On 15 September the prefect told the mayor that only after a pump had been purchased and a brigade established in the commune would he be able to ask the Conseil Général to award a grant. Thus cautioned, the commune council at Marchenoir itself became reticent about the idea, postponing any action. In August 1845 the council considered it again, deciding to explore the possibility of joining with the neighbouring commune of Saint-Léonard-en-Beauce, in order to share the costs involved. But there seems to have been no further action until 1852, when there was compiled a list of men willing to serve as *pompiers*; but Marchenoir still had no pump. At the council meeting on 8 May 1853 the mayor argued that there was an urgent need to acquire a pump for the brigade, but the council decided that it could not find any funds for a pump, given the other demands upon its resources. The mayor returned to the matter at a council meeting on 7 August 1853, arguing that a pump was absolutely essential in that agricultural locality; his case was accepted and the council agreed to allocate towards the purchasing of a pump the sum of 300 fr., which had recently been raised from selling some of the commune's land. But that was by no means the end of the matter, which was debated again by the council at its meeting on 18 February 1855 when the mayor explained that a pump would cost about 900 fr., that an appeal had raised 270 fr., and that a grant of 200 fr. might be expected from the department and also of 200 fr. from an insurance company, given that the commune's public buildings were insured with it against fire. The council, acknowledging the undoubted benefit of a pump, then agreed to charge the outstanding sum – of 250 fr. – to the commune's account. But it was not until 1859 that a pump was acquired and the brigade formally authorised by the prefect.[15] In the same year, a brigade was also established in the nearby commune of Josnes – its council had debated the idea at least fifteen years earlier, but took the matter no further when the prefect on 5 July 1844 informed the mayor that a request for a grant could not be considered by the department until after the commune council had taken a firm decision to purchase a pump and establish a brigade.[16] But it was not until 1865 that a brigade was established at Saint-Georges-sur-Cher, thirty years after the prefect had written to its mayor urging him to consider establishing one in his commune.[17]

Even when the prefect's efforts to promote the creation of fire brigades in the 1830s were specifically targeted upon particular communes, they were far from being very successful. More generally, probably most – and certainly many – of the department's communes discussed the circulars sent to them by the prefect or sub-prefect encouraging them to found fire brigades, but decided against doing so then, and brigades were not established in them until much later. Such discussion took place during the late 1830s, for example, in the following communes, but brigades were not established in them until the dates indicated: Saint-Dyé-sur-Loire (1847); Neung-sur-Beuvron (1851); Le Gault (1854); Sougé (1865); Villiers (1865); Lunay (1895); and Villedieu (1906). At Trôo no brigade was founded before 1914. The idea was most frequently rejected by commune councils because of the considerable costs involved in relation to their limited and already stretched resources, but also on occasions because the dispersed pattern of settlement in some communes meant that not all of its inhabitants would benefit equally from the services of a brigade.[18] By 1842 only twenty-one had been established in the department, despite massive efforts by the prefect and sub-prefects (although it must be admitted that those efforts were to some extent undermined by the time it often took for the central authorities to match their words with hard support in the form of grants). Enthusiasm (even when combined with promises of subsidies) on the part of the central authorities was by no means a sufficient condition for the creation of fire brigades by local communities: there had also to be local concern and engagement, both a willingness to raise the necessary funds primarily within the commune either by a public appeal or by a special tax, and a willingness on the part of sufficient able-bodied men to volunteer to serve in a brigade.

Popular pressures

The strength of any 'grass-roots' enthusiasm for forming *corps de sapeurs-pompiers* is difficult to assess. At Prénouvellon in 1846 the commune council decided to acquire a pump, acknowledging that it would be very useful and that the inhabitants of the commune had for a long time been wanting one; and at Monthou-sur-Bièvre the mayor called a meeting of the council in December 1893 to consider forming a brigade, as requested by several residents of the commune; the council of La Chapelle-Vendômoise decided to organise a *corps* during 1895 because it believed that one was wanted unanimously by the population of the commune; and that at Séris acted similarly during 1896 in response to the wishes of the population. Such popular pressure is rarely explicitly recorded but almost certainly lay behind many similar actions by commune councils, most especially in those cases where a council first acquired or decided to acquire a pump and to establish a *corps* and then, almost as an afterthought, asked the central administration how to achieve that objective.[19] It might also have been the case in those communes whose councils decided to create brigades and then reported their decisions to the prefect: but in such cases it is impossible to determine whether local or central, individual or public, pressure lay behind the creation of the brigades.

What the central authorities did provide and control was the structure within which local communities, and individual agents within them, could act.

Operating through his sub-prefects at Romorantin and Vendôme, through inspectors of fire services, and directly with the mayors of some 300 communes, prefects effectively controlled an information network and carefully monitored the structure, function and evolution of fire-fighting services within their administrative territory. At irregular intervals, the services were surveyed throughout the department (notably in 1810, 1820, 1835, 1842, 1865, 1887 and 1896). Almost constantly, prefects showed concern with the day-to-day operation of brigades in particular communes. They were keen to ensure that the pumps were properly maintained and *pompiers* properly trained to use them. They were responsible for approving the rules of emergent *corps*, for making sure that a brigade was of a particular strength and that its officers were appropriately selected and appointed, and its men rewarded for their voluntary services in ways permitted by law. Prefects were able to arrange financial support, both in the form of direct grants and by authorising commune councils to raise additional taxes to meet the capital expenditure initially incurred in acquiring a pump and equipment (such as helmets, uniforms or insignia, drums and bugles, as well as fire-fighting equipment and arms). Always attentive to the formation and functioning of the *corps*, prefects tried to monitor them particularly closely from 1831 to 1871, when so many were being founded and – more significantly in this context – when the *corps* were part and parcel of the Garde nationale (and so had the maintenance of law and order added to their fire-fighting role).

Corps de sapeurs-pompiers in nineteenth-century France were viewed by contemporaries as offering an exclusively public but essentially voluntary service: no brigade could be established without both the prior consent of a local (commune) council and the ultimate authorisation of central authority (as represented by the prefect), but commune councils were not compelled by law to establish brigades and recruits to each *corps* were volunteers, not conscripts. The prefect's role, therefore, was to encourage the creation of new brigades within as many localities of the department as possible and to ensure that all brigades functioned in accordance with a nationally determined legal framework. The State wished the opportunity to establish *fraternities* of firemen to be *equally* available throughout France while leaving to each commune the *liberty* to decide whether or not to form a fire brigade to serve the community. Clearly, within Loir-et-Cher not all communes accepted the advice and assistance offered by the prefects: the development of *corps de sapeurs-pompiers* was a markedly uneven process by contrast with the relatively uniform availability of that advice and assistance throughout the department. Any attempt to understand the historical geography of the fire-fighting services must therefore also be conducted at the scale of the locality and in relation to individual *corps*, within the context of their general form and functioning.

The structure of fire brigades

Membership

Founded in 1805, the *corps* at Vendôme had acquired a second pump by 1820 when it had a complement of 50 men: they were all artisans (such as roofers, carpenters and stone-masons) described by the inspector of fire services as possessing the skills needed to stop a fire spreading (by demolishing a building) and they were rewarded for their voluntary services by being exempted from serving in the National Guard and from providing lodgings for military personnel in time of war. Nominal lists of *pompiers*, detailing their ages and occupations, survive for almost a dozen *corps*.[20] In the 1850s and 1860s, brigades usually numbered between 30 and 40 men: their average age was thirty-five years, their ages ranged between twenty-two and fifty-four. Within agricultural communes the *corps* were composed mainly of *cultivateurs* (as at La Chapelle-Saint-Martin in 1867, Courbouzon in 1852, Maves in 1854 and Mulsans in 1862) or of *vignerons* (as at Avaray in 1852 and Lestiou in 1864), or of a mixture of the two occupations (as at Cour-sur-Loire in 1869). By contrast, men from a wide range of crafts and trades (including clog-makers and plasterers, shop-keepers and café-owners) formed the *corps* of cantonal centres (as at Mer in 1852 and Montoire-sur-le-Loir in 1866). Exceptionally, a *corps* included a professional man (like the notary at La Chapelle-Saint-Martin in 1867 and the schoolmaster at Maves in 1854). Virtually all *corps* included one or more carpenters, stone-masons and metal workers; quite a few included long-sawyers, roofers, blacksmiths, ploughwrights and rope-makers. There was, then, a significant skilled or craft composition to these *corps* which, combined with their distinctive age and sex structures and the apparently dangerous, public yet voluntary, nature of their service, gave them a separate identity and a prestigious role within their communities. Certainly in mid-century the *sapeurs-pompiers* of the brigades of rural Loir-et-Cher were – both officers and men – mainly farmers, *vignerons* and artisans. There is no evidence that they were being led by socially superior landowners or retired military men, as was sometimes the case elsewhere.[21] Even the brigade established in 1805 in the town of Romorantin, whose fourteen *pompiers* were all artisans, was commanded by a metal worker as its captain and by a roofer as its second captain.[22] It must, of course, be emphasised that although men volunteered to serve in a brigade there also operated a degree of selection, because (until 1871) brigades were part of the National Guard and so membership required the approval of the prefect. At Mulsans in August 1862, for example, the commander of the recently established brigade sent a list of the thirty-two men being nominated for membership, stating their ages, occupations and domiciles within the commune, and claiming that all of them were not only capable of fulfilling the tasks which would be given to them but were also loyal to the *pays* and to the Emperor.[23] The unique character of a *corps* – given this cluster of characteristics – was further accentuated by its reliance upon a neo-military code of conduct.

Discipline

During 1805–6 both fire prevention measures and fire-fighting services were carefully considered and established at Blois: detailed rules relating to them, approved by the mayor on 1 June 1806, ran to twenty-eight printed pages. The rules specified that the town's *compagnie de pompiers* was to comprise 1 captain, 1 lieutenant, 2 sub-lieutenants, 2 sergeants, 2 corporals and 32 men (thus totalling forty men to operate the brigade's two pumps). The captain, lieutenants and sub-lieutenants were appointed by, and received their authority directly from, the mayor; the four sergeants and corporals were appointed by the mayor, who selected them from a list of eight presented by the captain; the ordinary *pompiers* were nominated – in effect, appointed by the captain, with the mayor approving his list. Thus a clearly defined chain of authority and command was specified. The rules also provided for a disciplinary committee and for monthly training exercises to check and maintain the pumps and other equipment, and detailed the uniforms and insignia each member of the *corps* was entitled to wear and the duties expected of him both in training and in fighting an actual fire. The *compagnie* of firemen at Blois was clearly exceptional – by 1820 it comprised 84 men, equipped with three pumps – but its detailed rules were used by the prefect as a model for instructing other communes when they expressed interest in forming brigades.

Unfortunately, the rules of only a handful of *corps* appear to have survived, but they demonstrate clearly the significance of chains of command and codes of discipline, indeed of the generally militaristic character of the brigades.[24] The brigade at Saint-Dyé-sur-Loire (est. 1847) produced a set of printed rules (*c.* 1860) which described in detail the slightly different uniforms and equipment of the *sapeurs* and of the *pompiers*:

Une grande tenue seulement est obligatoire pour tous les hommes de la Compagnie; elle consiste pour les Pompiers: en habit d'uniforme avec épaulettes, casque avec chenille et aigrette, col noir, ceinturon avec giberne, sabre-poignard et fourreau de baïonnette, pantalon bleu, bottes ou souliers, gants blancs. Pour les Sapeurs: en bonnet à poil avec aigrette, barbe, tablier et hache. Les cols, ceinturons, sabres-poignards, habits, pantalons et chaussures ceux des Pompiers. Le tout de la plus grande propreté.[25]

The rules specified that officers, sub-officers and corporals were to be shown obedience and respect; fines would be imposed upon those who left their post without permission of the commanding officer or sub-officer and those who interrupted or shouted in the ranks; fines were also imposed upon those who presented themselves late for (or were absent without good reason from) meetings and training exercises of the brigade. On the first Sunday of each month, at 1.00 p.m., the brigade undertook its *manœuvres* of the pump, but for that purpose the brigade was divided into two sections and each section was required to present itself in alternate months. For such sessions with the pump the men had only to be *en petite tenue* rather than in their full uniform, and the wearing of clogs was tolerated during the winter months from November to April inclusive. It was the duty of the brigade's drummer to beat '*la retraite*' at the end of a brigade meeting, '*l'assemblée*' at 8.00 a.m. on the days of general meetings, and '*le rappel*' one hour before a meeting. At the exact time announced for a meeting or for an exercise with the pump, there would be a roll of the drum at the door of the *mairie* followed immediately by a roll-call, with men being fined for not responding or for not being properly attired.[26] The *sapeurs-pompiers* of the brigade at Mulsans (est. 1861), 'désirant mainte-nir la bonne harmonie qui doit exister dans la compagnie', produced in 1863 a detailed set of rules to be implemented by a disciplinary committee – which itself suggests the brigade might have experienced some particular disciplinary problems early in its existence.[27] The new rules drawn up in the same year for the brigade at Montoire (est. 1810) were more general but still made some very specific points: when there was a fire, the drummer was to beat '*le rappel*' (only the local authority had the right to beat '*la générale*' and to ring the church bell); when there was a fire-call, the most senior officer present would take command and the pump was not to be operated unless there was an officer present; when the course taken by a fire meant that the only way of checking it was to demolish the building, its destruction could only be ordered by the local authority not by the brigade; no fireman was to wear his uniform for normal working; men serving in the brigade would be provided by the municipality with 'une tunique, un casque, un fusil, un sabre et une giberne' (although the drummer was given a *caisse* and not a *fusil* or a *giberne*) and when they left the service they had to return them to the *mairie*. The rules of this brigade of some fifty men were enforced by a disciplinary committee of five, comprised of its commander (the sub-lieutenant), a sergeant, a corporal and two men selected by the commander.[28] In so many ways, then, the *corps de sapeurs-pompiers* had military characteristics, from their uniforms to their conduct. Although many of their members might have had the experience of military service, and although that might have given some of them a taste for military ways and even an acceptance of a rigid discipline, *sapeurs-pompiers* were not conscripts but volunteers.

Esprit de corps

Disciplinary codes were largely self-imposed: membership of a brigade (and so acceptance of its rules) was voluntary. Rules which constrained the behav-iour (and thus the *liberté*) of members of a *corps* were accepted because they applied to all of them (and thus conformed to the concept of *égalité*) and because they were acknowledged as being in the interest of the brigade as a whole (and thus in the *fraternité* of this voluntary association). Moreover, some of the rules were designed to promote *esprit de corps* within these brigades. The rules of the brigade at Mulsans were revised in 1863 with the explicit intention of promoting *la bonne harmonie* and *la vrai fraternité* among its members. But many of the rules and activities of brigades implicitly fostered fraternal feelings. It is notable, for example, that the first article of the regulations for the *corps* at Saint-Dyé-sur-Loire (1860) and at Montoire (1863) were concerned with the uniforms to be worn and equipment to be carried by each member. That considerable significance was attached to the appearance of *corps* is also evidenced in correspondence between mayors and prefects (mainly, of course, with mayors seeking uniform and equipment grants from prefects). Uniforms served to emphasise a unity of interest, as symbols of solidarity. *Esprit de corps* was promoted also by those festivities (usually culminating in a toast-filled dinner) which many brigades held annually on or near the day of their patron, Saint Barbara.[29] At Saint-Dyé-sur-Loire, for example, on the Sunday following Saint Barbara's day in the 1860s there was held a Mass and then at, 4.00 p.m., a banquet, and the brigade's members attended both in uniform (but did not carry their arms).[30] *Esprit de corps* was additionally privately cemented and publicly exhibited on those occasions when a brigade paraded on the commune's streets behind its flag (and often marching to the tunes of the brigade's *corps de musique*), as was the case at Oucques in 1865. Such manifestations of both fraternity and publicity were most usually seen on *le quatorze juillet* (when a local population acknowledged and celebrated its membership of a nation-state) and at funerals of comrades, of kith and kin of the *sapeurs-pompiers* themselves, and of prominent members, public figures of the local community. It was also witnessed on those occasions when a newly acquired pump was ceremoniously taken to church to be blessed by the priest and then celebrated by drinking, dining and dancing by most of the local community, as was the case at Villebarou in 1852 and at Villefranche in 1855.

Social and spatial tensions

While both the rules and practices of brigades were designed to encourage a sense of solidarity, that objective was not always achieved. The smooth functioning of brigades was disrupted by the social and spatial tensions latent within their organisation. The social cohesion of a brigade could be threatened by both personal disagreements and political discussions either among its members or between the brigade and the municipal council from which it derived its legal authority. One potential source of instability within such a voluntary organisation was its necessary chain of command and disciplinary structure. At Josnes in 1862 the long illness of the brigade's captain led to mass resignations (with no one willing to take on his responsibilities), and to disciplinary proceedings against some of the men. Not all officers were able to command the constant respect of their men; consequently, for some periods their orders were not totally heeded and the pumps were not properly exercised. Command problems of this kind, essentially of a personal nature, could simmer for years, seriously undermining the morale of a *corps*, as they did at Suèvres from 1840 to 1860, at Ménars in 1853, 1854 and 1860, and at Ouzouer-le-Doyen from 1865 to 1869. Sometimes the problem was not internal disagreement but external conflict with members of the commune council: for example, for at least eight months in 1857 the brigade a Ouzouer-le-Marché refused to exercise their equipment, declining to take orders from the local council, and in the autumn the brigade was dissolved by the mayor so that a new *corps* could be established; at Chitenay in 1883 and again the following year the commander of the brigade used the occasion of the celebration of Saint Barbara's Day to make speeches criticising some members of the commune council, and by 1886 the brigade was in abeyance.[31]

At times, personal conflicts combined with political rivalries to threaten the preparedness of brigades to fight fires effectively. At Fougères in 1884 the brigade refused to participate in the 14 July celebrations and some of its members disrupted them, shouting abuse at the mayor: this demonstration was, in fact, merely symptomatic of a long-standing rivalry between pro- and anti-republican factions within the commune (with, in this instance, the pro-republican brigade challenging the authority of the commune's reactionary council by disrupting the Republic's *fête nationale*!). This political squabble reverberated within Fougères for at least the next thirteen years, seriously interfering with the proper functioning of its brigade. Similar political conflicts embroiled the brigades of Nouan-sur-Loire in 1889, Prunay in 1896, and Prénouvellon in 1903.

Inter-communal brigade rivalries combined with spatial problems to produce a curious conflict involving the brigade from Fougères in November 1862. On the mayor's orders, the brigade attended a fire in the neighbouring commune of Ouchamps but the officer commanding the Ouchamps brigade (which was already fighting the fire) would not authorise the men from Fougères to operate their pump in his commune, allowing them instead merely to join the bucket-chain servicing the Fougères brigade's pump. Although the distance of the burning building from a source of water meant that only one pump could be properly supplied, it would have been more tactful and fraternal – and arguably more efficient – to have operated each of the two pumps alternately.

From the early nineteenth century communes which established fire brigades were expected to provide a service to neighbouring communes and some basic geographical issues – of distance friction, of centrality and settlement hierarchy, and even of settlement structures – clearly affected the historical development of fire-fighting services in Loir-et-Cher. In the 1830s, the prefect and sub-prefects were keen to see brigades established in Loir-et-Cher's cantonal centres, so that they could offer a fire-fighting service to their dependent communes. In May 1835 the prefect claimed that the department's fourteen brigades – ten of which were in cantonal centres – were able to service a total of 131 communes. In October 1837 the Inspecteur des compagnies de pompiers of Loir-et-Cher prepared a plan for improving the service in rural areas by encouraging the creation of *corps* based on several communes each of which would be expected to provide men for an inter-communal brigade. Leaving aside the already well-established and effective *corps* at Blois and at Mer, the inspector identified a further 29 communes around which specified communes could be grouped and in which an inter-communal brigade should be established. Of the 19, only one had a *corps* in operation in 1837 and only five more had set them up by the autumn of 1842, so the inspector's plan to improve the service centrally can hardly be said to have been successful: to be effective, *corps* relied upon local enthusiasm (a genuine fraternity) and local expertise rather than upon central encouragement and advice. In October 1842 the inspector was claiming that 19 communes had brigades and that a further 150 communes could be reached by them within one to three hours, but he was surprised that there remained some cantonal centres without brigades.

The prospect of having to provide a fire-fighting service for nearby communes was a special difficulty facing the creation of brigades in rural areas. In 1860 the mayor of Josnes petitioned the prefect for permission to enlarge his commune's *corps* beyond its authorised complement of thirty men, pointing out that the organisational problems of rural brigades were significantly different from those in towns: because the volunteers lived and worked throughout the territory of a commune, it took them longer to gather together as an operational team when the alarm was raised; furthermore, the men were expected to provide a service not only in Josnes itself but also in five other communes. At Montoire in 1842 the brigade of 32 men was reported as providing a service not only for that cantonal centre but also for seventeen other communes, the two most distant being 12 km away so that it took the brigade two hours to carry the pump there themselves and eighty-five minutes if the pump were horse-drawn. The operation of a pump beyond the boundaries of the brigade's own commune could lead to a further, financial, difficulty because in such cases the men were expected to be reimbursed for their loss of earnings by the commune in which the fire was located but such payments were not always made promptly. Even within a single commune, dispersed farmsteads or the existence of a number of hamlets in addition to the main *bourg* could militate against the proper functioning of a brigade and in a few communes efforts were sometimes made (and sometimes successfully) to have two pumps and teams of men, one in the *bourg* and another in a hamlet, as was the case at Saint-Claude-de-Diray in 1842 and at Ouzouer-le-Marché in 1857.

Social tensions and spatial problems undoubtedly affected the ways in which brigades were organised and functioned, and at times they could determine whether or not a particular brigade was operational at all in the short or medium term within a given commune. But the long-term evolution of brigades in Loir-et-Cher generally during the nineteenth century seems to have been influenced by the more fundamental issues of finance and recruitment.

The evolution of fire brigades

Almost all brigades had chequered histories, witnessing some periods when they thrived and were very active, and others when they languished. A few, for diverse reasons, failed to maintain their momentum and collapsed entirely for want of volunteers. Some – especially in the east of the department – had their operations seriously disrupted by the Prussian invasion of 1870–1, with their equipment being destroyed or confiscated, and recovery back to normal was sometimes difficult and slow.[32] But above and beyond these local and specific difficulties, almost all brigades confronted problems of funding and of recruitment with varying degrees both of seriousness and of success.

Financial resources

In the 1830s and 1840s, when the establishment of *corps* was being strenuously encouraged by central government, mayors of communes were often at pains to point out to the prefect that the costs involved in setting up a brigade – purchasing a pump and uniforms and equipment for the *pompiers* – could not easily be met at a time when communes were already burdened with heavy expenditure, especially on constructing and improving rural roads and on providing schooling. The mayors of Sargé and Prunay, for example, informed the prefect in the spring of 1836 that their councils had discussed the prefect's enthusiasm for having fire pumps in rural communes but had decided that it was not possible to establish brigades in their communes, because of a lack of funds.[33] Sometimes a commune council's decision to establish a brigade reflected its enthusiasm more than its determination, because there could be a delay of some years between the decision to found a *corps* and its practical and proper implementation. For example, at Cellettes in 1846 the brigade was still not properly equipped some seven years after its foundation because of the costs involved; attempts made at Herbault in 1839 to raise the funds needed to create a *corps* met with little response and one was not established there until fourteen years later; at Marchenoir a decision taken in 1842 could not be implemented, for lack of funds, until ten years later.

The main cost, of course, was the pump itself. In 1863 when the suggestion of establishing a *corps* was being debated at Mulsans the mayor estimated that the pump would cost about 500 fr., which was equivalent to
two-thirds of the amount being spent by the commune on running its
primary school and almost one-quarter of the sum it was spending on its
roads. The necessary funds could be raised from a variety of sources: local
taxation, public appeal, subventions from central government and grants
from insurance companies. Unsurprisingly, commune councils usually
endeavoured to maximise income from external sources in order to minimise
the additional charge which would have to be imposed internally on the
commune's tax-payers. Fund-raising efforts of this kind to some extent acted
as a brake upon initiatives to set up brigades. At times, after a disastrous fire
in a commune, the brake was dramatically released and funds readily became
available: for example, the commune council's reticence at Courbouzon about
establishing a brigade disappeared totally following a fire in the commune in
1842 and funds which had been allocated by the council for a poplar plantation (an integral part of the road improvement programme) were diverted
into meeting the start-up costs of a fire brigade. The early example set by 400
of the community conscious residents of Mer in 1810 in raising by public
appeal the entire cost of a pump (some 815 fr.) was not widely followed: even
in the immediate aftermath of a fire at Cour-Cheverny in 1846, a public
appeal raised only about half the cost of a pump, the balance of the expense
involved in setting up a *corps* being met by grants, from the department and
from an insurance company, and by a special allocation on the commune's
budget. An appeal at Chitenay in 1848 did raise 972 fr., enough to meet the
930 fr. which a pump had cost but not enough to meet the expenditure of a
further 700 fr. on helmets, buckets and other equipment; at Maves in 1854 an
appeal raised 1,025 fr., enough to pay for the pump and its accessories (930
fr.), so that other equipment was purchased with funds from two grants (of
200 fr. each) from the department and from an insurance company. Where
there was sufficient general support and sufficient prosperity in a commune,
some of the necessary funds could be raised by a hypothecated tax imposed
by the commune council, having first of all obtained the prefect's approval.
For example, a brigade had been established at Avaray in 1842 when a local
notable gave a pump to the commune, but its council wanted to raise 600 fr.
to buy helmets (needed no doubt in part for safety reasons but also, as the
mayor pointed out in his letter of 4 October 1842 to the prefect, because arm-
bands did not adequately identify *pompiers* in the tumultuous circumstances
of a fire): an insurance society made a grant of 200 fr. but the prefect gave
permission for the balance to be raised through an ear-marked tax, recognising that the commune had insufficient normal resources because of its
recent expenditure of 1,100 fr. on building a schoolhouse and of 200 fr. on
repairing the bell-tower of the church.[34]

That financial obstacles could check the initial establishment of brigades is beyond doubt. The start-up cost was, in terms of a commune's budget, considerable. But, once established, the costs incurred in operating a brigade were much less, being limited essentially to repairing and maintaining the equipment. Furthermore, once a brigade was established its benefit to the commune no doubt became clearer than its cost and there is no evidence that the functioning of an established *corps* was seriously endangered by a shortage of money. Paucity of funding was less of a threat to the operation of an established *corps* than periodic shortages of men.

Human resources

Service in a *corps* was voluntary: membership involved both costs, in time (on training exercises and in attending fires) and often in money (on clothing and equipment), and the benefits of *camaraderie* and social prestige within a community. There were officially no payments for service in a brigade, although compensation for loss of earnings in attending a fire was usual and some communes offered as incentives to men to join their brigades exemptions from the customary road-repairing duties which able-bodied males were expected to perform annually, or exemptions from the customary obligations to provide board and lodging for military personnel in wartime. Some *pompiers* were able to benefit from the *caisses communales de secours* established by some commune councils from the 1860s onwards from which payments were made to those injured while on duty and to the widows and under-age orphans of a fireman killed while on duty.[35] In part because of the hazards involved in serving in a brigade, *pompiers* individually and their *corps* collectively were generally accorded respect within their local communities and the prestige associated with being a fireman was aid to the recruitment of volunteers.[36] It might have been, however, that the perceived dangers were greater than the real hazards. In the five years 1891–5 in Loir-et-Cher, one fireman died (from the effects of smoke inhalation) and eleven were injured (the most serious accident rendering the fireman deaf, the least serious being an injury to the hand which prevented the fireman from working for six days): given that there were more than 4,200 volunteer firemen in the department during that period, the death-rate while on duty was about 1:21,000 and the accident rate about 1:2,000, odds which were probably at least as good and possibly better than those pertaining to the men while carrying out their normal work.[37]

A brigade's existence depended fundamentally upon the goodwill of its individual members and upon its collective *esprit de corps*, rather than upon financial or other inducements offered by the commune. Before the 1870s, only a few brigades encountered serious difficulties in recruiting *pompiers* and from those which did we can obtain insights into brigade *mentalité*. At Saint-Jean-Froidmental in 1858 a *corps* was established and a pump purchased but it took two years to assemble sufficient volunteers: local men hesitated to join because of the expense each had to meet in equipping himself. At Saint-Laurent-des-Eaux in 1859 the mayor reported that the brigade had been founded in 1847 and the men provided with uniforms at the commune's expense but the rise in the cost of living during the 1850s had led several *pompiers* to resign because they could no longer afford to look after their uniforms, so that the brigade was reduced to two-thirds of its authorised complement. Similar financial stresses during the 1850s were experienced at Avaray, whose mayor reported in October 1860 that limited response to a call for volunteers had meant that the pump had not been exercised for more than a year. At Pezou in 1865 the brigade could not find people willing to serve as its two officers, in part because of the additional costs they would incur in acquiring the uniforms and equipment of, respectively, a sub-lieutenant and a sergeant-major. At Saint-Claude-de-Diray in 1866 the '*esprit de concorde*' of the brigade was ruptured by resignations of *pompiers* objecting to a change in their terms of service: from 1832 each volunteer fireman had been exempted by the commune council from performing his two days of road maintenance annually but in 1861 the council had increased this general obligation to three days and so now expected each volunteer fireman to undertake one day's labouring on the commune's roads. Protests by the firemen led the mayor to exempt them from all road-mending duties, but a renewed attempt in the spring of 1866 to require each *pompier* to undertake one day's service led to renewed protests and resignations. A rather different recruitment problem was encountered at Ouzouer-le-Doyen from 1865 until 1868: the brigade there languished because of the commune's inability to find anyone willing to serve as its leader after the resignation of its sub-lieutenant because of his unwillingness to abide by the brigade's official regulations about compulsory attendance at training exercises and the imposition of fines for absenteeism.

Recruitment problems did affect some brigades before the early 1870s, but they were serious in only a handful of the almost 100 brigades which had been founded by then. Such problems acquired greater significance from the mid-1870s onwards, as a direct result of the decree of 29 December 1875 which required existing *corps* to be reconstituted by the end of 1876, required communes to underwrite their brigades financially for a five-year period and – more significantly – required volunteers to contract their services for a five-year period. This fundamental change in the firemen's terms of service certainly checked the flow of volunteers. From the late 1870s onwards, many brigades found it difficult to recruit men in the numbers they needed: there was a deep reluctance to sign on for five years at a time. At Chambon-sur-Cisse a request by the *pompiers* for them to be exempted from the obligation of military service in exchange for contracting into the *corps* for five years was rejected by the prefect because of the illegality of such an agreement. At Contres the brigade was reconstituted in April 1878 for a five-year period but in April 1883 its leader declined to renew his contract and the commune remained without a properly organised and functioning brigade for at least the next twenty years, relying on a team of twelve or thirteen men to operate the pump from time to time instead of the thirty disciplined and trained men needed for an official *corps*. Recruitment difficulties as a result of the 1875 decree led to a markedly slower rate of foundation of new *corps* in the final quarter of the nineteenth century than had been the case during the preceding forty or so years. Moreover, perhaps as many as ten *corps* founded before the 1875 decree came to be dissolved after it because of the adverse effect it had on their recruitment of volunteers. Recruitment problems certainly led to the dissolution of *corps* at both Marchenoir and Rilly in 1877 and at Chissay in 1897; the *corps* at Contres was not operational between 1883 and 1899 because of a shortage of recruits.

The *corps de sapeurs-pompiers* were voluntary associations: in those communes and localities where, for whatever reason, the *esprit associative* was not strong, they were unlikely to be established or, if established, to endure.

Although the State was involved at the outset in the creation of *corps de sapeurs-pompiers*, the role of fire brigades as voluntary associations dependent for their existence upon the goodwill of their individual members and upon a sense of social solidarity was equally clear. Especially within remote rural communes, *corps* were established and functioned in at least partial ignorance of the legal framework within which they were officially embedded. Gradually during the nineteenth century the State came to exercise a greater degree of control over fire brigades and their role as voluntary associations was correspondingly reduced. They had, of course, never been entirely free from state control: a brigade could only be established with the joint authority of a commune council and of the prefect, and it remained directly responsible to that authority. Unlike the position with respect to some other fraternal associations – such as (as will be shown) agricultural syndicates or musical societies[38] – there was never the possibility of the emergence of rival brigades in a single commune, each being promoted by groups with differing political or religious persuasions. The *corps de sapeurs-pompiers* were simultaneously fraternal, voluntary associations and – increasingly during the nineteenth century – part of the extending apparatus of the State. Indeed, they contributed one more component – the fire station – to the evolving public landscape of rural France during the nineteenth century: the fire station joined the school and other public buildings as symbols of the growing power of the State and of the declining power of the Church.[39] After all fires, once accepted as acts of God, were now being fought by forces of the State.

7

Anti-phylloxera syndicates

Une association syndicale est formée entre les sousignés à l'effet de défendre contre l'invasion du phylloxéra leur vignes situés dans la commune de Mer et climats limitrophes. Elle surveillera et traitera les vignes dans une circonférence qui aura Mer pour centre et pour rayon cinq kilometres.[1]

Article 1 of the statutes of the anti-phylloxera syndicate at Mer (1882)

The diffusion of a disaster

Between the mid-1860s and the mid-1890s a parasitic aphid accidentally introduced into Europe on vines imported by steam ship from North America devastated the European continent's vineyards, including those in France.[2] Phylloxera, a small, yellow root-feeding aphid, has had several scientific names: initially called *Phylloxera vastatrix* (the devastator) by the French scientist J-E. Planchon, it is now more correctly called *Dactylasphaera vitifoliae*.[3] In the 1860s and 1870s, when the phylloxera – as it remains more commonly known – was diffusing throughout France, vines and wheat were the country's two most important cultivated products. In 1875, a record year, wine production amounted to 84 million hectolitres and vineyards covered almost 2.5 million hectares: in economic terms, the value of wine production in France was second only to that of textiles. But in a remarkably short time the vineyards of France were ravaged by the phylloxera, whose larvae lived off the roots of vines and destroyed them.[4] As Augé-Laribé so starkly described this disaster, never before in the history of agriculture had a plant species been so suddenly and so completely destroyed.[5]

Brought to France on American vines, the phylloxera was first reported in the Midi, at Pujault (Gard) in 1863 and soon afterwards elsewhere in southern France; in 1869 it was identified in the Bordeaux region, although it is likely that the infestation occurred there several years beforehand. By 1878 a quarter of France's vineyards had been invaded by phylloxera, reducing wine production to 25 million hectolitres in 1879 and for many years it was less than

40 million. The livelihoods of those dependent upon viticulture were seriously threatened and the problem was recognised as being of national significance. But the battle against phylloxera proved to be difficult, long and costly. A prize of 300,000 fr. offered by the State in 1874 to anyone finding an effective way of combating it was never awarded. Unsurprisingly, given the growing connection between science and agriculture, some sought a solution in a chemical treatment: twenty years earlier applications of sulphur had conquered vine powdery mildew (*oidium*) and at the time copper sulphate was being increasingly and effectively used to control downy mildew (*mildiou*). But sprays and insecticides had little impact upon phylloxera while becoming increasingly labour-demanding. One effective solution was to flood vineyards, but such a measure could not be widely adopted given the location of many of them on slopes and terraced hillsides. Ultimately, the only effective but costly solution was to reconstitute the vineyards with vines grafted on to phylloxera-resistant American rootstock.[6]

Such a fundamental crisis in French viticulture was almost inevitably addressed by state intervention. A circular of 6 March 1876 from the Ministry of Agriculture to prefects stated that each viticultural department should establish a Comité d'études et de vigilance to co-ordinate the battle against the phylloxera on its territory. A circular of 9 February 1877 required the *comités* to report to the Ministry quarterly, but identification of the new pest proved to be too slow for the central authorities, so in April 1878 the Ministry undertook to send some representatives from the departments to the viticultural school at Montpellier to receive training in the identification and control of phylloxera. A law of 15 July 1878 aimed to check the spread of phylloxera by giving prefects powers both to control the movement of vine plants and of composted materials (in case they contained infected vines) and to treat affected vines using methods advised by the Ministry's Commission supérieure du phylloxéra. It also provided for the State to make a subsidy equal to that provided by the department or by a commune's municipal council to owners of vineyards who treated their vines by a method approved by the Commission supérieure. Under the law of 15 July 1878 prefects were able to authorise the creation of associations (*syndicats*) of *vignerons* for the purpose of fighting the phylloxera. A law of 26 July 1879 required affected departments to set up a Service du phylloxéra and a circular of 20 August 1879 set out the increased powers given to prefects through that agency to identify and to treat phylloxerated vines. Then, in 1887, prefects were authorised to order the setting-up of *syndicats de défense* in the affected communes of their departments. Syndicates created compulsorily represented, of course, the decentralisation of public powers rather than the actions of freely associating individuals. But independent associations of *vignerons* could also be established and such *syndicats* also received active encouragement and financial support from the State.[7]

Photograph 7.1 A *vigneron* treading grapes in the Sologne
Source: 3 Fi 6432 photo, Archives Départementales de Loir-et-Cher

In Loir-et-Cher, phylloxera was first identified in 1876, although it had probably been introduced into the area some ten years earlier, on vines brought from the Bordeaux region by a viticulturalist in the locality of Vendôme.[8] It was to devastate the department's vineyards but not before many *vignerons* had fought a battle against it, doing so in part at least by forming voluntary associations.[9] In the mid-nineteenth century, the value of wine production in Loir-et-Cher was second only to that of cereals: 20 per cent of the electorate were *vignerons* and in some communes and localities the proportion exceeded 50 per cent (photographs 7.1 and 7.2). Between the early 1850s and the late 1880s, and while the phylloxera was diffusing throughout the department, its vineyards expanded from about 26,000 to 45,000 ha, located mainly in the valleys of the Cher and the Loire, in the Petite Sologne (which formed a triangle in the approach to the confluence of those two rivers), and in the

Photograph 7.2 Grape picking in the Sologne
Source: 3 Fi 6387 photo, Archives Départementales de Loir-et-Cher

Loir valley, especially downstream from Vendôme (figure 7.1). But by 1902, the phylloxera having taken its toll, vineyards had retreated again to about 26,000 hectares. The viticultural landscape of Loir-et-Cher had been dramatically and rapidly reduced: by more than one-third within little more than a decade.[10]

In the summer of 1874, a Ministry of Agriculture circular describing the measures to be taken to combat phylloxera was distributed by the prefect of Loir-et-Cher to mayors in the department, coupled with a request that they report immediately to the administration should symptoms of phylloxera attack be observed in their communes. From January 1875 the prefect prohibited the importing into Loir-et-Cher of vine-plants coming from other departments or from abroad. No reports of phylloxera within Loir-et-Cher had been received by the prefect when in early September 1876, on the instruction of the Ministry of Agriculture, the department established a Comité d'études et de vigilance pour prévenir la propagation du phylloxéra. Later that month vineyards attacked by phylloxera were recognised in five communes: three in Vendômois (Vendôme, Rocé and Villetrun) and two just to the north of Blois (Saint-Denis-sur-Loire and Villebarou).[11] Initially the phylloxera – or, more precisely, its recognition – diffused slowly, being identified in fifty-seven communes by 1884, after which it spread quickly by 1894 to all communes with vineyards (figure 7.2). Most seriously affected by September 1882 was a group of thirty communes in the locality of Vendôme, focused on the middle

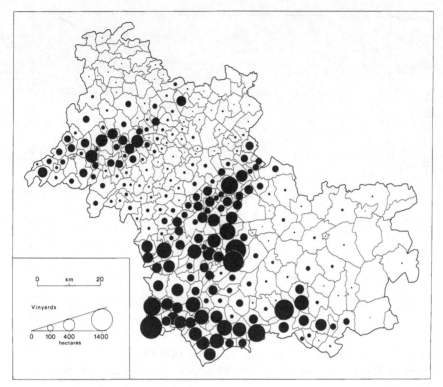

Fig. 7.1 Vineyards in Loir-et-Cher in 1878
Source: AD 7M 178 Etat indiquant par commune le nombre d'hectares de terrains
cultivés en vignes, d'après la statistique de 1878

Loir valley; by August 1884 a group of fifteen communes on the right bank of
the Loire, between Mer and Blois, had also been badly attacked, together with
a few communes in the Cher valley where it had been identified for the first
time in 1880. In the late 1880s and early 1890s the phylloxera diffused throughout the entire department (figure 7.3).

Crisis management by the central authorities

Detecting vineyards affected by phylloxera was, of course, a preliminary to
treating them and representatives of the State were closely involved in both
activities. Advice offered by the Ministry of Agriculture on ways of dealing
with affected vines – principally chemical treatment involving the injection of
carbon disulphide into the soil around their roots – was accepted by the
regional administration and in turn transmitted to local authorities: for
example, on 29 July 1877 the prefect sent a circular to the mayors of the

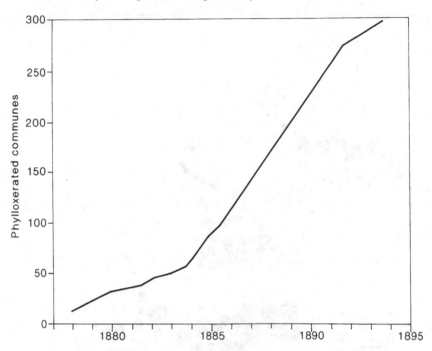

Fig. 7.2 Phylloxerated communes in Loir-et-Cher, 1876–95
Source: AD *Conseil Général: rapports du Préfet et procès-verbaux des délibérations 1876–1895*

department informing them of the various measures for treating phylloxera as recommended by the national Commission supérieure du phylloxéra.

The central authorities in Loir-et-Cher, alerted by national officials in Paris, were themselves aware of the potential crisis and took steps to manage it even before it was recognised as a reality in the department. What were deemed to be appropriate administrative structures were established early. In July 1876 the prefect announced the setting-up of a Comité d'étude et de vigilance pour prévenir la propagation de phylloxéra: its nine members were the president and the secretary of the *comice agricole* of the *arrondissement* of Blois, the vice-president of the *comice agricole* of Vendôme, the president and the secretary of the *comice agricole* of Romorantin, the president of the *comice agricole* of the canton of Lamotte-Beuvron, the mayor of Beauchêne (who was an *agriculteur*) and one *propriétaire-agriculteur* from the commune of Naveil. In September 1877 the prefect permitted the *comice agricole* of the *arrondissement* of Vendôme to establish its own Commission du phylloxéra, to study ways of combating the phylloxera in the locality where it had first been detected in the department. The Commission had as its members the officers

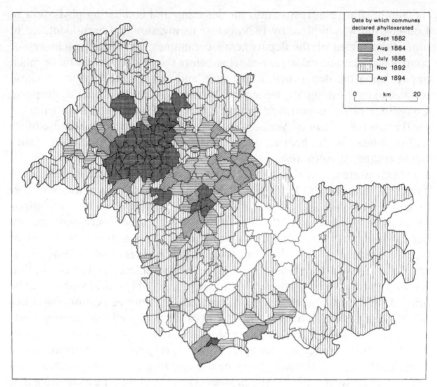

Fig. 7.3 The spread of phylloxera in Loir-et-Cher, by communes, September
1882–August 1894
Source: AD *Conseil Général: rapports du Préfet et procès-verbaux des délibérations
1876–1895*

of the *comice agricole* of the *arrondissement*, the mayors of its communes, and
three or four representatives from each of its cantons. In December 1877 the
Commission treated experimentally with carbon disulphide just 0.74 ha of
vines in the commune of Villetrun, both to test and – it was hoped – to demonstrate the chemical's efficacy. Responding to ministerial decrees, in the spring
of 1879 the department reorganised its Comité d'études et de vigilance into
one central committee (for the department) and three local committees (one
for each *arrondissement* – with the one for Vendôme coming to replace the
commission established by its *comice agricole*). Each committee had about
twelve members, including doctors, pharmacists and schoolteachers as well as
presidents of *comices agricoles*, mayors of viticultural communes, and some
vineyard owners. Later in 1879 the General Council of Loir-et-Cher also
agreed to the establishment of a Service du Phylloxéra to meet the crisis in the
middle Loire valley viticultural region on an inter-departmental basis. Thus

the administrative infrastructure for detecting and combating phylloxera in Loir-et-Cher was in place by 1879, before its invasion had been identified in more than fifteen of the department's communes: an official machinery of control was operational in Loir-et-Cher before the phylloxera had made much impact upon the department's vineyards and before it had begun to cause serious alarm among its *vignerons*. In early 1879 the General Purposes Committee of the department reported that the Commission du phylloxéra for the *arrondissement* of Vendôme was in difficulties not only because of its lack of funds but also because of the inertia and conservatism it was encountering among viticulturalists.

Official awareness of the seriousness of the problem preceded its perception by private individuals. Treatment of affected vineyards was initially undertaken by the public authorities rather than by individuals or by associations of individuals, but even the department's authorities were not entirely confident about their own decisions and actions. In treating a small area of vines in December 1877, the Commission du phylloxéra in Vendôme recognised that it would take some time to determine its efficacy and so in addition it set up a nursery for raising American vines on to which local vines could be grafted. The department's General Purposes Committee (Commission des affaires diverses) at its meeting on 21 August 1878 argued that treating vines with carbon disulphide or potassium chloride was expensive, about 1,300–1,400 francs per hectare spread over three years of treatment, and it considered whether or not it might be cheaper to grub up the affected vines and to replace them with American vines. The prefect told the committee that the department's Comité du phylloxéra was, by a large majority, opposed to the introduction of American vines (and one councillor on the General Purposes Committee was opposed to it on the grounds that the whole problem was itself a consequence of the introduction of American vines). Throughout 1877 and 1878 most individual *vignerons* were perceived by the authorities as exhibiting a basic conservatism, remaining sceptical both about the application of chemical insecticides and about replanting with American rootstock.

The seriousness of the situation was certainly grasped by the central authorities at an early stage. At its meeting on 25 August 1879 the Conseil Général of Loir-et-Cher noted that, following ministerial instruction, a Service du phylloxéra would be created for the department; and it was reminded by its finance committee (Commission des finances) of the importance of viticulture in the department (with its 31,026 ha of vines providing an annual income of approaching 30 million francs), of the gravity of the situation already in the Vendômois and of the rapidity with which phylloxera could spread throughout the department. The Service du phylloxéra undertook a major publicity campaign in the department's viticultural regions during 1879–80, organising a series of public lectures (many of them given by M. Tanviray, the department's recently appointed professor of agriculture) and engaging the assistance of public officials (including departmental and cantonal inspectors of roads, and schoolteachers). Not surprisingly, the Service discovered the phylloxera to be more widespread than had hitherto been supposed: by August 1880 it had been identified in twenty Vendômois communes and in four communes of the Val de Loire immediately upstream from Blois. Tanviray warned then that, although the total area affected by phylloxera was still small (only about 50 ha) and that although the initial invasion by the parasite had been slow (he himself had identified the introduction of phylloxera to the department fourteen years previously, in 1866, on American vines brought to Vendôme from Bordeaux), the spread of phylloxera was likely to progress geometrically. He argued strongly for action to be taken. On 15 November 1880 the prefect informed the Minister of Agriculture that of the approximately 38,000 ha of vines in the department phylloxera had attacked some 95 ha and totally destroyed 3 hectares. He further reported that the spread of phylloxera had slowed down the rate of expansion of vineyards in the department without yet having had any impact upon their valuation.

On 3 June 1880 the prefect authorised the treatment of affected vines by the department's Service du phylloxéra: applications of carbon disulphide were chosen on the grounds that it was cheaper, easier to use and less damaging to the vines themselves than some other chemicals available. That summer 70 ha of vines were so treated, with the costs being met by the department and the State. Commune councils were unwilling to provide grants in part because, as the prefect acknowledged in April 1881, they were already contributing significantly to the costs of other public services, notably local roads and primary schools; and he also reported to the Conseil Général that there was no sign that individual *vignerons* were willing to support financially the efforts of the administration to control the pest. Indeed, there was even opposition to the use of chemical insecticides. At first, those *vignerons* opposed to having their vines treated were ultimately persuaded by the authorities that such measures were in the common interest because of the persistent danger of further contagion from any untreated but infested vines. Gradually, however, opposition among the *vignerons* to the compulsory treatment of their vines mounted: on 17 September 1881 the prefect reported that the administration had to battle against a deep-seated opposition by some *vignerons* who, in their ignorance, believed that it was the carbon disulphide that was destroying the vines, confusing the remedy for the disaster with its cause. Strong opposition by *vignerons* to the treatment of their own, privately owned, vines by the public authorities led the latter to favour instead the creation of associations of *vignerons* – anti-phylloxera syndicates – which would take responsibility for treating the vines of their members. In the early 1880s the prefect of Loir-et-Cher, acting upon advice from the department's professor of agriculture, considered that the best means of fighting phylloxera was the formation of local anti-phylloxera syndicates, subsidised by the State and the department.[12]

Opposition by *vignerons* to having their vines treated with carbon disulphide by the Service du phylloxéra was by 1881 sufficiently strong and lively, even at times violent, to interfere with the proposed programme of treatment in that and the following year: the practice had to be stopped in the *arrondissement* of Blois and significantly curtailed in that of Vendôme. During 1881 only 11 ha and in 1882 only 24 ha of vines were treated by the Service (whereas 113 ha were identified as being attacked by phylloxera) and in both years the Service only treated vines when requested by their owners to do so, the prefect's request to the Minister of Agriculture for permission to treat affected vines compulsorily, by force, having been refused. Opposition to chemical treatment was no doubt based in part on suspicion of, even ignorance about, the method itself, but it was probably also based on pride in individual property and on prejudice against public authority. It was certainly fostered by inexpert applications of carbon disulphide which damaged healthy as well as diseased vines; by the fact that in addition to applying carbon disulphide the Service in 1881 and 1882 was testing five other ways of tackling phylloxera, thereby in effect admitting that it did not itself have total confidence in its methods; by rough estimations that treatment would cost 400 fr. per ha per annum for each of three successive years, and associated suggestions that the *vignerons* would have to contribute substantially to these costs when many considered that they were already paying heavily for public services; and by the apathetic attitude towards phylloxera held by many *vignerons* who attributed falling land values and a sluggish land market to a series of years with poor wine yields, not itself a new experience. It needs also to be recognised that the work of the Service du phylloxéra was constrained by the funds made available to it by the department and the State, which were never reported as being adequate for the size of its task.

During 1882 Jules Tanviray, the department's professor of agriculture, included phylloxera and its treatment in the lectures which he gave throughout the department to local audiences which ranged from 40 (at Salbris, on dairying) to 340 (at Villiers, on the fragmentation of farms – *morcellement* – and viticulture). In fact, fourteen of his twenty-eight lectures in 1882 were concerned with viticulture and the phylloxera problem. By the summer of 1882, Tanviray had concluded that carbon disulphide was the best insecticide to use but that compulsory treatment of vines was no longer appropriate, because it aroused too much local opposition, because it was too costly for the central authority, and because the rate of spread of phylloxera was too rapid, accelerating geometrically. Collaboration between the department and individual *vignerons* – with the former providing the carbon disulphide and necessary equipment, the latter the labour – reduced the costs but was nonetheless an inadequate solution, treating some infested vineyards but leaving others as sources of contagion. So, by the end of 1882, Tanviray was strongly advocating the formation of associations of *vignerons*, of anti-phylloxera syndicates, to undertake treating vines themselves but with financial support from the central authorities. On 12 November he lectured on anti-phylloxera syndicates to 300 *vignerons* at Mer and on 31 December to 110 at Marolles (just to the north of Blois): while an anti-phylloxera syndicate was indeed established at Mer, on 14 December 1882, none was ever created at Marolles – the professor's advice was clearly not uniformly accepted.[13]

By the autumn of 1882 the prefect reported that his administration's efforts to manage the crisis had not been successful and there was little hope of stopping the diffusion of phylloxera: far from being under control, phylloxera was still spreading. It had by then been identified in thirty-seven communes: thirty-one in Vendômois, five in the vicinity of Blois (mainly in the Loire valley) and one in the Cher valley. During 1882 the Service had only treated 24 ha of vines. At its meeting in September 1882, the Conseil Général considered whether or not to establish a nursery for growing American phylloxera-resistant vines (on to which French vines could be grafted) in case the attempt to bring the pest under control with insecticides should fail. On that occasion the Conseil Général decided not to proceed, but at its meeting in April 1883 it did decide to establish such a nursery near Romorantin, far from the then phylloxerated areas of the department, using seeds of American vines in order to avoid bringing in yet more phylloxera on vine plants. Other nurseries were in due course created near to Blois and to Vendôme.

For his part, the department's professor of agriculture, Tanviray, had already concluded by mid-1882 that the surest way of arresting the spread of phylloxera in the long term would be to grub-up affected vines on a large scale and he was studying the implications of, and the ways and means of taking, such a drastic step. Tanviray recognised that the application of chemical insecticides was a medium-term palliative rather than a long-term solution. It was at his suggestion that the Conseil Général had decided in the spring of 1883 to set up a nursery for the cultivation of American vines on to which local vines could be grafted. In the medium term he advocated the continued application of carbon disulphide not in general to all vineyards but specifically to the small patches of infested vines, and then only to vines whose owners gave permission for such treatment, with the Service du phylloxéra providing the insecticide and the equipment needed to inject it around the roots of vine plants and with the *vignerons* themselves providing their own labour. Tanviray's tactical and strategic views were shared by his successor, Trouard-Riolle, whose lectures in the late 1880s focused upon two topics, the benefits to agriculture in general of using chemical fertilisers and the difficulties of fighting phylloxera. Between October 1887 and July 1888, Riolle gave twenty-four lectures throughout the department to audiences which averaged 145 people: half of his lectures dealt with the phylloxera crisis, and it seems that although he covered the application of chemical insecticides by anti-phylloxera syndicates he also advocated strongly the adoption of American rootstock. Letters from Trouard-Riolle to the prefect during 1887–8 make it clear that, in many of the communes in which he lectured about the phylloxera crisis, he encountered apathy, indifference and even resistance to his arguments.[14] Only a minority of the communes with phylloxerated vines which he visited during that period subsequently established anti-phylloxera syndicates. From the mid-1880s, it seems, the department's agricultural advisors were becoming somewhat less keen than hitherto about anti-phylloxera syndicates, not because of the resistance to them demonstrated by some *vignerons* but because the chemical treatment of vines came to be seen only as a short- or medium-term palliative, not as a long-term solution. Paradoxically, it was not until the mid-1880s that the *vignerons* themselves began to look more favourably upon the chemical control of phylloxera.

By the summer of 1883, with phylloxera rapidly spreading and by then identified in forty-seven communes, it seems that the opposition of many *vignerons* to having their vines treated by the administration was waning in the face of the growing scale of the crisis and the Service du phylloxéra was by then receiving many requests from vineyard owners in the Vendômois to have their vines treated by the public authorities. Tanviray continued to advise that the best means of fighting the pest was the creation of anti-phylloxera associations or syndicates which would be locally organised (in cantons or communes) but centrally subsidised and advised, to apply chemical insecticides to their vineyards. In August 1885 the prefect reported to the department's General Council that 'la gravité de la situation a enfin ému les vignerons' ('the seriousness of the situation has finally affected the vine-growers'). Recognising that treating vines was costly both in terms of chemicals and of labour, the prefect argued that syndicates provided the best way of finding the necessary resources and that, whereas only one anti-phylloxera syndicate had been in existence in the department at the beginning of August 1884, a further nineteen had been created since then. A new phase in the war against phylloxera had begun.

The spread of anti-phylloxera syndicates

In the summer of 1879, the prefect of Loir-et-Cher – responding to instructions from the Ministry of Agriculture – had authorised the creation of anti-phylloxera syndicates in the department but initially he did little to promote them. In September 1881 the Commission des affaires diverses of the department noted that because of opposition from individual *vignerons* to the compulsory treatment of their diseased vines, the creation of syndicates was being encouraged, for this allowed the *vignerons* to treat their own vines but to do so under the eyes of the administration. But it was not until December 1882 that the first such syndicate was established, in the commune of Mer on the right bank of the Loire. Tanviray had lectured there the previous month on the organisation of anti-phylloxera syndicates and it was not until he devoted more of his limited resources to the promotion of such syndicates during 1883 and 1884 that others were founded. In August 1884 the prefect reported that his administration had decided to abandon its practice of treating vines itself in order to concentrate instead on the promotion of syndicates to undertake the task. Both Tanviray and his successor from 1887, Trouard-Riolle, were convinced that the safeguarding of viticulture in Loir-et-Cher in the medium term lay in collective action by *vignerons* to combat the phylloxera. Their conviction was shared by the prefect, who instructed Trouard-Riolle to redouble his efforts in explaining the advantages of association, adding that he was counting upon his zeal and devotion to promote as much as possible the creation of such syndicates. This he certainly did by intensive lecture tours throughout the department and by articles in newspapers and agricultural journals, an attempt to overcome what was regarded as peasant apathy based upon ignorance.

Success of an immediate but limited kind ensued (figure 7.4). Between August 1884 and April 1885 nineteen syndicates were established, with a total of 1,655 members working 1,636 ha of vines. Subsequently, the creation of syndicates was much slower. The prefect reported in August 1885 that many *vignerons* were inexplicably opposed to the treatment of their vines, and in August 1886 that many incredulous and indifferent *vignerons* remained obstinately opposed to progress. Only another thirteen syndicates were founded after April 1885, the last one being founded in February 1895. The final total of thirty-three syndicates represented only 11 per cent of the potential adoption field of all viticultural communes in the department and their temporal diffusion took the form of a truncated logistic curve with only an origin phase but without widespread adoption or saturation (figure 7.5). The spatial spread of the syndicates reveals their early development along the right bank of the Loire, in and around the canton of Mer, rather than along the Loir, in and around the canton of Vendôme, which was the initial core from which the phylloxera itself diffused (figure 7.6). The creation of syndicates was not a simple response to the single stimulus of the phylloxera infestation: the canton of Mer and its environs appears to have been particularly receptive to the idea of association and collective action, whereas *vignerons* of the Vendômois were for some while resistant both to having their vines treated and to establishing anti-phylloxera syndicates. Considering the department as whole, anti-phylloxera syndicates were relatively few and also short-lived. The number of syndicates in operation grew rapidly from just one by early August 1884 to a peak of twenty-two in mid-1886, declining almost as quickly to only ten in 1888 and thereafter more slowly to only one in 1900 (figure 7.4).[15] By mid-1887 the prefect was expressing his regret that several syndicates had ceased operations and the newly appointed professor of agriculture, Trouard-Riolle, was commenting upon apathetic *vignerons*, some of whom simply left their vines to be killed off by phylloxera.

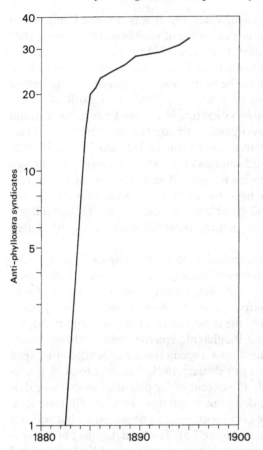

Fig. 7.4 Anti-phylloxera syndicates in Loir-et-Cher, 1880–1900
Sources: AD 7M 175–88

From as early as 1884 the official view of the administration was that while chemical treatment could not be expected to destroy the phylloxera completely it could successfully prolong the life of afflicted vines, and the administration continued to advocate such treatment even after 1890, when it recognised that in the long term the reconstitution of the vineyards with French vines grafted on to American phylloxera-tolerant vines was a better way of adjusting to the disaster.

By the mid-1890s, opposition to chemical applications had been eliminated: partly because the techniques of treatment had gradually been improved and the hazard of being burned while applying the chemical greatly reduced; partly because of the defeat of ignorance by good teaching and example on the part of the Service du phylloxéra and the department's professors of agri-

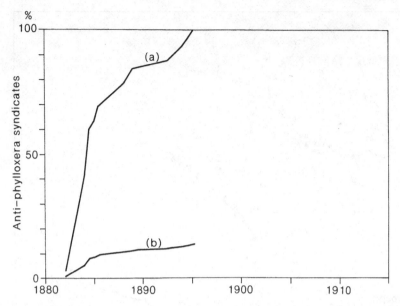

Fig. 7.5 Anti-phylloxera syndicates in Loir-et-Cher, 1880–1915, (a) as a cumulative percentage of all such syndicates and (b) as a cumulative percentage of the number of phylloxerated communes
Sources: AD 7M 175–88

culture; partly because the ravages of the phylloxera had by the mid-1890s dramatically spread throughout the entire department, and partly because conservative *vignerons* were anxious to preserve their own vines rather than reconstitute their vineyards on American rootstock, in order to avoid producing the unpleasantly 'foxy' wines which the latter were considered to yield. By the mid-1890s the public authorities had convinced many viticulturalists of the merits of chemical treatment but they had failed to convince most of them of the benefits of associating in syndicates in order to carry out applications. Instead, many *petits vignerons* treated their own vines using equipment borrowed without charge from the department. For example, in 1900 the professor of agriculture reported that during the previous year there had been 226 borrowings of the department's sixty-seven injection tubes, with each tube being borrowed for an average of five days and – he calculated – being loaned by a *vigneron* to at least five of his neighbours, so that more than 1,000 people had used the department's injecting equipment during 1899. Even so, it must be recognised that less than 600 ha of vines – a very small proportion of the remaining total of about 26,000 ha of French vines in the department – were chemically treated in that year. Environmentally, phylloxera had been brought under a degree of control but it had certainly not been conquered.

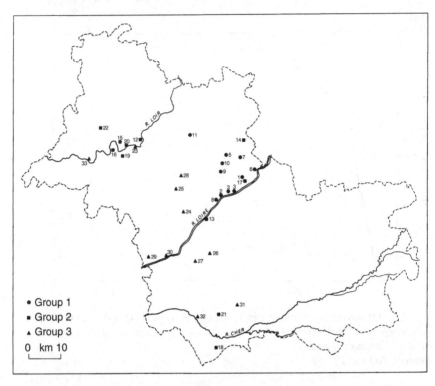

Fig. 7.6 The spatial spread of anti-phylloxera syndicates in Loir-et-Cher, 1882–95
Sources: AD 7M 175–88
The individual syndicates are shown as members of three equally sized and
chronologically ordered groups of the total population of thirty-three syndicates, as
follows:
Group 1 1 Mer (14 December 1882); 2 Cour-sur-Loire (15 August 1884); 3 = Suèvres
(16 August 1884); 3 = Suèvres (hamlet of Fleury) (16 August 1884); 5 Talcy (24
August 1884); 6 Avaray (1 September 1884); 7 Séris (7 September 1884); 8 Ménars
(11 September 1884); 9 La Chapelle-Saint-Martin (14 September 1884); 10
Villexanton (28 September 1884); 11 Oucques (5 October 1884)
Group 2 12 Saint-Ouen (26 November 1884); 13 La Chaussée-Saint-Victor (21
December 1884); 14 Josnes (25 December 1884); 15 Villiers-sur-Loir (5 January
1885); 16 Thoré-La-Rochette (6 January 1885); 17 Mer (21 January 1885); 18
Châteauvieux (31 January 1885); 19 Marcilly-en-Beauce (29 March 1885); 20 Naveil
(12 April 1885); 21 Saint-Romain-sur-Cher (25 December 1885); 22 Fortan (2
February 1886)
Group 3 23 Vendôme (9 April 1886); 24 Saint-Sulpice (15 July 1888); 25 La
Chapelle-Vendômoise (24 August 1888); 26 Chitenay (25 November 1888); 27
Ouchamps (1 March 1889); 28 Champigny-en-Beauce (5 September 1889); 29
Monteaux (21 September 1892); 30 Onzain (18 March 1894); 31 Chémery (1 June
1894); 32 Thézée (31 July 1894); 33 Montoire-sur-le-Loir (3 February 1895)

Psychologically, it might be argued, peasant individualism prevailed and only a few anti-phylloxera syndicates were established. But any such conclusion needs to be qualified. Given both that the principal stimulus for setting up such associations was external to the communities of *vignerons,* and that the chemical treatment of vines was not an obviously and completely successful method of treating affected vines, the limited enthusiasm of many *vignerons* for anti-phylloxera syndicates is understandable.

Disillusion and dissolution

Anti-phylloxera syndicates, promoted by the government and its agents, failed in their immediate intention: the extent to which the dissolution of these syndicates carried with it a long-term disillusion with the local authority's technical advisers and with its republican politicians is not easily determined. But it must have depended to some extent on the reasons for the failure of the syndicates as perceived by their members.

Historians have tended to argue that these syndicates failed because of their high operational costs despite the fact that one of their purposes was to achieve economies of scale in the purchase of necessary materials and expertise.[16] Analysis of the accounts of the syndicates in Loir-et-Cher suggests that many *vignerons* might not have considered the costs to them as being excessive. Each syndicate was entitled to receive a subvention from the Ministry of Agriculture equal to any grant made to it by the department, which itself tended to award a sum equal to the subscriptions paid by a syndicate's members. The syndicates' activities were, therefore, heavily subsidised from public funds and their operational costs were commonly kept down by use of their members' own labour (except for the most skilled operations involved in applying the insecticide). The cost to an individual *vigneron* might in some instances have been as low as 3 per cent of his annual income; but costs varied considerably and theoretically might have been as high as 18 per cent if *all* of the treatment were undertaken by paid labour (an improbable, at most exceptional, circumstance). There is remarkably little evidence that members considered their subscriptions to be too onerous. The president of the syndicate at Châteauvieux wrote to the prefect in December 1885 that its members would not agree to increase the rate of subscription established for the first year until chemical treatment was seen to be successful; while at Suèvres in October 1887 the syndicate was unable to settle its bills because, after two years of its operations, disillusioned members were withholding their subscriptions for its third year. The efficacy – or inefficacy – of the treatment was of greater concern to viticulturalists than was its cost. In 1897, the newly appointed professor of agriculture, M. Vezin, explicitly stated in his first report that a *petit vigneron* was not worried about the cost of the inexpensive insecticide; nor was a *vigneron* likely to have been unduly concerned about the cost of the labour needed, because he undertook most of the work himself. But Vezin did emphasise that what worried *vignerons* was the cost of the necessary equipment, of the *pal-injecteur*, and for that reason the department had a stock of some sixty or so such injecting tubes which could be borrowed without charge by *vignerons*. There is no clear indication that the costs of chemical treatment were generally considered by most *vignerons* to have been excessive.

There is, by contrast, abundant evidence that syndicates thought that government support for their activities was inadequate, that promises made were not fulfilled. The need for public accountability meant that the syndicates were required to present to the prefect detailed financial accounts of their operations: many of the syndicates' officers became irritated by the bureaucratic procedures which they were required to follow to the letter, with failure to do so embroiling them in seemingly interminable correspondence with the prefecture. Many syndicates were angered by the long delays before subsidies promised by the minister and by the prefect were actually paid and some, such as that at Séris in 1886, found themselves consequently without adequate funds to cover the costs of treatment undertaken, as they saw it, with certain official approval and promised financial support. Relations between the syndicates and the administrators of the department were undoubtedly marred on occasions by the latter's slow response to – or even total neglect of – requests for assistance: thus the mayor of Châteauvieux wrote to the prefect in March 1885 explaining that the latter's inattention to a request submitted the previous August for founding a syndicate within the commune had paralysed the goodwill which had existed there; and in the same month the president of the syndicate at Saint-Ouen, no doubt irritated that it had taken since the previous November to complete the bureaucratic procedures for setting up a syndicate, wrote to the prefect that he was impatient to receive approval of its estimates for 1885 so that the treasurer could collect subscriptions and work on treating the vines could begin. Instances of this kind marred relations between peripheral peasants and central committees: they could hardly have helped anti-phylloxera syndicates to flourish.

But a more important problem, contributing most significantly of all to the failure of these syndicates, was the limited effectiveness of the remedy which they were promulgating. This was well exemplified by the syndicate at Josnes, which had 190 members working 105 ha of vines. Two treatments were applied in 1885 and three in 1886; in the summer of 1887 the syndicate's president reported that its executive committee acknowledged that vines treated with insecticide had not been invaded as rapidly by phylloxera as had untreated vines; the committee agreed, with only one dissenting voice, that two further treatments should be undertaken because they would check the spread of phylloxera even if they would not eliminate it entirely. But by 1888 isolated dissent had become widespread disillusion, for the committee then concluded that the phylloxera had gained a stranglehold which it was impossible to loosen and the syndicate at Josnes was dissolved. Many similar instances can be cited of syndicates dissolved because chemical treatments had failed to produce satisfactory results. For example, at La Chaussée-Saint-Victor a syndicate established in the autumn of 1884 operated only for one year: its president explained in a letter of 3 April 1886 to the prefect that although its members believed in the principle of treating vines they had not observed in practice any improvement in the condition of their vines, and so the majority of them had decided to dissolve the syndicate. At Suèvres in February 1887, members of the syndicate which had been founded in September 1884 decided to dissolve it because the chemical treatments had not produced the expected results and because the greater part of their vineyards had been destroyed by phylloxera. They decided that their resources would be better spent on establishing a nursery of American vines, grown from seed, to enable them to reconstitute their devastated vineyards. Similarly, the syndicate created at Talcy in September 1884 was dissolved by its members in May 1887 because treating the vines had been ineffective, half of the vines of its members had by then been destroyed and most of the remaining vines were infested. At Châteauvieux, the syndicate established in the spring of 1885 was dissolved by its members at the end of 1887 because three years of chemical treatments had not been a success, producing results which were so disappointing that its members wanted nothing more to do with it. Similarly, the syndicate established at Vendôme in April 1886 for five years was soon dissolved, at the end of 1887, because the treatment of affected vines had had a negative result. It was, it seems, less a matter of costs being too high than one of benefits being too low. In a letter of 7 March 1889 the prefect of Loir-et-Cher wrote to his counterpart in the department of Sarthe that most of the anti-phylloxera syndicates, although supported financially both by the department and by the State, were dissolving themselves because they were unable effectively to combat the affliction.

Anti-phylloxera syndicates proved not to be the salvation of the department's viticulture as the authorities had promised they would be. As early as August 1887, the prefect reported that some syndicates had ceased their operations and in August 1889 he stated that many syndicates had dissolved themselves on the expiry of the period of three years for which they had been established: the *vignerons* had clearly not been convinced that such syndicates provided an appropriate solution to the phylloxera problem. From the early 1890s, the reconstruction of vineyards with French vines grafted on to American phylloxera-resistant rootstock was coming to be seen not only by the central authorities but also by the viticulturalists as the preferred long-term solution. In September 1893 the Finance Committee of the General Council reported that combating phylloxera by treating vines chemically had been abandoned almost everywhere and that instead use was being made of phylloxera-resistant stock. That claim was undoubtedly premature, because the professor of agriculture continued to report annually throughout the 1890s and in the early 1900s that vineyards were still being treated with carbon disulphide, prolonging the lives of vineyards until they could be reconstituted by grafting French vines on to American rootstock.

Politicisation of the peasantry?

The impact of the failure of the anti-phylloxera syndicates on the *mentalités* of the peasantry is difficult to discern because its specific consequences were indivisible from those stemming from general causes, including the wide range of national agricultural policies and the transformation of the world's agricultural systems during the second half of the nineteenth century. That the syndicates had a political role is indisputable. Local struggles for control of syndicates no doubt at times reflected personal animosities, provoking resignations of the kind witnessed at Séris in the summer of 1885, but they were also underpinned by political rivalries which occasionally surfaced as they had in the adjacent commune of Mer earlier that year. In February 1885 the mayor of Mer, a fervent republican, wrote to the prefect about plans to establish in the commune a second anti-phylloxera syndicate, plans to which he was himself opposed. The mayor argued, with some intelligence, that the real need was to expand membership of the existing syndicate so that it could treat all of the vines in the commune, whether already afflicted or still healthy. Indeed, he envisaged a syndicate not restricted to the commune but one extending its collective and so effective defence of the vines throughout the entire canton. He deplored the fact that most *vignerons* refused, out of ignorance and incredulity, to accept this view. But during the winter months of 1884–5 the mayor had worked energetically to persuade the *petits vignerons* throughout the commune to associate in a single syndicate. This splendid union, as the mayor described it, was crowned by having as its president an ardent republican. But its executive committee included M. Bidault, a veterinary doctor and – more importantly – an inveterate enemy of the Republic. Disturbed by the possible political consequences of an enlarged association of *vignerons*, he set about to undermine it by supporting proposals for the foundation of a second syndicate, feeding upon long-standing rivalries between *vignerons* in northern and southern parts of the commune of Mer. While the prefect, for technical reasons, deferred approving these proposals, he was unable to do so for long and by the spring of 1885 Mer had two anti-phylloxera syndicates of opposing political persuasions.

The role of the anti-phylloxera syndicates in the politicisation of the peasantry was likely to have been undermined by their own ultimate lack of success: failure in this in particular might have checked the growth of agricultural syndicalism in general in the department, and even of republicanism, since the syndicates were, for the most part, closely linked with the government. The rapid expansion of agricultural syndicalism in Loir-et-Cher during the 1880s was certainly arrested during the 1890s (see chapter 8 below). Examination of the relationship between the failure of anti-phylloxera syndicates and the development of livestock insurance societies is also illuminating: only 7 per cent of viticultural communes which established anti-phylloxera syndicates between 1883 and 1895 also created livestock insurance societies between 1883 and 1905, whereas 15 per cent of viticultural communes without such syndicates acquired such societies. Furthermore, of the twenty-five viticultural communes which did establish anti-phylloxera syndicates in the period 1883–95, eight (32%) had not founded livestock insurance societies by 1914; and the seventeen communes which did set up such societies did not do so, on average, until twenty years after such a syndicate had been established. The creation and dissolution of anti-phylloxera syndicates appears to have acted as a brake upon the development and spread of other forms of agricultural syndicalism. Although the battle against phylloxera did involve a good deal of concerted action, because the battle was lost it also seems ultimately to have given the *vignerons* of Loir-et-Cher a certain distaste for such action, even if, as Clapham claimed, it initially gave them a renewed taste for it.[17]

It probably also led in many cases to a reinforcement of the traditional peasant *mentalité* which considered struggle against the whims of government as inevitable as that against the vagaries of nature. Certainly by the early 1880s the Catholic and right-wing thrice-weekly newspaper of Loir-et-Cher, *L'Avenir*, was blaming the agricultural crisis on natural disasters and on the government's policy of taxing agriculture rather than protecting it against foreign competition. On the other hand, measures taken by the administration to support the agricultural sector – such as the improvement of communications and of education – were seen by *L'Avenir* as attempts to buy support for republicanism. Of course, its views were countered by its lay and left-wing opposite number, *L'Indépendant*. But the electorate expressed its view in the ballot box. Moderate left-wing republicanism was triumphant in the department in the early 1880s but as the general agricultural crisis deepened from the mid-1880s so criticism of its programme mounted, aided and abetted by its particular failure to win the war against phylloxera. In the cantonal elections in the department in the autumn of 1885, left-wing candidates lost considerable ground to candidates on the right and this trend was accentuated in the elections of 1889. Many of the gains of the right were made in the cantons along the valleys of the Loire and the Cher, which included the major viticultural districts of the department, although some *vignerons* – like those of the cantons of Vendôme and Montoire along the Loir – chose to express their discontent by supporting candidates of the extreme left.[18] But it seems that moderate republicanism lost support among the *vignerons* of Loir-et-Cher during the phylloxera crisis. The battle for the minds of the peasantry, like that against the phylloxera, had not been won.

8

Agricultural associations

Les Syndicats sont des associations de personnes ayant une cause commune à
défendre. L'union fait la force.

Jules Tanviray, in his brochure *Syndicat des Agriculteurs de Loir-et-Cher (projet de
création)* (1883)

...offrir à ses membres un centre d'appui et de confraternité.

Article 1 of the Statutes of the Syndicat des Viticulteurs de Cellettes (1906)

Context

Referring to Western Europe in general rather than to France in particular,
Michel Augé-Laribé claimed that the practice of association brought to the
social fabric of agriculture a change almost as great as that engendered by the
railway network in the economic sphere.[1] Within France, at least one in six of
all communes had an agricultural syndicate by 1914, when there were almost
7,000 such associations throughout the country with a total of more than 1
million members.[2] It has been claimed – admittedly, by a committed Loir-et-
Cherien agricultural activist – that by the 1920s there were few departments in
France where the impact of agricultural associations and organisations was as
complete as it was in Loir-et-Cher.[3] But even a more detached observer, Augé-
Laribé, remarked that it was in Loir-et-Cher that the first, best known and
most imitated agricultural syndicate was established in 1883, prior to the law
passed in 1884 which was to provide the legal framework for such associations.[4]

The legislation of March 1884 was intended primarily to permit the formation of professional trade unions, but it was also to provide the legal context
for the development of agricultural associations nationally, regionally and
locally. The law of 1884 allowed groups to form unions or syndicates in order
to defend their interests and under its umbrella agricultural syndicates were
soon established to provide their members with goods and services at dis-

counted prices and of guaranteed quality, and later to process and to market the products of their members, with all such activities being undertaken by syndicates in the defence of their members' interests (and especially so during the decades of depression at the end of the nineteenth century). The precise origins, structures, functions, aims and evolution of agricultural syndicates have been shown to have varied considerably from place to place and from period to period within France.[5] Local syndicates often reflected national political rivalries, between right and left, between clerical and anti-clerical forces; some were promoted by traditional power-brokers from the local *notables*, others were the creations of active farmers, while still others owed their existence in large measure to the efforts of the local state, in the form of a department's prefect and agricultural advisors. While this study of agricultural associations will need to bear in mind these complexities, the key aspect to be considered here will be the light which such associations can throw upon the question of the peasant fraternalism, on the issue of individual and collective attitudes in the countryside.

As agricultural associations, syndicates were not without precedents. Livestock insurance societies and anti-phylloxera syndicates, considered in earlier chapters, were components of what might be seen as a broad transition from individualism to associationism in agriculture. But other, better known even if locally restricted, cases can also be cited, such as the cheese-making co-operatives which had existed in the Jura mountains since the Middle Ages and the *confréries* (confraternities) of *vignerons* in Touraine during the early modern period. But such associations were not well known in their times and so were not widely imitated.[6] Nor were they direct, functional precursors of the agricultural syndicates of the late nineteenth century. Closer in time at least were the agricultural societies founded in the eighteenth and early nineteenth centuries.[7] Such societies tended to be associations of learned men with an interest in agriculture – whether that interest was intellectual, social or more directly commercial. As part of the Enlightenment, they sought to promote the study of agriculture and to diffuse knowledge about agricultural experiments, innovations and what was deemed to be best practice. In addition, they addressed what were considered to be the social and economic problems of rural society. They were, in effect, study groups or *académies*: each one tended to have a small, restricted membership comprised principally of *notables* (leading citizens) and wealthy landowners. They also had an important social role, as a club, as a networking medium. As such, they were somewhat removed from the mass of the farming populations of their localities – and, indeed, they were often better connected with members of other, distant, agricultural societies. Somewhat closer to local populations were the *comices agricoles* established sometimes by these societies but more often by the prefects of departments. These local agricultural committees had better links to their communities, from whom they recruited members and supporters. They

encouraged agricultural improvement, most importantly by organising practical demonstrations and competitions at (normally annual) fairs at which
public officials (and politicians) awarded prizes and delivered speeches about
how best to improve agriculture specifically and the quality of rural life generally. The famous portrayal by Gustave Flaubert in *Madame Bovary* of such a
fair reveals their character in lively detail but with an irony and even a sarcasm
that makes one doubt the efficacy of such events as part of the programme of
agricultural improvement.[8]

The Society of Agriculture and the *comices agricoles*

The first Society of Agriculture

The Society of Agriculture of Loir-et-Cher had a discontinuous history from
its initial foundation at the end of the eighteenth century. The first society was
founded in January 1799 by the department's administrators, acting upon a
suggestion from the Minister of the Interior. It was in existence for only two
years, during which period it held about forty meetings in the former bishop's
palace at Blois. This was a centrally constructed society, not a spontaneously
established association of agriculturalists. It was to comprise twenty-five 'resident' members from the department, and an unlimited number of associated
corresponding members. Its president was M. Camereau, who was the department's principal administrator. Among its 'resident' members were the department's secretary, the librarian of Blois, members of the local judiciary and the
pépiniériste (head gardener) from the national garden at Blois, as well as a
number of *notables* and large property owners. The Society promoted discussion of a wide range of issues – including how to maintain a plantation of
chestnut trees and how to preserve and make good use of dove-cotes falling
into disrepair because of the new law permitting pigeons to be taken as game
– and also established some trials, for example of wheat seeds from Siberia and
Poland. Its first fête, on 28 June 1799, involved a public procession to and cer-
emony in the former cathedral of Blois, converted by the revolutionary regime
into a *temple des réunions décadaires* (chamber for public monthly meetings).
Speeches were followed by the singing of patriotic songs and the distribution
of bouquets to those *citoyens* who had distinguished themselves in terms of
their agricultural productivity. Then the procession went out of town to a field
where the president ploughed two furrows with the decorated plough team
which had been leading the procession, doing so to alternate cries of 'Vive l'a-
griculture' and 'Vive la République'. Clearly, this first society was about more
than agriculture. In 1801 the new prefect of the department, M. Corbigny,
informed the Minister that he wanted to dissolve the Society and establish a
new one, because the society's members knew little about the rural economy.
The Minister replied that, unless the Society concerned itself with political

questions, the prefect could not dissolve the Society but he could, of course, establish another one. It seems that the first society ceased to exist but it was not until four years later that a new one was founded.[9]

The second Society of Agriculture

In February 1805, Prefect Corbigny established the Société Libre d'Agriculture et d'Economie Rurale du département de Loir-et-Cher.[10] The Society was to comprise fifty members residing in the department, as well as associate corresponding members from within it and outside it. To be a member it was necessary to be a landowner or a farmer, and to have an interest in experimental cultivation, or to have special knowledge of some aspect of the rural economy: thus its founding members included not only agriculturalists but also a justice of the peace, tax collectors, a forestry official, a pharmacist, a doctor, and a priest. Members were to pay an annual subscription of 12 francs and it was assumed that the Society would be able to call upon funds set aside by the government for the advancement of agriculture. Each year the Society was to hold six private meetings, and also one public meeting at which prizes were to be distributed and a lecture given on an approved topic of general interest. The Society established from among its membership eight commissions, each with the remit to investigate and from time to time report upon a specific aspect of the rural economy, such as ploughs and artificial meadows, viticulture, silviculture and horticulture. The associate corresponding members were also invited to report at least once a year to the Society's committee any useful information coming from their own experiences or researches.

The Society thus encouraged agricultural improvement through the diffusion of information at its meetings and in its publications (which included the *mémoires* written by its members, and a bulletin), by practical demonstrations in the field, and by awarding prizes for what it judged to be good practice or worthwhile experimentation. In its early years the Society was promoting, for example, improved rotations and especially the cultivation of artificial meadows, sugar beet and potatoes, and improvements to the quality of local sheep by cross-breeding them with merinos, but it also embraced wider concerns, such as encouraging the vaccination of rural populations against smallpox.

As an association, the Society was socially exclusive and geographically constrained: its nominated members came from the *notables* and the bourgeoisie, and most of them lived in the town of Blois. In 1815 the Society had 47 members, 39 of whom were living in Blois – and seven of the others were living less than 20 km from the town. The Society's fourteen corresponding members were, for the most part, residing much further away from Blois (notably in the Sologne, a *pays* of large landowners) and so maintained

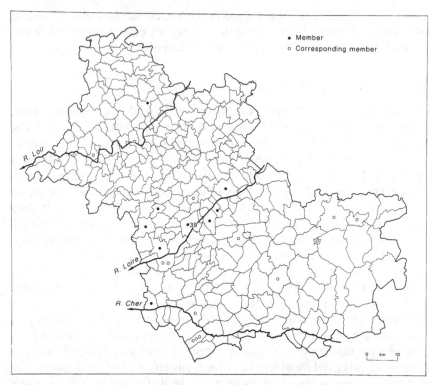

Fig. 8.1 Membership of the Society of Agriculture of Loir-et-Cher, 1815
Source: Annuaire de Loir-et-Cher (1815)

contact with the Society literally by correspondence rather than by attending its meetings, held in Blois itself, often in a room of the department's *prefecture* (figure 8.1).

Addressing directly the problem of geographical – but not of social – distance, in 1819 the Society was reorganised: its corresponding members in the *arrondissements* of Vendôme and Romorantin were persuaded by the department's authorities and the Society's committee to establish separate agricultural societies in each of those two districts, while the former society focused its concerns upon the *arrondissement* of Blois. In 1819, the Society of Blois had 45 members, of whom 39 resided in Blois and the other six lived elsewhere in the *arrondissement* of Blois. The Society of Vendôme had 17 members, that of Romorantin 13 (figure 8.2).[11] The two latter societies came from the early 1820s to be referred to as *comices agricoles*. While that of Romorantin became firmly established and soon established local sections within some of its cantons, that of Vendôme had an uncertain beginning and seems not to have been definitively established until 1837.

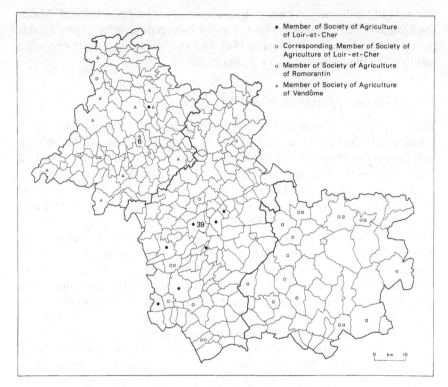

Fig. 8.2 Membership of the Societies of Agriculture for the *arrondissements* of Blois, Romorantin and Vendôme, 1819
Source: Annuaire de Loir-et-Cher (1819)

The central authorities – in the form of the prefect and his two sub-prefects – were heavily involved in the formation and functioning of these societies, but such involvement became a source of conflict in 1839 and was to lead to the reorganisation of the Society. On 12 January the prefect wrote to the Society's secretary, Guérin d'Ogonnière, pointing out that detailed accounts should be provided of the Society's use of grants made to it by the government, because of the need to ensure that such funds had been employed for the purposes intended. Offended by the tone and implication of the letter, d'Ogonnière replied on 3 February, pointing out that the Society was operated according to its rules and that its activities and accounts were made public. He objected to such interference by the authorities, both in principle – having devoted himself for nineteen years to the work of the Society – and in practice – because it would be difficult to provide detailed accounts of the kind requested by the prefect. The prefect remained insistent and in due course, but not until 30 June 1840, d'Ogonnière informed the prefect that he had resigned as secretary of the Society. Although a new secretary was appointed – M. Marin Desbrosses – the Society was clearly thrown into confusion for a considerable period by this episode and during 1841 and 1842 was reorganised into what was in effect the third Society of Agriculture.

The third Society of Agriculture and the comices agricoles

The rules of the reorganised Society were approved by the minister on 18 January 1843; its newly constituted committee included Edouard Malingié, a well-known agricultural improver and director of the farm school at Pontlevoy,[12] as the Society's president. On 17 July 1844 the society's secretary, de Souvigny, submitted to the prefect a report on *la situation morale* of the Society: he emphasised that although the reorganisation during the first part of 1843 had not been without difficulty, by June the Society had 50 members, 35 of whom had also belonged to the previous society. Unlike its predecessors, membership of the new society was not limited: de Souvigny claimed that consequently the Society could now include among its members the department's most distinguished practising agriculturalists, its most enlightened theoreticians, and its most worthy tenant farmers and owner-occupiers. The Society raised its annual subscription to 20 fr., to enable it to have its own funds as well as grants from the government for awarding prizes, but such a rate must have acted as a constraint upon its membership. By 1846 the Society had 116 members and 135 by 1847, after which it settled down at around 120 members. The new society operated over the whole of the department, but unsurprisingly most of its members resided in the *arrondissement* of Blois, especially in the Beauce plateau, the Val de Loire and the Petite Sologne (figure 8.3).

Although far from being inclusive, this third society was much less exclusive than had been its predecessors and it explicitly worked to raise agricultural standards in the public interest, diffusing knowledge about best agricultural practices throughout the department and rewarding such practice (whether by its own members or by others). The Society's efforts attracted national attention and commendation. In 1845 a government inspector of agriculture, M. Royer, spent ten days visiting farms in Loir-et-Cher and was very impressed with the activities and influence of the Society. In a letter of 11 August written from Paris to the prefect, Royer praised the 'generous and enlightened' individuals who had 'planted the flag of progress' in the department, tackling both practical problems in agriculture and social problems in the wider rural economy. But, Royer stressed, such individual efforts would be fruitless if they remained unknown, so he attached considerable importance to the role of the Society of Agriculture at Blois in diffusing information about them: he heaped praise upon the 'zeal and devotion' of the Society's officers and believed that the future of agriculture must be great in a *pays* where such public-spirited men gave such an example of progress.

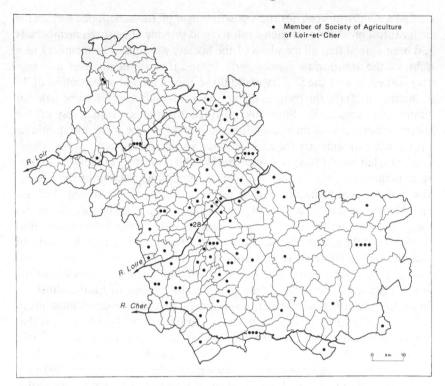

Fig. 8.3 Membership of the Society of Agriculture of Loir-et-Cher, 1846
Source: Annuaire de Loir-et-Cher (1846)

The Society's secretary understandably used such a glowing report to persuade the prefect to obtain more central funds for it to use in the encouragement of agriculture. Royer's visit was to have even deeper repercussions, for he had advised the Society to organise its activities geographically on the basis of cantons. In the early 1840s the Society had an ambivalent geographical basis: in theory, it operated over the entire department but in practice it operated principally in the *arrondissement* of Blois, for which it in effect acted as a *comice agricole* alongside the two other *comices agricoles* for Romorantin and Vendôme. In the summer of 1845 officers of the Society came to realise, presumably as a result of Royer's visit, that more central funds were likely to be allocated to it if were to become the *comice agricole* for the *arrondissement* of Blois – not least because grants to *comices agricoles* by departmental administrations were automatically matched (or more than matched) by grants from the Ministry.

Accordingly, on 10 August 1845 de Souvigny, the Society's secretary, informed the prefect that its committee had decided that there should be created a *comice agricole* for the *arrondissement* of Blois, with one section for each canton provided that each canton could provide at least ten members. It had been agreed that all members of the Society would also be members, as a right, of the appropriate *comice agricole*; and that all members of a *comice* who wished to join the Society could do so for an annual subscription of 15 fr. instead of 20 fr., the balance of 5 fr. being the subscription to be paid for joining the *comice*. De Souvigny provided the prefect with a list of the Society's members, so that he could choose from among them and nominate a president and secretary for each cantonal section of the new *comice*. Those so nominated would then be empowered to call a meeting in their cantons of other members of the Society, in order to constitute a section of the *comice*. The prefect acted promptly, writing to his nominees on 21 August 1845. By early December the prefect was so pleased with the progress being made in creating cantonal sections of the *comice agricole* in the *arrondissement* of Blois that he urged the sub-prefects of Romorantin and Vendôme to follow suit and promote such a cantonal organisation for the already existing *comices agricoles* of their *arrondissements*. By contrast, Malingié, the first president of the *comice agricole* of Blois, was so disappointed by the lack of local enthusiasm for creating cantonal sections (and specifically by the absence of most presidents and secretaries of sections from a general meeting of such officers) that in June 1846 he told the prefect that he was inclined to resign his office. Clearly, the 'official' view differed from that of the individual most closely involved. Nonetheless, although *comices agricoles* were slower to be developed than the progressive Malingié would have wished they were to become a very visible part of the agricultural scene of Loir-et Cher.

From the mid-1840s the *comices agricoles* of Loir-et-Cher, operating at the level both of its *arrondissement* and of its cantons, increasingly took over the role of the Society of Agriculture in promoting good farming practices, and they made much more effort than had the Society to reach a wider range of the farming community within the department. They did so through meetings and publications but most especially through arranging fêtes (annually or almost annually in the case of the *arrondissement* and on a less frequent, rotational basis in the case of the cantons) at which speeches were made, *concours* (competitions) held and prizes awarded. Before each such fête or *concours*, visits were arranged to farms of the most progressive agriculturalists of the locality in which it was being held – such visits were seen as a practical teaching exercise and, as the secretary of the Society of Agriculture described it to the prefect in a letter of 16 August 1845, undoubtedly the most useful role of the *comices agricoles*.

The role of the Society of Agriculture was gradually eroded by the *comices agricoles*: its membership fell to 102 in 1847 and to 91 in 1848. Part of the reason for the decline might well have been the growing reluctance on the part of the government to support such societies with grants and its developing preference for targeting its support for agriculture through what it perceived to be 'official' organisations, such as the *comices agricoles*. The role of the Society was certainly further undermined by the foundation in October 1850 of an official Chambre Consultative d'Agriculture for Loir-et-Cher. Members of the Society rapidly joined the new consultative body, so that the Society lost much of the reason for its existence. In September 1852 the prefect informed the society's secretary that from the following 1 January the Society would have to vacate the building loaned to it by the authorities in the courtyard of the *mairie*, that the council of the department had decided not to allocate any funds to them for 1853, and that in his view the Society would probably dissolve itself.

The Society of Agriculture had, it seems, operated within its own organisation what it – but not the government – considered to be a *comice agricole* for the *arrondissement* of Blois. When it collapsed in the early 1850s, therefore, there was no longer any mechanism for distributing government funds for the encouragement of agriculture within the *arrondissement*. Consequently, in November 1854, the prefect set up for that purpose a Commission Consultative d'Agriculture. Allocations of funds by the government and by the department for the five years from 1853 to 1857 were to be distributed by this body in the form of grants and prizes. But such a way of proceeding was considered by the prefect to be unsatisfactory, so he informed the Minister in June 1855 that he was endeavouring to reconstitute a society of agriculture. In fact, what came to be formally established was – more appropriately – a *comice agricole* for the *arrondissement* of Blois. A group of thirty landowners and farmers from the *arrondissement* met on 5 December 1857, at the invitation of the president of the Chambre Consultative, M. Rousel. He, like at least sixteen of the thirty, had been a member of the defunct Society of Agriculture. They agreed that since the Society had ceased to function, farmers within the Blois *arrondissement* had come to be disadvantaged by comparison with those in Romorantin and Vendôme. So they decided to found a *comice agricole* for Blois: it was to be divided into sections according to soil and farming type, rather than by cantons, with its activities being held successively in different cantons. The foundation of the new *comice* was approved by the prefect early in 1858 – indeed, he claimed credit for its creation when informing the Minister of it. The *comice agricole* of Blois had as its first president the Marquis de Vibraye; it operated as two sections, one for territories on the left bank of the Loire and one for those of the right bank.

It was, then, only from the late 1850s that *comices agricoles* functioned effectively throughout Loir-et-Cher. They then did so during the rest of the century, following the same policies and practices of their predecessors. Their principal sources of income were grants from central and local government, and donations from a few members, rather than subscriptions from the general membership. Their principal expenditures were on cash prizes, medals

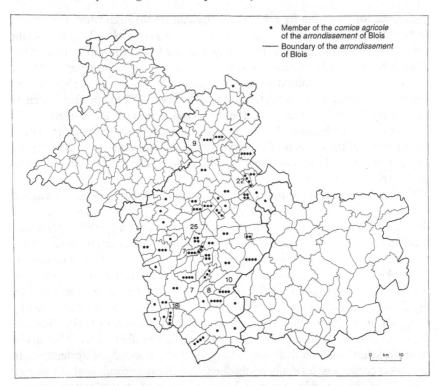

Fig. 8.4 Membership of the *comice agricole* for the *arrondissement* of Blois, 1883
Source: Bulletin Annuel: Société d'Agriculture de Loir-et-Cher (Comice de l'arrondissement de Blois) (1883)

and printing. Their principal activities were organising farm visits and fêtes, and producing publications on agricultural innovation and improvement. They had larger and geographically more widespread memberships than did the agricultural societies; for example, the *comice agricole* of Blois in 1883 had 220 members distributed among fifty-nine communes (figure 8.4). Many communes had no members and many others had only a few. Directly, the *comices agricoles* involved only a small proportion of the agricultural population of Loir-et-Cher; indirectly, they impacted upon many more of those engaged in agriculture but the extent of their influence is impossible to measure precisely.

It is clear, however, that memberships of the societies of agriculture and of the *comices agricoles* were largely confined to the rural elite: the large landowners, progressive farmers and enlightened, educated and often professional members of rural society. While from time to time these associations penetrated into the lives of the mass of the rural population (doing so in often spectacular ways at agricultural fairs, competitions and demonstrations), they were not part of that daily life world. Furthermore, while the societies were to a considerable degree and the *comices agricoles* were to some extent elitist academies, the latter were much more closely linked to the machinery of government. While societies and *comices* might have served as models for peasant associations, the extent to which they did so in Loir-et-Cher was probably very limited. The very close links between societies of agriculture and the formation of peasant associations, which have been claimed for some other departments of France, seem not to have been so strong in Loir-et-Cher.[13] The societies of agriculture and *comices agricoles* of Loir-et-Cher significantly promoted the expansion and diffusion of knowledge about agricultural practices, but – like those in some other parts of France[14] – did not engage peasants as active agents or themselves become directly responsible for the creation of other professional agricultural associations. It is, therefore, now necessary to turn more directly to those agricultural associations in which peasants themselves were very closely involved.

Agricultural syndicates

Although it has often been noted that Loir-et-Cher was the site of France's first agricultural syndicate and that it was the largest and most active of the departmental syndicates, the role of that and of other agricultural associations in the department has received limited attention from historians and historical geographers.[15] But in his classic regional monograph on the Val de Loire, published in 1934, Roger Dion made two significant observations: first, that within the Loire valley it was in the section between Orléans and Blois that the spirit of association in agriculture had its most precocious awakening: the earliest such association he found throughout the entire Val de Loire was one established in 1878 at Saint-Claude-de-Diray, close to Blois on the left bank of the Loire, for the collective purchase and use of a threshing machine; and second, that agricultural syndicalism appeared soonest in viticultural (and formerly viticultural) *pays* for reasons linked functionally to their dependence on chemical fertilisers, which syndicates were set up to supply to their members. He noted that the initial objective of the Syndicat des Agriculteurs de Loir-et-Cher, founded at Blois in 1883, was to obtain for its members quality-assured fertilisers at reasonable prices.[16] A brief account of the work of the Syndicat provided in 1950 by its president, Eugène Nivault, concentrated on its activities from the 1920s onwards.[17] A brief note by D. Braux in 1954 on the co-operative movement in Loir-et-Cher commented upon its considerable growth during the interwar years and claimed that the Syndicat des Agriculteurs de Loir-et-Cher was for forty years the only grouping of farmers in the department.[18] Georges Dupeux, in his classic monograph on the social and political history of Loir-et-Cher, published in 1962, considered agricultural associations only cursorily. He referred briefly to the department's

Société d'Agriculture, to its *comices agricoles*, and to the Syndicat des Agriculteurs de Loir-et-Cher (whose almost exclusive role, Dupeux argued, was the provision of fertilisers). His main general claim was that a new beginning of association in agriculture was detectable in the department in the early twentieth century, with a recourse to collective organisation in order to reduce the costs of farming: it was not until the early twentieth century, according to Dupeux, that association began to affect peasant ways of behaving.[19] In J. Vassort's recent admirable survey of Loir-et-Cher from the Revolution of 1789 to the eve of the Second World War, the department's agricultural syndicates are virtually ignored and there are but brief references to its Société d'Agriculture.[20]

The historical geography of agricultural associations in Loir-et-Cher before 1914 remains a largely uncultivated field. Fortunately, there are some contemporary publications and very considerable contemporary unpublished material relating directly or indirectly to the department's agricultural syndicates before 1914. Their growth and significance was reported and commented upon in the local press, such as *L'Avenir* and *L'Indépendant*, and in the agricultural press, such as *L'Agriculture Pratique du Centre*, and the principal syndicate regularly published its own *Bulletin*. The monitoring of such associations by the local and central authorities produced a mass of unpublished documentation on the syndicates, while the prefect's reports to the department's general council were published as part of its proceedings. From these sources in combination can be reconstructed the historical geography of the department's agricultural associations before 1914.[21]

Historical development

The development and achievements of agricultural syndicates must be set against their historical background. During the 1880s and 1890s many farmers in Loir-et-Cher were feeling increasingly the impact of foreign competition while viticulturalists saw their livelihoods being undermined by the invasion of the phylloxera. But the State, in the form of central as well as local authorities, was keen to support the peasantry for political reasons and saw the promotion of agricultural syndicates as one way of offering effective aid, channelled through its officially appointed professors of agriculture.[22]

Jules Tanviray was Loir-et-Cher's first professor of agriculture, taking up his post on 1 April 1879. Coming from a farming family, Tanviray was one of the tiny minority of peasants who received a formal education in agricultural methods, first at the farm school in the department, at La Charmoise, near Pontlevoy, and subsequently at a national farm school. He then worked his own farm from 1871 until appointed, after an open competition, professor of agriculture for Loir-et-Cher.[23] During his first four years in office Tanviray was principally concerned with encouraging *vignerons* to combat phylloxera and with educating peasants into applying more chemical fertilisers in order to increase crop yields.[24] Somewhat disappointed by the results of his numerous lectures throughout the department, Tanviray concluded by mid-1883 that the best means of fighting phylloxera was the creation of local anti-phylloxera syndicates and that the best way of promoting the use of chemical fertilisers was the foundation of a central agricultural syndicate which would combat fraudulent commerce in fertilisers and also reduce their price to farmers. In March 1883 Tanviray called a meeting of about fifteen people to consider the statutes which he had drafted for an agricultural syndicate; copies of the amended statutes were then sent to the almost 300 mayors of the department, seventy of whom replied, submitting the names of more than 200 people who had indicated their support for such a syndicate. The first formal meeting of the Syndicat des Agriculteurs de Loir-et-Cher (hereafter referred to as SALC) approved its statutes on 7 July 1883; by the end of that year the association had 345 members.[25] Tanviray left Loir-et-Cher three years later, to direct a farm school elsewhere in France. During his seven years as professor of agriculture, he had worked energetically and effectively as a government official whose primary aim was agricultural improvement. Tanviray's example of close co-operation between the local authority and the farming community – and especially his promotion of their joint interests in SALC – was followed both by M. Trouard-Riolle, who succeeded him as professor of agriculture in 1886, and by M. Vezin, who moved into the post in 1897.

The Syndicat des Agriculteurs de Loir-et-Cher was to prove to be such a great success in the department – and, indeed, it came to be emulated elsewhere in France and to be noticed by agronomists in England[26] – that it needs to be considered in detail, separately from the department's other agricultural syndicates. Such consideration will, however, be deferred until after the general picture of agricultural syndicalism in Loir-et-Cher has been drawn: for the moment it is sufficient to note that SALC had about 350 members by the end of 1883, more than 4,000 by 1893, more than 6,000 by 1903, and approaching 17,000 by 1913. To some extent this remarkable growth was achieved by the eclipse of other, smaller and less effective, syndicates.

The Syndicat des Agriculteurs de Loir-et-Cher was founded in 1883, just before the Waldeck Rousseau law of March 1884 which was to be the legal framework for subsequently established syndicates. The Syndicat was, in so many ways, ahead of its time and was to remain the only agricultural syndicate in the department for three years. But for the moment attention will be focused upon the department's other general purpose agricultural syndicates. During the fifty years between 1883 and 1913, more than forty such syndicates were founded in Loir-et-Cher, although not all of them survived to the end of that period. After the creation of SALC in 1883, the next agricultural syndicate – the Syndicat Agricole et Viticole de Thenay – was not founded until 1886. During the following three years there was a mushrooming of syndicates, with

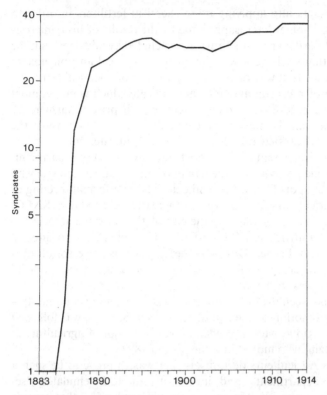

Fig. 8.5 Agricultural syndicates in Loir-et-Cher, 1883–1914
Sources: AD 7M 244 and 10M 55–9

twenty-one new ones being established by the end of 1889. Between 1890 and 1905 the pace was much slower, with only one or two – or even no – new syndicates being set up each year. Thereafter it quickened again (although not to the rate of the late-1880s), with eight syndicates being founded between 1906 and 1911.

This general picture of growth must, however, be modified because from the mid-1890s onwards some syndicates went out of operation. For example, three stopped functioning in 1896 and there were also other dissolutions of syndicates between then and 1914. A first peak of functioning syndicates was reached in the mid-1890s, when there were more than thirty in operation. That number then declined to twenty-six by 1902, after which it grew again, passing thirty in 1906 and reaching thirty-six by 1914 (figure 8.5). By 1895, when SALC had about 4,000 members, the other thirty syndicates had about 5,500 members. By 1908, when SALC had about 11,000 members, the other thirty-one syndicates had about 6,500 members. Unfortunately, data on membership numbers is not sufficiently robust to determine precisely when the membership of SALC came to exceed that of all of the other general purposes syndicates of the department, but it most probably happened during the opening years of the 1900s: in 1903 SALC had 6,300 members, and by 1904 it had 7,375, whereas the total membership of all other agricultural syndicates seems to have peaked at around 6,500 in 1908.

Most farmers of Loir-et-Cher came to be members of an agricultural syndicate. The population census of 1891 records for the department a total of just over 33,000 heads of agricultural holdings (owner-occupiers, tenant farmers, nurserymen and market-gardeners).[27] In 1892 SALC had almost 4,000 members and other agricultural syndicates had about 5,000. So, just over one in four 'farmers' of the department were members of an agricultural syndicate in the early 1890s, with almost as many of them in SALC alone as in all of the other syndicates. By 1908 more than half of 'farmers' were members of a syndicate and by 1914 the figure had reached almost three-quarters. By the former date about one in three 'farmers' of the department were members of SALC; by the latter date the proportion was more than one in two. Although these figures are approximate, the general picture which they provide is undoubtedly one of a high level of agricultural syndicalism, increasingly dominated by SALC.

Geographical distribution

The geographical distribution of general purpose agricultural syndicates within Loir-et-Cher was very specific (figure 8.6). The earliest ones to be established – at Thenay (1886), Soings (1887) and Chitenay (1887) – were all located in the Petite Sologne and this distinct *pays*, within the angle formed by the confluence of the Loire and the Cher rivers, was to acquire about one in four of all such syndicates created in the department by 1914. A second concentration developed within the Loire valley itself: syndicates were founded in 1887 at Saint-Gervais-la-Fôret and at Blois (for the canton of Blois-East), and in 1888 at Blois (for the canton of Blois-West), Onzain, and Saint-Gervais-la-Fôret. The Val de Loire was also to acquire almost one-quarter of the syndicates established by 1914. But the largest geographical concentration of syndicates was in the valley of the Cher, where two were founded in 1887 at Montrichard and Châtillon-sur-Cher, six in 1889 at Selles-sur-Cher, Saint-Georges-sur-Cher, Seigy (2), Châtillon-sur-Cher and Billy, and a further five were set up between 1890 and 1893. In all, fifteen syndicates – almost half of the total for the department – were established in the Cher valley by 1914.

A fourth, but much less marked and historically later, concentration of syndicates was located in Vendômois, where they were created at Selommes (1887), Vendôme (1899), Montoire-sur-le-Loir (1904), Coulommiers (1905) and Thoré-la-Rochette (1911). Elsewhere in the department, it is the paucity

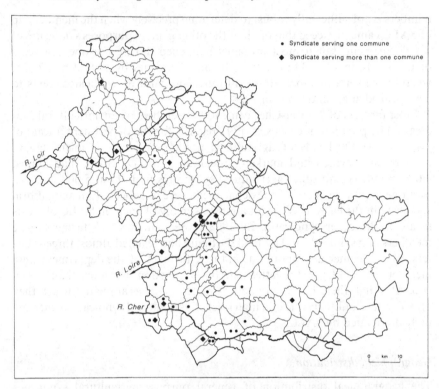

- Syndicate serving one commune
- Syndicate serving more than one commune

Fig. 8.6 The geographical distribution of agricultural syndicates in Loir-et-Cher, 1883–1914
Sources: AD 7M 244 and 10M 55–9

or total absence of syndicates which is striking. In Perche, just one was established, at Mondoubleau (1887), but it was intended to serve the whole of its canton; in the whole of the Grande Sologne, just three syndicates were set up, at Chambord (1888), Romorantin (1889) and Pierrefitte-sur-Sauldre (1906); in Beauce and in the Gâtine tourangelle there were none.

The distinctive patterns thus far identified in the history and geography of Loir-et-Cher's agricultural syndicates now need to be investigated in terms of the processes which underpinned them. The most appropriate point of entry into this problem is an examination of the characteristics of the syndicates themselves.

Characteristics

Aims and activities

The objectives of syndicates can be determined in part from their titles but more comprehensively from their aims as stated in their statutes and from their activities as described in a variety of sources. A syndicate usually operated under one of three sets of titles: *syndicat agricole* (or *syndicat des agriculteurs*), *syndicat agricole et viticole* (or *syndicat des agriculteurs et vignerons*), or *syndicat viticole* (or *syndicat des viticulteurs*). Recognising that some syndicates changed their names (or at least were not always recorded as having only one name), it can be said that approximately one-quarter of the syndicates of Loir-et-Cher came into the first category, about one-half into the second, and about one-quarter into the third. It is certainly the case that more than two-thirds of the department's agricultural syndicates were specifically for the benefit, in whole or in part, of *vignerons*.

The most broadly based syndicates, like the *syndicat des agriculteurs* established at Billy in 1889, had as their principal, explicitly stated, aims the collective purchasing of farming supplies in order to obtain them for their members at the keenest prices and in order to protect them, especially in the case of chemical fertilisers, against fraud. Some also offered advice to their members about the best materials to use for particular purposes, seeing it is their role to enlighten the farming community. The costs of syndicates were met either by annual subscriptions collected from members or by a small percentage premium on the cost of materials supplied to each member.

Exceptionally, a few of these generally titled syndicates had explicitly stated broader aims. Those established at Coulommiers (1905) and at Pierrefitte (1906), for example, adopted the model statutes supplied by the Union Centrale des Syndicats des Agriculteurs de France. While working for the material benefit of their members, these syndicates were also explicitly concerned with their moral, intellectual and professional development as a means by which rural populations could be encouraged to remain in the countryside. Thus in addition to providing members with agricultural supplies, these syndicates also to some extent made arrangements for mutual aid and for livestock insurance.

When a syndicate claimed in its title specifically to be not only *agricole* but also *viticole*, it clearly covered as wide a range of the farming community as did the more generally titled syndicate, but the explicit reference to viticulture may be taken in such cases to be a recognition of the greater importance of *vignerons* in the locality concerned. The statutes of such syndicates, like that established at Chitenay in 1887, referred not only in general terms to their role in supplying a range of farming materials (such as seeds, fertilisers and tools) for their members but also specifically to those materials directed to viticulture, for example for fighting mildew and phylloxera. Those associations with the most restricted title of *syndicat viticole* or *syndicat des viticulteurs* aimed, of course, to defend the particular interests of their specialist memberships. Their objectives included not only supplying quality-assured and keenly priced materials but also, for example, providing information on the best means of combating mildew and phylloxera and monitoring the movement of vine cuttings, to ensure that phylloxerated vines were not brought into the locality. Some of these syndicates, such as that at Saint-Georges-sur-Cher (1892), also aimed to assist with the reconstitution of vineyards destroyed by phylloxera, by replanting them with vines grafted on to American phylloxera-resistant rootstock and by setting up nurseries for the propagation of such stock. Furthermore, a few syndicates – like the syndicat agricole et viticole created at Thenay in 1886 – also aimed to provide replacement labour when a member was unable to work because of illness, injury or infirmity.

At least four agricultural syndicates had distinctive objectives. The Syndicat de Saint-Fiacre was founded at Saint-Gervais-la-Fôret in 1888 for its market-gardeners, who cultivated holdings in that commune on the old flood-plain on the left bank of the Loire, although according to the mayor of Saint-Gervais the syndicate's members resided just across the river in the town of Blois. The Union Vinicole de Loir-et-Cher (1894) aimed to publicise the wines of the department as widely as possible, both through advertising and through participation in exhibitions and competitions. Members of this syndicate undertook to supply for its approval and support only unadulterated wines produced in Loir-et-Cher. The Syndicat des Viticulteurs at Cellettes (1906) was established with the aim of finding new outlets for the wines of the commune (while not actually selling them itself) and, more importantly, of defending their good name against any fraudulent activity (such as diluting them with water). The Union Syndicale de Vineuil-Saint-Claude (1911) was open to all producers of asparagus and other agricultural products whose farms lay in the communes of Montlivault, Maslives, Huisseau, Vineuil or Saint-Claude. It aimed to inform its members of the best farming practices, to supply them with the materials they needed to carry them out, to publicise their products, and to provide a certification for asparagus produced in the five communes so that it could be sold with a guarantee of its having been pro-duced in 'Vineuil-Saint-Claude'.

Membership and organisation

Most agricultural syndicates in Loir-et-Cher had small memberships. In 1895, for example, the average size of the thirty existing syndicates (excluding SALC) was 183; in 1908 the average size of the thirty-one existing syndicates (excluding SALC) was 210. But sizes of syndicates varied considerably, for example, in 1895 from fewer than 50 to more than 1,000, and in 1908 from fewer than 50 to more than 2,000. Generally speaking, the wider the geo-graphical area served by a syndicate the larger its membership was likely to be.

Most of the department's agricultural syndicates served only single com-munes, but the more important minority served two or more communes, a canton, an *arrondissement* or the entire department. The syndicate founded at Onzain in 1888 admitted to its membership farmers and *vignerons* not only from that commune but also from eight others nearby. Its initial membership of 119 increased to a peak of 185 in 1893 and thereafter declined to 160 by 1908. The syndicate at Saint-Georges-sur-Cher (1892) admitted vineyard owners living in the commune and also those living outside who owned at least 0.25 ha of vines in the commune. It normally had sixty to seventy members. The syndicate at Thoré-la-Rochette (1911) was open to landowners, farmers and *vignerons* living in the commune or in its neighbouring communes. The Union Syndicale de Vineuil-Saint-Claude (1911) accepted farmers not only from those two but also from three neighbouring communes.

A number of syndicates whose geographical reach extended over a canton were founded in 1887. A syndicate was created that year for the canton of Blois-East and one for the canton of Blois-West in 1888: membership of the former peaked at 269 at the end of 1890, after which it declined to only 74 by 1894 and it ceased operating two years later; membership of the latter peaked at 180 in 1892, after which it declined and ceased operating in 1896. The syndicate established at Montrichard in 1887 was for farmers throughout its canton. It had 376 members at the end of 1887 but that figure seems to have represented its peak: it thereafter fell slowly to 340 by 1894 and then more quickly to 180 by 1902. The cantonal syndicate founded at Contres in 1887 had 250 members by the end of that year and 300 by the end of the following year, after which it seems to have stabilised for a long period before declining in numbers during the early 1900s. A *syndicat agricole régionale* founded at Selommes in 1887 recruited from throughout its canton: its membership rose from 70 in that year to 115 two years later, after which it declined to 98 by 1908. The syndicate created at Mondoubleau, in Perche, in 1887 was for farmers not only in its own canton but also for those in neighbouring cantons. Its initial membership of 190 increased to 525 by 1894, after which it gradually declined to 300 by 1908.

The Syndicat Agricole et Viticole du Vendômois (1899) was open to farmers throughout the *arrondissement* of Vendôme, but it appears never to have had more than 200 members. By contrast, the syndicate established at Romorantin in 1889 for farmers in its *arrondissement* came to be the second largest syndicate in Loir-et-Cher: with only 30 founding members, it had increased to 250 by the end of 1889, to 1,100 by 1899, and to 2,115 by 1908. As for the department as a whole, SALC was the only truly agricultural syndicate with such a wide geographical catchment and sphere of influence: its role will be considered in some detail shortly. One other agriculturally related syndicate did operate throughout the entire department. The Union Vinicole de Loir-et-Cher established at Blois in 1894 seems not to have had a very successful beginning, for it had no more than 50 members until the early 1900s; it then seems to have undergone a reorganisation and changed its name to the Association des Viticulteurs de Loir-et-Cher, and to have very considerably increased its membership to more than 600 by 1908.

Economic logic favoured larger associations with a wide geographical reach. While it must have been difficult in such societies to maintain a genuine spirit of local fraternity, based in part upon face-to-face contact or at least personal knowledge, they could instead be seen as having contributed to a sense of class solidarity, based upon the perception of common interests by individuals who were not known personally to each other.

Fraternal association and political tensions

Agricultural syndicates were clearly materialist institutions and no doubt many peasants joined them for the economic advantages which they could obtain by becoming members. But behind the associations lay the spirit of fraternity, more readily detectable – given the nature of the sources – among their founders than among the mass of their members. Such idealist motivations have left only few traces in the records, in part because by law syndicates were not permitted to engage in political and religious discussions or activities. But such gleanings do provide hints of a richer harvest of fraternity as well as of political tensions. For example, the statutes of the syndicate of viticulturalists at Cellettes (1906) stated that one of its aims was to provide for its members 'un centre d'appui et de confraternité' ('a fraternal support network').

Political and religious ideologies underpinned the development of agricultural syndicates in various parts of France and Loir-et-Cher was no exception. The most successful syndicate in the department was the Syndicat des Agriculteurs de Loir-et-Cher, established by Loir-et-Cher's republican professor of agriculture and strongly supported by its republican administration. Although not permitted to take explicitly political positions on economic or social issues, it clearly became common knowledge that SALC's founders and officials were on the left of the political spectrum. On 10 January 1890 the Police Commissioner reported to the mayor of Blois that SALC was operating within the democratic spirit of the law and that it was appealing much more to small farmers than to large landowners; he knew that it already had, and confirmed that it deserved, the support of the department's administration. On 18 January 1893 the Police Commissioner's approving report on SALC stated that it had always been led by republicans, that it was having a beneficial influence throughout the department, that in ten years it had grown from 50 to 4,000 members, and that it had among its leaders men like Alphonse Riverain, whom the Commissioner described as 'un véritable apostolat auprès de la petite culture' ('a real enthusiast for small farmers').

During the 1880s a few other republican syndicates were established in particular localities within the department, such as that at Châtillon-sur-Cher (1889). But it is principally against the background of the growing success of the republican-spirited SALC throughout the department that the emergence of other agricultural syndicates, promoted by the Church and by individuals with right-wing opinions, has to be seen. Although in other departments of

France, such as the Aveyron, the Church was a very significant agent in the promotion of agricultural syndicates,[28] in Loir-et-Cher the Church's role was a minor one. Few instances of its significant involvement with agricultural syndicates can be cited. The clearest is in relation to the syndicate established at Chitenay in 1887, which had as its most general aim the union of its members in 'une pensée de paix sociale, de confraternité chrétienne et d'assistance réciproque' ('a spirit of social harmony, Christian brotherhood and mutual aid'). This syndicate was established under the patronage of Nôtre-Dame and its members were obliged to participate in the patron's annual fête, as well in those of Sainte Cécile and of Saint Vincent. All were also expected to attend the religious service for the funeral of a member, to provide help for a member who was ill, and to give assistance as soon as possible when a member's property caught fire.

While the Church seems to have played only a minor role directly in the creation of syndicates, evidence for the involvement of individuals with right-wing views is unambiguous. Some of the evidence is indirect, but there is also considerable direct evidence. Some syndicates revealed their political leanings through the support they received from, or their own affiliation to, larger institutions. The two syndicates for the cantons of Blois-East (1887) and Blois-West (1888) were clearly established as right-wing alternatives to SALC. On 10 January 1890 the Police Commissioner informed the mayor of Blois that these two syndicates only existed because they had support from the national Société des Agriculteurs de France, a conservative and royalist grouping of large landowners and aristocrats; on January 1893 he reported that the two syndicates had been created by conservative politicians in order to counter the influence of the republican SALC – he correctly claimed that their now amalgamated membership was not increasing. Within three years they had ceased to function. The *syndicat viticole* of the canton of Montrichard (1887) included in its aims the diffusion of information to its members about best viticultural practice and for that purpose it decided, when it was being established, to affiliate to the Union Centrale des Syndicats des Agriculteurs de France, an anti-republican, right-wing conservative and paternalist national organisation which was linked to, but independent from, the Société des Agriculteurs de France. Its primary aim was to stem the flow of migration away from the countryside by promoting agricultural associations which in turn would promote a range of collective, co-operative and mutualist activities as a way of improving the quality of life in rural districts.[29] The syndicates at Coulommiers (1905) and Pierrefitte (1906) were also both affiliated to the Union Centrale des Syndicats des Agriculteurs de France.

Other agricultural syndicates in Loir-et-Cher had – or were perceived by contemporaries to have – conservative, even anti-republican, tendencies without being associated with larger institutions. For example, the first agricultural syndicate to be established after SALC – that at Thenay in 1886 – was founded, according to the commune's mayor in a letter to the prefect dated 8 January 1890, in opposition to the mutual aid society created by the municipality for the commune's *vignerons*: the mayor claimed that although the syndicate's explicit aim was the supply of fertilisers, its real intention was to form a nucleus of opposition to the (republican) council, an opposition which had become clear in all the elections. But by January 1891 the principal members of this syndicate had also joined SALC and by January 1895 the mayor was able to report that the syndicate had become much less political. A different and distinctive example is that of the syndicate at Chambord (1888). The prefect told the Minister of Agriculture in his letter of 28 February 1889 that the syndicate at Chambord was likely to have beneath its ostensible aims some reactionary intentions, given the very distinctive nature of that commune, dominated by the vast, largely wooded estate of a single landowner, M. le Duc de Parme, on whom most of the people of the commune depended for their livelihoods.

Although one of the syndicates founded during the late 1880s and early 1890s – that at Châtillon-sur-Cher (1887) – was considered to be devoted to republican institutions, many of the new ones were on the right politically and were established in opposition to the growing influence of the republican SALC. Those at Soings (1887), Billy (1889), Romorantin (1889) and Selles-sur-Cher (1889) were described in January 1890 by the sub-prefect of Romorantin as having organisers who were reactionary and anti-republican. At the same time, the sub-prefect of Vendôme reported that the cantonal syndicate at Mondoubleau had been established by reactionaries, clearly with a political purpose, many of them being *châtelains* (large landowners). That at Selommes (1887) was said to concern itself only with purchasing fertilisers for its members, but it was noted that it had a reactionary president. Also in January 1890 the mayor of Contres reported to the prefect that the cantonal syndicate established there in 1887 under the pretext of supplying materials to its members had above all a political aim: it had been organised in direct opposition to the republican SALC and it had reactionaries among its members, including its president. In March 1896 the prefect described the syndicate at Cour-Cheverny (1887) as having a militant monarchist as its president.

In short, at least fourteen (almost two-thirds) of the twenty-two syndicates founded between 1886 and 1889 can be identified as having been set up in opposition to SALC (whose membership almost trebled during that period from 902 to 2,680) and/or as having anti-republican sentiments. This identification rests upon the evidence of republican witnesses which might have been exaggerated in detail, but the general picture is incontestable. The administration's report to the Minister of Commerce in March 1892 emphasised that while many of the syndicates had been set up by reactionaries, in order to bring together people with similar political views, their memberships were not growing but falling. The report claimed that some of these reactionary syndicates might be closing down for lack of support. In fact, none was to do so until four years later, when the syndicates for the two cantons of Blois stopped functioning. Indeed, right-wing syndicates continued to be newly established. On 6 January 1894 the mayor of Chissay reported to the prefect that a syndicate had been founded his commune in November 1893 in opposition to the municipality and to the republic, and he claimed that admission to membership was being restricted to those calling themselves conservatives.

By January 1896 the prefect was reporting to the Minister of Commerce that while most of the agricultural syndicates in the department had been established by the reactionary party, their total membership was falling; but he recognised that even these associations were serving their members well in obtaining fertilisers for them on conditions which they would not be able to obtain for themselves as individuals. It is probable both that the local administration was becoming less anxious about the political undercurrents within syndicates and that the syndicates themselves were becoming confident about the economic advantages which they could offer to their members. Even the syndicate at Romorantin, described in February 1896 by the sub-prefect of the *arrondissement* as having only reactionaries on its central committee, was acknowledged by the same authority two years later to be not only providing its members with a range of farming supplies but also concerning itself more and more with the marketing of their products, so that green beans and asparagus were being sent directly to the *halles* of Paris with the help of the syndicate.

Syndicates certainly operated as fraternal associations: for example, members were not permitted to place orders for supplies on behalf of non-members. But there was a practical limit to their spirit of fraternity: purchases were made on behalf of members, so that the syndicates themselves were not liable to agricultural suppliers for the debts of their members. For example, the statutes of the syndicate at Châtillon-sur-Cher (1889) stated quite clearly that 'il n'existe aucune solidarité entre les associés et chacun n'est tenu que pour ce qu'il a demandé' ('there is no solidarity [collective financial responsibility] among its members and that each individual is only responsible for the orders which he has placed').

Failures and successes

About one-third (30 per cent) of the agricultural syndicates established in Loir-et-Cher between 1883 and 1914 had ceased operating by the latter date. By contrast, only two of them – and one in particular – went almost uninterruptedly from strength to strength. The reasons for these varying fortunes of syndicates now need to be considered.

Given that the principal economic rationale of syndicates was the association of farmers in order to obtain economies of scale in the purchasing of agricultural supplies, it would not have been logical for two syndicates providing very similar services to have developed in a single commune, competing with each other for members within the same limited pool of farmers. That point was certainly appreciated, for example, by the mayor of Montrichard: in a letter of 11 January 1892 to the prefect, he pointed out that when there were two syndicates competing with each other for members in the same locality they divided the farmers into two groups rather than uniting them in one, so that the bargaining power of each with the suppliers of merchandise was lower, and hence the prices of the merchandise less reduced, than would potentially be the case where there was only one syndicate operating. It could be argued that that point was widely appreciated, because there were few instances in Loir-et-Cher of two competing syndicates founded within a single commune. It might specifically have been the case that appreciation of that problem prevented plans for a second syndicate from coming to fruition in the communes of Cour-Cheverny and Mondoubleau.

Where there were two syndicates in a single commune, they were not always directly in competition. This was the case at Vineuil, where there was a general purpose syndicate (1895) and a specialist one for promoting the locality's asparagus, and at Saint-Gervais-la-Forêt, where there was a syndicate for *vignerons* (1887) and one for market gardeners (1888). It was in effect also the case at Saint-Georges-sur-Cher, where the general purpose syndicate created in 1889 for both farmers and *vignerons* was supplemented for the latter by the founding in 1892 of a syndicate for supplying vine cuttings and American rootstock to enable the commune's vineyards to be reconstituted after being devastated by phylloxera. Only in three instances – at Châtillon-sur-Cher, Seigy, and Selles-sur-Cher – were two syndicates established in the same commune which could be seen as being directly in competition with each other for members, but in all of those cases the syndicates established were still functioning in 1914. The only possible case of competition leading to the closure of a syndicate was at Montrichard, where the opening in 1910 of a new Syndicat de Défense des Viticulteurs des Côtes du Cher was followed in 1914 by the closure of the Syndicat Agricole et Viticole du Canton de Montrichard.

A syndicate which recruited its members from a single commune might also, of course, have found itself in competition with another syndicate also recruiting locally but doing so from a neighbouring commune, from its cantonal centre or from the principal town in its *arrondissement*. It would also, almost certainly, have found itself competing for members with SALC, which recruited throughout the entire department. Small syndicates were potentially vulnerable. Membership numbers fluctuated from year to year but, taking the most favourable case of their peak memberships, the mean size of the thirteen syndicates which were dissolved by 1914 was 143. The societies which collapsed (for whatever reason) were smaller than the average size of all agricultural syndicates.[30]

Some syndicates, as has been shown, were promoted by conservatives for clear political reasons. It might be that some of these politically led syndicates simply did not receive sufficient grass-roots support to sustain them. Such right-wing syndicates accounted for about 37 per cent of all of the syndicates established in Loir-et-Cher before 1914, but they represented also 54 per cent of all of those which collapsed by then. Many of their members were not politically faithful and there is evidence that members of a right-wing syndicate could simultaneously be members of the republican Syndicat des Agriculteurs de Loir-et-Cher: such was the case, for example, at Cour-Cheverny, according to a letter of 15 January 1894 from a councillor of the department to the prefect. The success of SALC must have been, to some extent, at the expense of other syndicates. Experience had shown that smaller syndicates were unable to compete effectively, and those that survived after the Great War were combined in 1921 into a single Union des Associations Agricoles that was able to offer itself as a serious alternative to SALC.[31]

Within this general picture of the department's agricultural syndicates, the paramount importance of the Syndicat des Agriculture de Loir-et-Cher has already become clear. This syndicate merits special attention, but it can be treated here relatively briefly by simply drawing upon my previously published essay.[32]

The Syndicat des Agriculteurs de Loir-et-Cher

Historical development and geographical distribution

The significance of the Syndicat des Agriculteurs de Loir-et-Cher needs to be viewed in terms both of its general historical context and of the particular individuals whose charisma provided the necessary link between structure and action.

The success of SALC owed much to the enthusiasms and energies of its first two presidents, Jules Tanviray and Alphonse Riverain. That SALC and the local authorities worked hand-in-hand was acknowledged by Alphonse Riverain, who in 1886 succeeded Tanviray as president of SALC. He held that office until his death in 1929. Like Tanviray, Riverain was a practising farmer who had been a pupil at the farm school of La Charmoise. On his large farm at Areines, near Vendôme, Riverain carried out various improvements – draining, marling, cultivating artificial meadows, applying chemical fertilisers – and maintained a passionate interest in practical farming even though he became increasingly involved in the ramifying activities of SALC.[33]

Both Riverain and Tanviray recognised that the advantages which SALC offered to farmers were fundamentally material in character but both harboured republican views which embraced associations as practical manifestations of the concept of fraternity. Riverain especially conceived a moral and

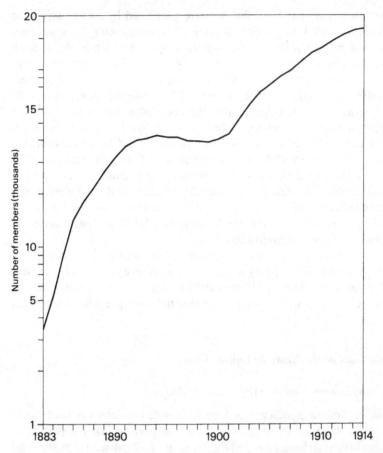

Fig. 8.7 Membership of the Syndicat des Agriculteurs de Loir-et-Cher, 1883–1914
Source: Berthonneau (1905) 5–6 and Anon. (1920) 18

social role for SALC, but both he and Tanviray considered that in order not to prejudice its growth their own political opinions had to be suppressed or at least restrained. In public, SALC made claims to 'complete political neutrality'; but that it was led by republicans was certainly known to the Police Commissioners of Blois in the 1890s and was very probably recognised by those joining SALC.[34]

From its birth in mid-1883, SALC steadily increased its membership to reach a total of almost 4,000 by the end of 1892. It remained at about that level until 1901, when growth was renewed and continued uninterruptedly to reach a figure of almost 17,000 by 1914 (figure 8.7). Growth in SALC's membership was almost exactly paralleled by an expansion of the volume of business it conducted (figure 8.8). The hiatus in the development of SALC

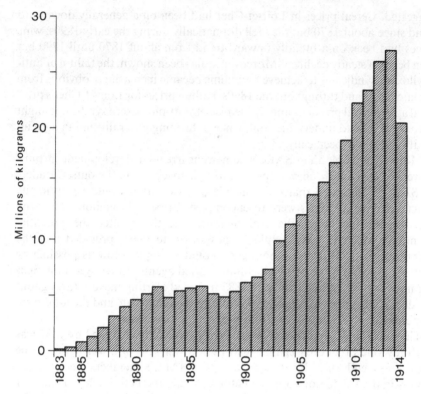

Fig. 8.8 Goods supplied to its members by the Syndicat des Agriculteurs de Loir-et-Cher, 1883–1914
Source: Berthonneau (1905) 5–6 and Anon. (1920) 18

during the 1890s has been explained erroneously by its leaders as being a consequence of the competition which SALC felt from the creation of 'numerous communal or cantonal syndicates' in the department.[35] By the end of 1892 there were indeed, as has been shown, twenty-five other general purpose agricultural syndicates operating in Loir-et-Cher, but by 1901 that number had only increased by two. By comparison with SALC, the other syndicates were small; and their total memberships grew hardly at all during the 1890s, from about 5,100 in 1892 to about 5,200 in 1900. The check to the growth of SALC during the 1890s cannot be attributed to the competition for new members provided by other agricultural syndicates. Indeed, many of them were right-wing associations and might not have appealed to those who preferred to join a syndicate which was ostensibly neutral politically but which was known to be run by left-wing republican sympathisers. All agricultural syndicates suffered arrested development during the 1890s. Two reasons for this may be suggested. Cereal prices in Loir-et-Cher had been on a generally downward trend since about 1870 but they fell dramatically during the early 1890s; wine prices had been on a broadly upward trend from about 1870 until 1890 but then began a steady decline.[36] Moreover, as has been shown, the failure of anti-phylloxera syndicates to achieve their aims became increasingly obvious from the late 1880s and throughout the 1890s. Falling prices for Loir-et-Cher's principal agricultural products and the demise of anti-phylloxera syndicates might have combined to undermine confidence in farming generally and in agricultural syndicates specifically.

More remarkable than SALC's somewhat irregular development in time, however, was its very uneven spread through space. From the outset, leaders of SALC appreciated that the principle of association would be optimally practised if its activities were to encompass farmers throughout the entire department: the larger the real membership, the greater the potential economies of scale which could be passed on to them, provided that the administrative and distributional costs could be kept as low as possible by establishing a network of local, commissioned agents. Tanviray's campaigning initiative during the spring of 1883 included alerting mayors throughout the department about his proposal but both the immediate and the long-term responses to his stimulus varied from one *pays* to another.

The geographical pattern of membership which had emerged by 1887 was not only one of uneven spatial distribution but also one which was to be remarkably enduring. By the end of 1887, SALC's members were located mainly in the Val de Loire and the Petite Sologne, and in the valley of the Loir and in Perche; there were relatively few in the Grande Sologne (figure 8.9). By 1892, towards the end of its initial phase of rapid growth, this overall pattern had intensified rather than altered in outline (figure 8.10). Even the massive expansion of SALC from 1901 onwards did not significantly change the basic distributional pattern of its members: by 1910 most were concentrated in the Val de Loire, the Petite Sologne and the valley of the Loir (figure 8.11). From its inception, SALC relied to a considerable extent upon local agents to recruit and provision members, and from the early 1910s it additionally established local depots for supplying and servicing agricultural machinery. The networks of such agents and depots which SALC had established by 1910 broadly reflected but did not mirror exactly the distribution of its members in that year (figures 8.12 and 8.13).[37] The distribution of SALC's membership was not only a product of factors related directly to SALC itself but also of some factors external to it. For example, membership was negatively related to physical distances from Blois and positively related to literacy levels: a clear tendency has been demonstrated for peasants in cantons closer to Blois to join SALC in significantly larger numbers than those in more remote cantons, and also for peasants in cantons with higher male literacy levels to join SALC in significantly larger numbers than those in less literate cantons (it being

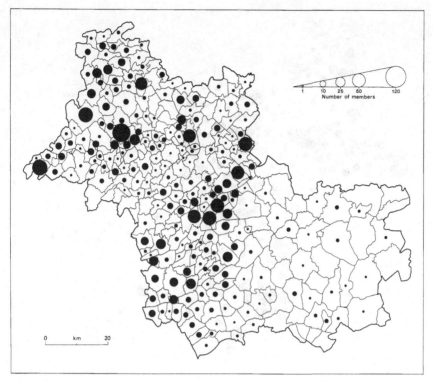

Fig. 8.9 Membership of the Syndicat des Agriculteurs de Loir-et-Cher, by communes, 1887
Source: Supplément au Bulletin du Syndicat des Agriculteurs de Loir-et-Cher (15 January 1888)

inferred that the more literate cantons included the better educated and more adaptable sections of the peasantry).[38] Additionally, the spatial diffusion of SALC's membership needs to be set within its agrarian context, taking particularly into account factors of farm size and type. SALC acquired members most readily in those *pays* characterised by small farms (often but not exclusively owner-occupied) and especially by *vignerons*.[39] In his report to the Minister of Commerce on 15 January 1891, the Police Commissioner for Blois said that *les petits agriculteurs* had flocked to SALC: it had, he claimed, hardly any large landowners among its members and only few owners of medium-sized farms, while three-quarters of its members were small farmers democratic in spirit and supporters of the republican regime's institutions. The largest concentrations of members were to be found among the *vignerons* of the Val de Loire and the Petite Sologne and of the valley of the Loir. Not usually possessing more than one or two cows, these *vignerons* had limited

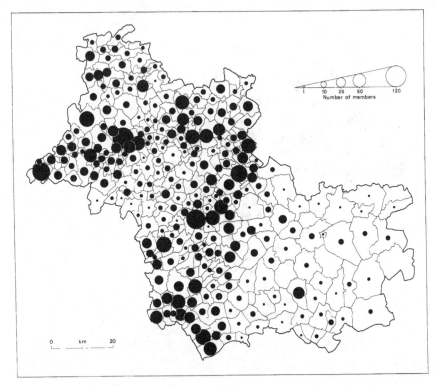

Fig. 8.10 Membership of the Syndicat des Agriculteurs de Loir-et-Cher, by communes, 1892
Source: Supplément au Bulletin du Syndicat des Agriculteurs de Loir-et-Cher (15 January 1893)

supplies of manure and therefore turned to SALC for fertilisers (and then also for other supplies). Furthermore, Dion has linked the growth of agricultural syndicalism in the Loire valley with the decline of viticulture: when vineyards were devastated by phylloxera many *vignerons*, as an alternative to expensively reconstituting their vineyards with vines grafted on to American phylloxera-resistant rootstock, turned instead to cultivating other crops which dramatically increased their demand for fertilisers and considerably added to production costs.[40] Membership of SALC, as a source of good-quality and reasonably priced fertilisers, came to be seen by many small farmers as an integral component of their survival algorithm. With a growing membership, SALC in turn diversified its activities well beyond supplying chemical fertilisers to include other agricultural materials, as well as information and credit.[41]

The undoubted success of SALC between 1883 and 1914 was based upon its ability to engage effectively in the process of agricultural improvement

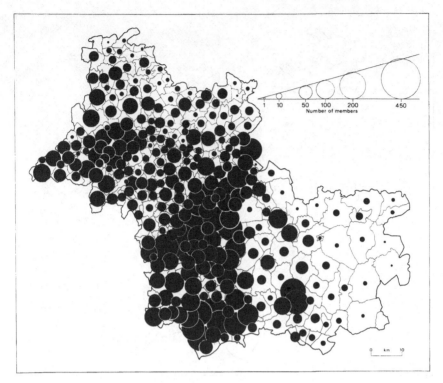

Fig. 8.11 Membership of the Syndicat des Agriculteurs de Loir-et-Cher, by communes, 1910
Source: Printed membership list consulted at the offices of Groupe Franciade, Blois

which was under way before the birth of the syndicate itself. SALC promoted agronomic changes as a means of ensuring the continuity of the rural social fabric. It acted in the defence of farming interests, reducing the costs and increasing the productivity of farmers. It helped in particular to create the circumstances in which small farmers might face the growing economic pressures upon them and survive. By 1914 perhaps one-third of the farmers of Loir-et-Cher were members of SALC. It could be argued that many of them had joined the association in order to protect their individual freedom, as part of their own survival algorithm. Most, it may be inferred, had done so as materialists rather than as outright idealists because the collectivism of the syndicate nonetheless left many individuals as owners of the means of production.

What is most striking, of course, is the geography of SALC's membership: although the syndicate operated throughout the department, its members were very unevenly distributed within it. While a spirit of fraternity among the founders of SALC might have underpinned its general development, the

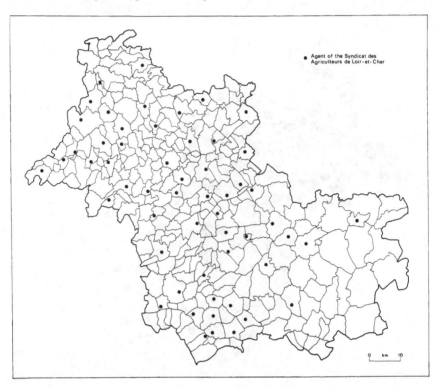

Fig. 8.12 Agents of the Syndicat des Agriculteurs de Loir-et-Cher, 1910
Source: Journal de l'Agriculture Pratique du Centre (January 1910)

specific levels of response varied from place to place within the department. Membership of SALC had a local geography: that was, of course, a characteristic which it shared with other associations. That the general concept of fraternity had a local specificity becomes very evident in the case of threshing syndicates.

Threshing associations

Lower cereal prices and higher labour costs were becoming a feature of farming in France generally towards the end of the nineteenth century, and so endeavours to increase crop yields, notably by making more and more use of chemical fertilisers, were coupled with efforts to decrease labour costs, especially by making more and more use of labour-saving machinery. One particularly significant development was the adoption of steam-powered threshing machines, the use of which could result in a five- or six-fold increase in productivity per unit of labour, even though such machines required an operating

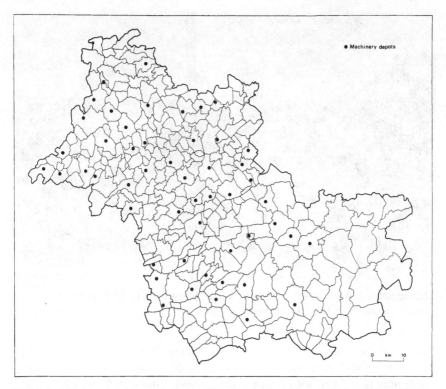

Fig. 8.13 Machinery depots of the Syndicat des Agriculteurs de Loir-et-Cher, 1910
Source: Journal de l'Agriculture Pratique du Centre (April 1910)

team of between fifteen and twenty men (photograph 8.1). The machines were very expensive, indeed too expensive for individual farmers (and especially small farmers) to afford. New enterprises therefore developed, threshing companies, which hired out both their equipment and their labour on a contract basis to undertake a given amount of threshing.[42] An alternative was for farmers to form associations for the collective purchase and operation of steam-powered threshing machines. Such syndicates could operate, it was calculated, at about 70 per cent of the cost of the threshing companies, with the purchasing costs of the machines covered in seven to ten years.[43] More than twenty such syndicates were established in Loir-et-Cher before 1914. They were set up as co-operatives, as *sociétés civiles particulières*, regulated by article 32 and the following articles of the *Code Civil*, and then some were established under the law of 29 December 1906.

Although the recording by the local authorities of the existence of threshing syndicates was not always consistent, by combing through the mass of unpublished data – surveys, reports and correspondence – it is possible to

Photograph 8.1 A steam-driven threshing machine at work in the Sologne
Source: private collection, the author

overcome the deficiencies of any one source and to construct a credible picture from such a combination of sources. In total, twenty-two threshing syndicates have been identified as having been established in Loir-et-Cher before 1914 (figure 8.14). The first was founded in 1879 and four others in the early 1880s. It was not until 1897 that the next syndicate was established, but shortly afterwards there followed the most active period of their creation: during the eleven years between 1902 and 1912 fifteen new syndicates were established, representing 68 per cent of the total founded in the department by 1914.

Their geographical distribution within the department was very striking (figure 8.15). The first five to be established were all located within the two adjacent communes on the left bank of the Val de Loire, two in Saint-Claude-de-Diray (1879 and 1883) and three in Vineuil (1880, 1883 and 1884). In all, the Val de Loire was to acquire ten threshing syndicates: one more at Saint-Claude-de-Diray (1911) and one each at Saint-Dyé-sur-Loire (1906), Montlivault (1907), Avaray (1909) and Chailles (1909). With two others just to the west of Blois – at Marolles (1906) and Saint-Sulpice (1912) – the Val de Loire had the largest concentration of threshing syndicates in the department. A second group was located within the Petite Sologne, where there were syndicates at Monthou-sur-Bièvre (1902), Cheverny (1904), Sambin (1909), Chémery (1909) and Feings (c. 1910). Nearby, in the valley of the Cher, two were established at Chissay in 1897. A third, smaller, group of syndicates were

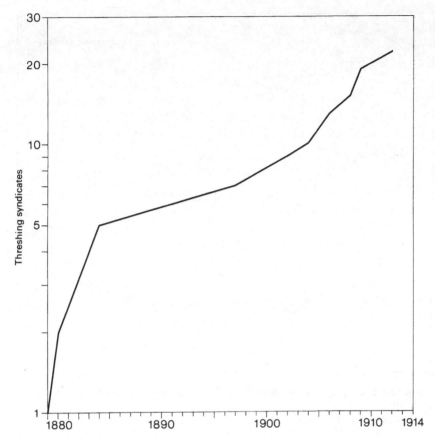

Fig. 8.14 The historical development of threshing syndicates in Loir-et-Cher, 1880–1914
Sources: AD 7M 244 and 10M 55–9

located on the margin of Beauce, but quite close to the Val de Loire, with two at Josnes (1904 and 1906) and one at Séris (1906). The absence of threshing syndicates from large parts of the department is very striking. So too, however, is their presence in such numbers in the Val de Loire and in the Petite Sologne. These were not the principal cereal producing areas of the department: they were, however, localities characterised by small farms (on which the purchase individually of threshing machines would not have been an economically sensible course of action) and, as has already been seen, a strong sense of fraternity. The way in which that principle was expressed in practice can be seen in the way in which the syndicates functioned.

The operations of syndicates are recorded in their statutes, copies of which have survived for a handful of the associations, namely two in Vineuil, two in

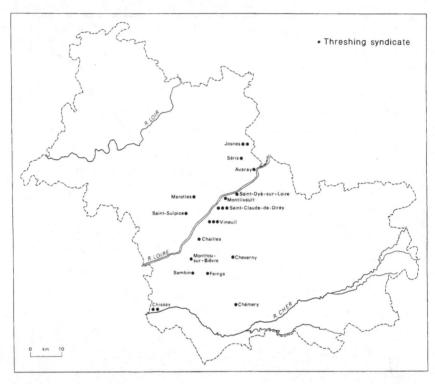

Fig. 8.15 The geographical distribution of threshing syndicates in Loir-et-Cher, 1880–1914
Sources: AD 7M 244 and 10M 55–9

Josnes and one in Chémery. Co-operative societies for the threshing of grain were founded at Vineuil in February 1880 (the Société Co-opérative de Vineuil) and in January 1883 for one of its hamlets, Noëls (the Société Co-opérative des Noëls-Vineuil). Their aim was to provide for their members the means of threshing grain in the best conditions and at the lowest possible costs. As co-operative societies, they renounced any idea of making a profit and any form of commercial speculation. The capital needed for purchasing a threshing machine for each society was raised from members, each of whom purchased a 100-franc share, the total number of shares available being the number required to raise the sum necessary to cover the capital purchase. If the number of individuals wishing to purchase a share was not equal to the number of shares which needed to be sold, then an individual could buy multiple shares but they would have to be sold on to new members when they came along. At a general meeting held ten days before threshing was due to commence, each association appointed a twelve-man committee to run it for the year. At a second general meeting, held on the second Sunday of December, each association approved the annual accounts and determined the charge to be levied for each hectolitre of corn threshed. Each association's expenses included, of course, payment to those operating the machine, the cost of fuel and oil for running it as well as the cost of maintenance and repairs, and the costs of housing and insuring it. In addition, the expenses included an amortisation payment of 5 per cent to each shareholder who had had his grain threshed by the association during the year.

Entitlement to use a society's threshing machine was extended to the father, the son, and father-in-law (and also to the mother, the daughter, and mother-in-law) of each shareholder on condition that the cereals to be threshed were all in the same building and could be threshed at the same time, without the machine having to be stopped. Each society's committee would usually meet on Trinity Monday (in late May) to agree all of the purchases necessary for cleaning and heating its machine and then again ten days before the beginning of threshing, to finalise the details of the threshing programme and the route which the machine and its team of operators would follow.

In the case of both of the societies at Vineuil, priority in the programme was given to threshing for those individuals who were required for military exercises, to those who were moving house and to those who had to make repairs either to their house or to their wine press, provided that they had notified their society's committee at least one month before the beginning of threshing. Also, those operating the machine could claim to have their cereals threshed before those of other members of their society. The itinerary to be followed by the threshing machine for other members of each society was set out in their statutes (figure 8.16). The society established in 1880 at Vineuil had a four-year itinerary for its machine: in the first year, threshing began in the *bourg* of Vineuil itself and then proceeded on a clockwise course throughout the commune; in the second year, it again started in the *bourg* but then proceeded on an anti-clockwise course; in the third year, threshing began at what was approximately the mid-point of the course, at Roche, and then proceeded clockwise; in the fourth year, threshing again began at Roche but then proceeded in an anti-clockwise direction. Such an arrangement indicates that not only the principle of fraternity but also that of equality mattered to the association. By contrast, the second society established at Vineuil, that of Noëls-Vineuil in 1883, had a much shorter, linear rather than circular, route which was followed each year in the same direction, from La Haute Rue to Léry. Each member was required by his society to be on hand when the threshing machine was being brought to his farm and to guide it to his farm along a local road. If it were not possible for the machine to have such access, so that it became necessary for threshing to be undertaken on a public road or even at a distance from a road less than that permitted by local regulations, then the member rather than the society was required to assume any third-party

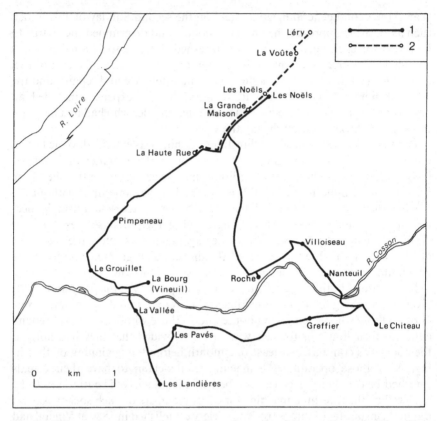

Fig. 8.16 The routes taken by the machines of two threshing syndicates in Vineuil in the 1880s
Source: AD 10M 59

responsibility not covered by the society's insurance. Under no circumstances were the machines to be left on a public road overnight. Shareholders were expected to participate in the threshing work itself.

With accumulated experience, some changes were made to societies' statutes. By 1900, for example, the society at Vineuil (1880) ruled that any member making a false claim to a privileged place in the threshing programme would be fined 50 francs for the first offence and expelled from the society for a second offence. Also, three days had come to be designated as days on which the threshing team and the shareholders would not be required to work: 15 August, the Sunday of the fair at Blois, and the Sunday of the schools' prize-giving.

Two threshing associations were founded at Josnes, as at Vineuil, initially one based upon the *bourg* (the Syndicat de battage de Josnes, established in

June 1904) and subsequently one based upon a hamlet (the Syndicat de battage de Josnes, the Amicale d'Issy, established in April 1908). In general terms, these two associations operated as did those at Vineuil, but there were some specific and significant differences.

For the 1904 association, for each share a member was required to pay 100 fr. but an individual could purchase as many shares as he would require half-days of threshing each year. If the capital needed to purchase a machine could not be raised by selling sufficient shares, then the Syndicat des Agriculteurs de Loir-et-Cher undertook to loan the difference, on condition that the first dividends of the association would be used exclusively to repay that advance. If in any given year, a member needed to have more time for threshing than his share ownership allowed, then he would be able to have the additional threshing done but at a price which would be 2 fr. 50 c. more than the normal charge for each half-day's threshing. This association was run by a small committee, of only seven members. The charge for each half-day's threshing was fixed at a level which would cover the payments to be made to the operating team, the maintenance, repair and running costs of the machine, the premiums for fire and accident insurance, the costs of fuel and oil, and not only pro rata payments to shareholders to amortise their capital but also payments into a reserve fund. After the cereals of all of the members of the association had been threshed, the machine could be used for threshing on the farms of members of SALC who had made such a request: the rate for them was to be, for each half-day, 2 fr. 50 c. more than the charge to members of the threshing association itself.

For the Amicale d'Issy in Josnes, the price of each share was 50 fr. and each member could purchase as many shares as the number of quarter-days of threshing that he would need annually. A shareholder could pay for additional threshing above his entitlement, but at a rate per half-day which was 2 fr. 50 c. above the normal price. The Amicale was run by a committee of only five people. It was clearly an association of small farmers, whose needs for threshing were limited. Information supplied by the association to the prefect at its foundation indicated that it had only twenty-two members, who had collectively purchased fifty-four shares: but eight of them each owned only one share, three had purchased two shares, five had purchased three and another five had purchased four, while one farmer had bought five shares.

The statutes of the two associations at Josnes make it clear that a member who left the locality would be reimbursed for the outstanding value of his share(s); that shares could be transferred to another person, but only if the association's general meeting approved the transferee's becoming a member; and that in the event of the death of a member, the threshing rights attached to a share could not be divided among his heirs and that only one person, designated by them, would accede to that right (if there were no heirs remaining in the region, then their share of the capital would be reimbursed to them).

While the Amicale d'Issy had only twenty-two members, that at Noëls-Vineuil had sixty. From an inquiry conducted in 1919, it seems that membership of threshing associations ranged from about ten to about seventy. While membership was open to both women and men, without exception men comprised more than 90 per cent and many associations had no women at all. That established at Chémery in 1909 is recorded as having twenty-three members in 1925. The Syndicat de battage de Chémery had somewhat wider aims than the others considered so far: its objective was to purchase and make available to its members agricultural equipment and especially threshing machines. Such equipment and machines were hired out to members on a rental tariff agreed by the association's four-man committee. The income thus raised was used to meet the costs of insurance premiums, repairing and maintaining equipment, payments to employees, payments to shareholders, interest on loans raised, and contributions to a reserve fund. The operations of the association, which included arranging insurances, hiring and firing employees, and determining the sequence in which the requests from members would be met, were carried out by its committee.

All of the threshing syndicates identified here as having been founded by 1914 were still operating during the 1920s and, in some cases, during the 1930s when additional associations were also established. They formed a significant and distinctive set of associations.

Other associations

In an endeavour to complete this picture of agricultural associations in Loir-et-Cher between 1815 and 1914 brief reference needs also to be made to a few other rural, agriculturally related, syndicates, even though the evidence relating to them is slender and imperfect.

Most of the agricultural syndicates considered above were concerned with supplying farmers with materials needed for agricultural production, only a few were concerned with the any aspect of the marketing of products and none with the processing of products. The development and diffusion of marketing and processing co-operatives were to be an important feature of the 1920s and 1930s in Loir-et-Cher: before then there were very few. A law of 29 December 1906 favoured the development of co-operative societies not only for production but also for processing and marketing of agricultural products. A co-operative distillery seems to have been established at Soings in 1907. Dairying co-operatives for butter production were reported in 1909 as having been recently established at Sougé and at Saint-Georges-sur-Cher as a result of lectures on agricultural co-operation given by M. Vezin, the department's professor of agriculture.[44] Vezin himself reported in 1910 that the department had one dairying co-operative at Saint-Georges-sur-Cher and a creamery at Pontlevoy, and that a co-operative for the sale of milk was being established at Mer and considered at Saint-Aignan.[45] But it was not until after the Great War that the movement to establish processing and marketing co-operatives really took off.

During the 1890s the Ministry of Agriculture was actively promoting the founding of agricultural credit societies and a law of 5 November 1894 provided a framework for their operation. There is evidence of three such societies having been established in Loir-et-Cher before 1914: at Chitenay (1894), Pontlevoy (1897) and Saint-Laurent-des-Eaux (1897). These societies did not have their own capital but borrowed in order to make loans. Each was founded with only three members. According to the mayor of Chitenay, the society there ceased functioning in mid-1898. The society at Saint-Laurent-des-Eaux had thirty-seven members 1901, when that at Pontlevoy had forty members. The activities of these societies were soon to be eclipsed by those of SALC, which established in 1902 at Blois a branch of the Caisse Régionale de Crédit Agricole de la Beauce et du Perche. By November 1903 it had 167 members, whereas the memberships of the societies at Pontlevoy and Saint-Georges-sur-Cher had not increased beyond their levels in 1901. By the early 1920s, and possibly sooner, there were no agricultural credit societies in Loir-et-Cher which were not affiliated to the Caisse Régionale.[46]

In addition, one can note some syndicates whose concerns were with water management – for example, with drainage (Selles-sur-Cher, 1911), irrigation (Naveil, 1907) or river control more generally (Saint-Viâtre, 1879 and 1882; Saint-Dyé-sur-Loire, 1913).[47] Management of the environment by such syndicates deserves quite separate study, but is neither necessary nor appropriate here.

In sum, agricultural associations – and especially the agricultural syndicates and most particularly SALC – constituted an important part of the general picture of fraternal associations in Loir-et-Cher between 1815 and 1914. It is now time to return to that wider perspective.

9

Synthesis: conclusions, comparisons and conjectures

L'association [est] un phénomène de solidarité dans l'espace, comme le crédit est un phénomène de solidarité dans le temps.

Eugène Fournière *L'individu, l'association et l'Etat* (1907) 54

La routine paysanne?

Many contemporary observers of farming practices in Loir-et-Cher during the nineteenth century commented upon the reluctance of the peasants to change their ways and identified *la routine paysanne* as a major obstacle to agricultural improvement. As Dupeux pointed out, 'il n'est guère de rapport de préfet ou de sous-préfet qui ne la signale' ('there is hardly a report of a prefect or sub-prefect which does not refer to it').[1] Dupeux was probably exaggerating for effect, but it was not only the higher officials who made such remarks, in part perhaps out of frustration with – as they perceived it – the slow response of the peasantry to adopt new practices in agriculture and more generally in the rural social economy.

On 11 April 1811 the mayor of Ouchamps wrote to the sub-prefect of Blois in scathing terms about the peasantry of the canton of Contres: 'Rien peut amener les colons de nos campagnes à abandonner leurs méthodes vicieuses. "Nos anciens le faisaient comme ça, nous voulons le faire de même." En un mot, l'agriculture qui se perfectionne partout est encore, en ce moment dans le canton, ce qu'elle était du temps de Charlemagne' ('Nothing can persuade the inhabitants of our countrysides to abandon their faulty practices. "Our forefathers did it this way, we want to do it the same way." In short, while agri- culture is being improved everywhere it is still in this canton today just as it was at the time of Charlemagne').[2] On 2 September 1819 the secretary of the Society of Agriculture of Loir-et-Cher, in his response to the prefect's enquiry about how to improve the condition of horned animals in the department, argued that 'l'opposition la plus réelle est la routine. Nos pères ont fait ainsi; nous faisons de même' ('the most solid obstacle is [peasant] routine. Our fathers did it this way, we do it the same way').[3] Responses from communes to the decennial agricultural inquiry of 1852 paint a similar picture of conservatism, of a cautious attitude towards innovations and even of an almost superstitious reliance upon traditional beliefs and practices. For example, they revealed in the canton of Savigny a continuing belief in the existence of propitious and non-propitious days for certain agricultural activities, irrespective of local soil and climatic conditions: thus one respondent stated that sowing normally began on 21 September, Saint Matthew's Day, but others preferred to sow on 4 October, the day of Saint Francis of Assisi; but on 6 October, Saint Bruno's Day, nobody sowed because farmers believed that corn sown on that day would not do well. In reply to the inquiry's question about which fertilisers were being used, the mayor (or secretary) of Fossé in the canton of Blois-East said that, despite the advocacy of chemical fertilisers by the agricultural societies, farmers in the commune were content to use organic (cattle, horse and sheep) manures: 'ils n'ont pas de récoltes étonnantes, mais ils engrangent toujours: cette vieille méthode leur suffit; ils ne sont pas disposés à en changer' ('they do not lead to remarkable harvests, but they always produce: this old practice satisfies them; they are not inclined to change it').[4] In that phrase is encapsulated the peasant farmer as a 'satisficer' rather than a 'maximiser' but it also hints at the peasant as someone unwilling or unable to experiment, to take risks.

Representations of the peasantry of Loir-et-Cher, like those of many other parts of France, as routinised individuals have relied very considerably upon contemporary anecdotal evidence of the kind just cited. Such an image can be countered in a variety of ways. First, it is based upon what could be interpreted as the possibly non-comprehending remarks of non-peasant observers, often those of *notables* and local government officials committed to an Enlightenment model of progress based upon scientific reason.[5] In such a vision the 'real' rural world was always likely to be lagging behind the 'theoretical', improved and even utopian, world which some thought it could become. Second, as has been pointed out above and as has been amply demonstrated by a number of authors, notably Dupeux, farming and the rural social economy in Loir-et-Cher did change considerably during the nineteenth century: they were far from being static.[6] Third, to the extent that the image of a routinised peasantry is correct, such conservatism can be rationalised as a way of managing risk. An empathetic view of the peasantry sees it as 'a class of survivors' which recognises a world of scarcity rather than of surplus, a world of uncertainties, risks and dangers which only some survive. As John Berger has expressed it, 'the path through the future ambushes is a continuation of the path by which the survivors from the past have come . . . The path is tradition handed down by instructions, example and commentary . . . To a peasant the future is this narrow path across an indeterminate expanse of known and unknown risks.'[7] In short, routine made sense to the peasants themselves and should therefore have made sense to their contemporaries and should make sense to us today.

Co-operation had long been part of the peasant's survival algorithm. What is now clear, however, is that during the nineteenth century new strategies of co-operation were adopted by some of the French peasantry as they fought increasingly for their survival. In rural Loir-et-Cher, as has been shown, there developed between 1815 and 1914 a vast number and great array of instrumental voluntary associations: this study has identified about 600 associations (an average of two per commune in the department as a whole), many of which were created before the beginning of the Third Republic in 1871. In this particular respect, Dupeux's admirable monograph on the social and political history of Loir-et-Cher was clearly mistaken, for he argued that it was not until the beginning of the twentieth century that 'l'association commence à pénétrer dans les moeurs paysannes' ('association began to enter into peasant ways of behaving').[8] A separate study of the agricultural syndicates of four-teen communes in the Cisse valley revealed considerable activity there before the end of the nineteenth century: membership of the Syndicat des Agriculteurs de Loir-et-Cher in those communes increased from 91 in 1887, to 177 in 1897 and to 505 by 1907; the Syndicat Agricole et Viticole established in the commune of Onzain in 1888 had 180 members by 1891; 4 anti-phylloxera syndicates were established in the valley between 1888 and 1894, and 10 livestock insurance societies were created there between 1900 and 1914.[9] Agricultural associations flourished in the Cisse valley from the 1880s. But the wider scope of this present study has indicated that rural fraternal associations were also active throughout most of Loir-et-Cher and had even earlier origins. An attempt will now be made to synthesise and evaluate the main findings of this study, first in relation to the department and then in relation to France as a whole.

Fraternal associations in rural Loir-et-Cher 1815–1914

The preceding analyses of separate sets of instrumental voluntary associations now need to be brought into a comparative synthesis. Each of the five sets of associations has been considered discretely, in absolute terms; now they will be examined comparatively, in relative terms. In this way some of the significant generalities will be brought into the light, to stand alongside the many specificities already noted.

The timing of voluntary associations

In order to compare the historical development of voluntary associations between 1815 and 1914, the growth of each set has been plotted over time as a cumulative percentage of the total number that came to be established within

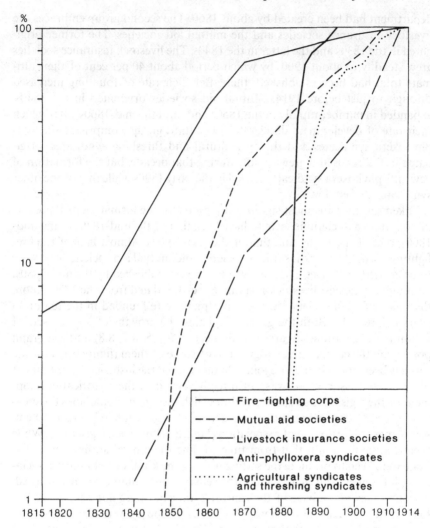

Fig. 9.1 The comparative historical development of fire-fighting corps, mutual aid societies, livestock insurance societies, anti-phylloxera syndicates, and agricultural and threshing syndicates in Loir-et-Cher between 1815 and 1914

that set during that period (figure 9.1). In terms of their timing, the five sets of associations appear to fall into three groups. The fire-fighting corps alone form the first 'group': they originated before 1815, they had an origin phase which continued until about 1830 and a main diffusion phase from then until about 1870, after which their rate of formation slowed but continued through to 1914. Fifty per cent of the corps which came to be established in the department had been created by about 1860. The second group embraces the livestock insurance societies and the mutual aid societies. The former originated in the 1830s and the latter in the 1840s. The livestock insurance societies grew steadily to about 1900, by which period about 40 per cent of their ultimate total had been established; thereafter their rate of founding increased through to just before 1914. Mutual aid societies originated in the 1840s, expanded in number rapidly in the 1850s and until the mid-1860s, after which their rate of development decelerated. The third group, comprising the anti-phylloxera syndicates and the agricultural and threshing syndicates, originated in the 1880s and grew rapidly during that decade, but the formation of new anti-phylloxera syndicates ceased in the early 1890s while that of the other syndicates slowed down.

Taken together, the most significant periods for the formation of these sets of voluntary associations seem to have been that of the mid-1840s to the mid-1860s, that of the 1880s, and that of the early 1900s. Almost half of the fire-fighting corps, livestock insurance societies and mutual aid societies that were to be created in Loir-et-Cher by 1914 had been established by the mid-1880s; although the earliest livestock insurance societies dated from the 1830s, more than half of such societies in the department were founded in the ten or so years before 1914. Both the graphs depicting the growths of specific sets of voluntary associations (figures 4.1, 5.1, 6.1, 7.5, 8.5, 8.7, 8.8) and the graph portraying the comparative progression of some of them (figure 9.1) indicate – as has been noted in the foregoing chapters – that the historical development of voluntary associations was not directly related to the historical development of the legislative structures within which they grew. Sometimes the creation of voluntary associations ran ahead of legislation relating to them, sometimes the formation of associations lagged behind such legislation. While there can be no doubt that from time to time the central authorities of the State were keen to promote the setting up of some kinds of voluntary associations, it is not evident that their efforts were directly or immediately rewarded. The advocacy and advice of the central authorities were not always heeded, for a variety of reasons: promises of financial support were not always kept (or perceived by potential recipients to have been kept); when grants were awarded, their payment was often made only after considerable delays; and the advice and intentions of the central authorities were often viewed with suspicion. Explanations of the timing of the historical development of voluntary associations need to look beyond their legal and official contexts.

All of these voluntary associations were forms of risk management and it is therefore of interest and significance to note the historical sequence in which they developed. The earliest to be created were the fire-fighting corps, intended to provide protection against the risk to property from fire damage: a peasant's house and farm buildings were likely to be his largest capital asset and could also contain other valuable property, such as stored cereals and animal fodder.

The second group of associations (the livestock insurance societies and the mutual aid societies) were concerned with protection against risks to labour, be it animal power or human effort, on both of which the running of peasants' farms depended; and, of course, a peasant's beasts – not only his draught animals – were likely to be his second or third largest capital asset, after his land and buildings. The third group comprised associations (the anti-phylloxera syndicates and the agricultural and threshing syndicates) whose objectives were to provide protection against risks to the crops on the land which a peasant cultivated, against the dangers for example of insect pests and of impoverished soils and, consequently, of poor crop yields. This ordering of the emergence of voluntary associations reflected to some extent the relative importance which peasants attached to the different components of their livelihoods: their farmhouse and outbuildings, their own labour and that of draught beasts, their other animals, and their crops.

The spacing of voluntary associations

In previous chapters the geographical distributions of different sets of associations were mapped and considered for specific dates between 1815 and 1914. But those distributions can also be considered for the period as a whole, providing an overall view of the geography of each set of associations. Such an approach emphasises the extent to which voluntary associations within Loir-et-Cher did indeed have local geographies. Communes which acquired at least one livestock insurance society during this period were most numerous in the Val de Loire, the Petite Sologne, the Petite Beauce and the valley of the Loir (figure 9.2). Those with at least one mutual aid society were concentrated principally in the Val de Loire, the Petite Sologne, and the Cher valley (figure 9.3). Much more widely distributed were those communes which had a fire-fighting corps at some time during the period 1815–1914, but even these associations were unevenly spaced throughout the department, with concentrations in the Val de Loire and the Petite Sologne (figure 9.4). Anti-phylloxera syndicates had a much more restricted distribution, mainly in two localities: around Mer, both in the Val de Loire and on the adjacent plateau of the Petite Beauce; and around Vendôme, in the valley of the Loir (figure 9.5). Communes which witnessed the formation of at least one general purpose agricultural syndicate and/or at least one threshing syndicate were located mainly in the Val de Loire, the Petite Sologne and the valley of the Cher (figure 9.6).

Within the geographical distributions of these individual sets of associations, the recurrent importance of some *pays* in particular stands out: especially the Val de Loire and the Petite Sologne, but also the valley of the Cher, parts of the Petite Beauce, and the valley of the Loir. This general impression can be checked more rigorously in relation not to *pays* but to cantons. For each canton has been calculated the percentage of its constituent communes (its

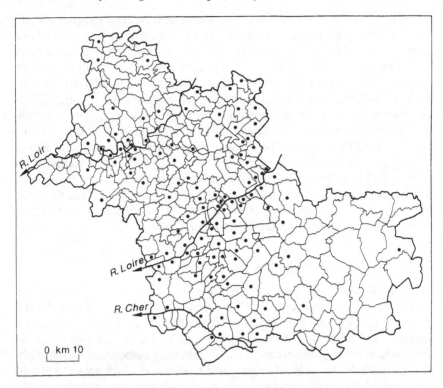

Fig. 9.2 Communes in Loir-et-Cher with at least one livestock
insurance society between 1815 and 1914

'field' of communes) which acquired a fire-fighting corps and/or a livestock
insurance society and/or a mutual aid society (hereafter referred to as 'indicator' associations) at some time within the long nineteenth century. As each
commune had the opportunity to acquire three different kinds of association,
the 'field' of communes within a canton was three times the actual number of
communes. Anti-phylloxera syndicates have been excluded from this analysis
because they could not potentially be established throughout the department
but only in those communes in which viticulture was practised. Similarly, agricultural syndicates (and so also threshing syndicates) have been omitted
because of the clear dominance of the Syndicat des Agriculteurs de Loir-et-
Cher in the domain of agricultural syndicalism. But the livestock insurance
societies, the mutual aid societies and the fire-fighting corps together numbered more than 500 so that they represented a very considerable proportion
of the grand total of almost 600 voluntary associations embraced by this study.

This synthesis of the geographical distribution of 'indicator' voluntary
associations by cantons confirms the impression already obtained (figure 9.7).

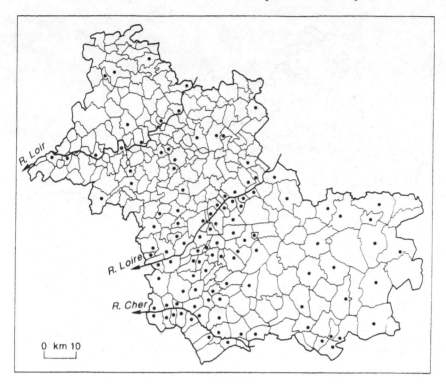

Fig. 9.3 Communes in Loir-et-Cher with at least one mutual aid society between 1815 and 1914

There was a core of six contiguous cantons in which more than 60 per cent of their 'fields' of communes acquired at least one 'indicator' association during the long nineteenth century: they were Blois-East (85.7%); Contres (70.6%); Blois-West (70.4%); Mer (66.7%); Bracieux (64.3%); and Montrichard (61.5 %). Collectively, this core of cantons with the highest levels of 'indicator' associations straddled the Val de Loire but also included part of the Petite Beauce in the north-east and of the Petite Sologne in the south-west. The significance of this core was enhanced by its extension into four of its adjacent cantons with high proportions (more than 40%) of 'indicator' communes: these were Ouzouer-le-Marché (42.9%) and Marchenoir (40.7%) in the Petite Beauce, Saint-Aignan (53.3%) in the Petite Sologne and the Cher valley, and Herbault (54.0%) which extended from the Val de Loire a little way into the Gâtine tourangelle. Away from that enlarged central core only two other cantons had 'indicator' communes in excess of 40 per cent – Lamotte-Beuvron (52.4%) and Salbris (55.6%). Associations were thinnest on the ground in the north-west, in the four adjacent cantons of Droué (11.1%), Mondoubleau

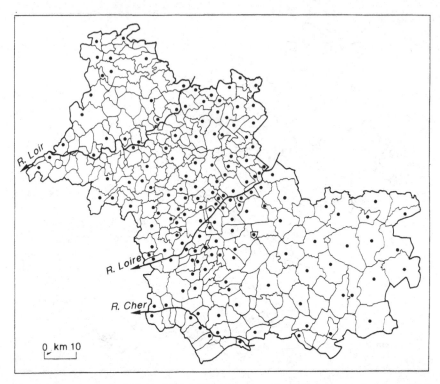

Fig. 9.4 Communes in Loir-et-Cher with at least one fire-fighting corps between 1815 and 1914

(26.2%), Savigny-sur-Braye (25.0%) and Montoire-sur-le-Loir (17.5%) – in effect, in that part of the *pays* of Perche which lay within Loir-et-Cher. In sum, the geographical concentration of voluntary associations upon the Val de Loire itself and upon its adjacent *pays* of the Petite Beauce and the Petite Sologne is very striking.

The foregoing chapters have taken some steps towards an understanding of the geography of voluntary associations in Loir-et-Cher, but in doing so they have also caused the enquiry to stop for reflection. Apparently obvious explanations of the geographical patterns uncovered have soon been found wanting. The distribution of livestock insurance societies did not reflect that of livestock within the department; the distribution of mutual aid societies did not reflect that of the overall population who could all have benefited from them; the distribution of fire-fighting brigades was not related to the incidence or risk of fires; the distribution of anti-phylloxera syndicates was not directly related to the diffusion of phylloxera throughout the department; agricultural syndicates (other than SALC) were focused largely upon one *pays* rather than

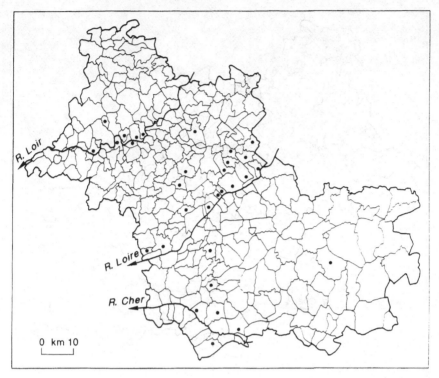

Fig. 9.5 Communes in Loir-et-Cher with at least one anti-phylloxera syndicate between 1815 and 1914

spread throughout the whole of what was a dominantly agricultural department; and although the membership of SALC was distributed throughout the department it was greater in some *pays* than in others. Similarly, the threshing syndicates were concentrated in the Val de Loire and the Petite Sologne, although they could also have played a useful role in other *pays*. Voluntary associations in Loir-et-Cher had specific geographies which require closer examination: they were by no means always to be found in the numbers and in the places that a straightforward functionalist, fundamentally economic, interpretation might suggest.

A closer analysis can begin by attempting to identify some of the factors which might have lain behind the geographies of associations. A fuller discussion of the processes which might have underpinned the patterns will be essayed in due course, as part of a wider consideration of the social significance of these associations. For the time being the problem will be addressed in terms of a simple spatial analysis of the geography of the 'indicator' associations (figure 9.7). From the work presented here on five sets of

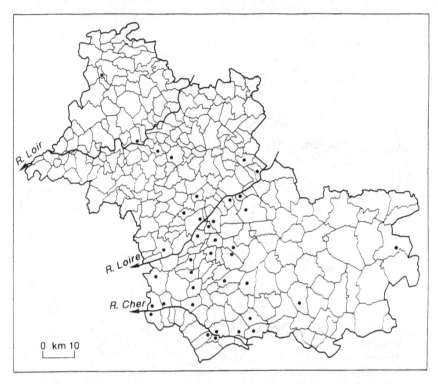

Fig. 9.6 Communes in Loir-et-Cher with at least one agricultural syndicate or threshing syndicate between 1815 and 1914

voluntary associations as well as from more general works, it could be argued, for example, that the distribution of voluntary associations in rural Loir-et-Cher during the nineteenth century was likely to reflect the role of Blois in their promotion, the role of larger cantonal centres and settlements in their spread, and the role of small farmers, especially *vignerons*, in their adoption. For each of the twenty-four cantons of the department have therefore been compiled the following data: the distance of the *chef-lieu* of the canton from the town of Blois; the percentage of the population of each canton living in the *chefs-lieux* of its constituent communes in 1881; the percentage of farms of less than five hectares in 1862; and the percentage of the non-built-up area occupied by vineyards in 1879.[10] The percentage of 'indicator' associations in each of the cantons has then been correlated in turn with those data. Correlating 'indicator' associations and distance from Blois produced a coefficient of −0.67, significant at the 1 per cent level: cantons closer to Blois had significantly higher percentages of communes with 'indicator' associations than did cantons more distant from the *chef-lieu* of the

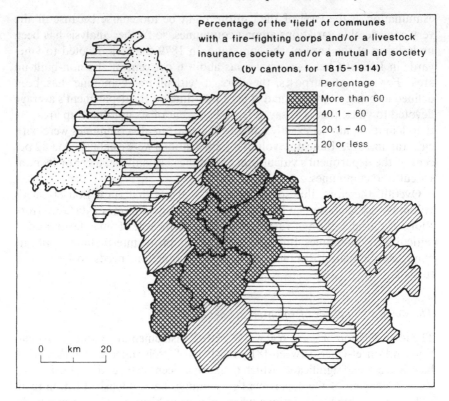

Fig. 9.7 The comparative geographical distribution of three sets of fraternal associations in Loir-et-Cher, by cantons, between 1815 and 1914

department. Similarly, correlating 'indicator' associations with the degree of nucleation of populations in the communes of the cantons produced a positive coefficient of 0.47, significant at the 5 per cent level: communes and cantons with the higher levels of nucleation were more likely to see such associations established than were those with the lower levels of nucleation. But now the picture becomes less clear. Voluntary associations and small farms had a correlation coefficient of 0.37, which was almost but not significant at the 5 per cent level; while their correlation with vineyards produced a coefficient of 0.23, which was not near to being significant at the 5 per cent level. Of course, taking the area under vineyards as a proxy for the numerical importance of *vignerons* within a locality is not entirely satisfactory: it has been used because the occupational data given in census documents and in electoral registers is so unreliable that it cannot be employed here (the terms *vignerons*, *cultivateurs-vignerons* and *cultivateurs* were used inconsistently, not rigorously, in these records).[11] But in relation to vineyards, examination at the level of the canton might be too coarse because of the very considerable variations among communes, so further analysis has been undertaken at the level of the commune. In 1879 the area devoted to vine-yards in Loir-et-Cher as a whole was about 6 per cent of the non-built-up area. For present purposes, therefore, a 'viticultural commune' has been defined as being one which had at least one-third of the department's average devoted to vines, that is to say at least 2 per cent of its non-built-up area. By that definition, almost exactly half of the department's communes were 'viti-cultural' in 1879. 'Indicator' voluntary associations were established in 82 per cent of the department's viticultural communes but only in 58 per of its non-viticultural communes.

Overall, therefore, this spatial analysis suggests that rural Loir-et-Cher's voluntary associations can be understood partially in terms of distance from Blois, of the nucleation of rural populations, of the role of small farms and of 'viticultural' communes. But this somewhat simplistic interpretation, with its emphasis on spatial distance and economic function, needs to be probed further.

The social significance of voluntary associations

The identification of some 600 work-related, instrumental voluntary associa-tions in Loir-et-Cher between 1815 and 1914 of itself suggests that they must have had a social significance which to date has been vastly undervalued. More than 70 per cent of the department's communes had established within them at least one of the associations which are the subject of this present study. They were to be found throughout the whole of the department, although their distribution within it was uneven; and they were being established throughout the whole of the long nineteenth century, although some periods within it saw much more activity than did others. That these voluntary associations were numerous, as well as widely spread through both time and space, has now been established beyond doubt. But it remains to assess their social significance.

Urban and rural, bourgeois and peasant

Many voluntary associations in Loir-et-Cher were to be found in its three prin-cipal towns and some were located in all of its cantonal centres. These were, of course, the largest and best connected settlements in the department. To the extent that the formation of voluntary associations depended upon a 'critical mass' of people and upon a 'network' of information, then it comes as no sur-prise to discover that many of those established in Loir-et-Cher during the nineteenth century were located in its main towns and larger rural centres of population. What is surprising, however, is the depth to which the practice of forming voluntary associations penetrated into the countryside, into small and remote villages and hamlets, into thinly settled and poorly connected communes.

Voluntary associations were by no means a uniquely urban phenomenon, but it might be that many of the ideas and activists underpinning them came from the towns or at least emerged out of urban contacts and experiences. To a considerable extent the flow of official information about associations was structured by the nested hierarchy of administration, from Minister in Paris to prefect in Blois, then from him to his two sub-prefects at Romorantin and Vendôme, and finally from each of those three to the approximately 100 mayors of communes in each of the department's three *arrondissements*. Such central control of official information combined with the settlement hierarchy of the department to structure the evolution of associations to some degree. But the existence of information about associations and of a critical mass of people as potential members of them were not in themselves sufficient conditions for the development of associations, however necessary they might be considered to be. The geographical structure of information and of settlements did not carry with it the formation of voluntary associations as an historical inevitability. There had also to be significant agents working within that structure. Many of the local agents active in the formation and running of voluntary associations in rural localities were men connected into urban ideas and practices: they were themselves bourgeois. Among the promoters and organisers of associations were often to be found, for example, notaries, doctors, vets, pharmacists, and schoolteachers. They soon came to be more important than the traditional power brokers within rural society, the notables and the priests. But this was by no means the full picture, for there were other significant agents. Artisans and craftsmen often played key roles, particularly in the formation of mutual aid societies. So too were farmers and *vignerons* important in the creation of associations, and not only of agricultural associations.

To be successful, associations had to have enthusiastic and committed leaders at their head but also a body of supportive and satisfied followers. The idea of association might have been advanced by a few but it had to be put into practice by many. In that sense the membership at large was as crucial to the functioning of an association as was its organising committee. Peasants mattered to these associations, sometimes as their officers and always as their members. If members in general lost faith in an association, then its organising committee faced an up-hill task and often never made it to the top: such was the general case most strikingly with anti-phylloxera syndicates but other specific cases were to be seen in other kinds of association. In addition, whether or not peasants were behind the formation of associations, they were often closely involved in running them. In particular, livestock insurance societies and mutual aid societies required for their effective operation the close and local monitoring of claims for benefits made by their members. Such policing was undertaken at the 'grass-roots' by responsible local members in the *quartiers* and *sections* of communes. Similarly, the massive activities of the Syndicat des Agriculteurs de Loir-et-Cher depended upon an extensive network of local agents to collect orders and payments for agricultural supplies and to arrange for their delivery to depots and collection by members. In varying ways and to varying degrees, peasants were engaged with Loir-et-Cher's fraternal associations.

Manifest and latent agendas

Voluntary associations in nineteenth-century France were obliged by law both to set out their aims and objectives very explicitly in their statutes and regulations and to have them approved by the central authorities. Their manifest agendas are readily recoverable from the records which they were required to construct and maintain. But behind that facade can also at times be glimpsed other, latent agendas.

The associations considered here have all been of the 'instrumental' rather than 'expressive' kind. They were not created just to satisfy or express specific interests which members had in relation to each other: some such voluntary associations – like musical societies and other recreational associations – did exist in Loir-et-Cher during the nineteenth century but have been excluded from this present study.[12] The associations considered here targeted their activities upon the wider rural social economy in order to bring about a situation which would be of benefit to their members (or, in the case of the fire-fighting brigades, to their communities), would assist them in their struggle for survival. What the associations offered was a variety of benefits in kind or in cash, or in both. Some provided support during an immediate crisis (such as when a house or haystack caught fire, or when a horse died or was unable to work because of an accident); others offered a means of dealing with long-term problems (such as the constant need for effective and cheap chemical fertilisers). Members combined in order better to defend their individual material interests. Such associations offered protection against hostile outside forces, both natural and cultural. That was the common purpose of these fraternal associations and that was, no doubt, why many individuals joined them.

But many associations also reflected the idealism, in its varied guises, of their promoters. Although by law voluntary associations were not allowed to involve themselves in either political or religious matters, many of them managed to do so to some extent and many of them were implicitly involved in what has been termed 'the moralisation of the masses'.[13] Voluntary associations were engaged in the battle for the minds of the peasantry. While not addressing political or religious questions directly at their meetings or in their operations, some associations nonetheless were able to make it clear – from their slogans, from the composition of their committees, from their lists of honorary members, or from the nature of their public celebrations – that they were committed to a particular ideology. Many associations represented materialism in the service of idealism. In Loir-et-Cher, as in other departments, the formation of voluntary associations was one of the ways in which the national contest between Church and State was waged regionally and even locally. In Loir-et-Cher, associations of the right were fewer in number and generally less successful than those of the left. No doubt the founders and organisers of associations were more ideologically aware than were most of their members and no doubt many of those joining associations did so for the material benefits and the sociability that they could thereby obtain, but the existence of other, more idealistic agendas has also to be acknowledged, as does also the potential for conflict both between and within these associations.

Social cohesion and social conflict

Voluntary associations in Loir-et-Cher during the nineteenth century might have been based upon the principle of fraternity and the commonsense practice of co-operation, but they did not always and everywhere enjoy a harmonious existence. An association of individuals with common sectional or geographical interests might have been expected to have promoted a sense of social cohesion among its members and often did so. But such an association could, in defending the interests of its own members, come into conflict with another association or with local or regional authority. Many did so. Additionally, sometimes relations among individual members of the same society broke down as a result of the surfacing of personal or other tensions. What were established as institutions to promote social cohesion sometimes exhibited social conflict.

Common objectives were the initial basis for individuals grouping themselves into an association. The successful delivery of those objectives was probably the best means of ensuring that the distinctive social identity of the group was sustained, but other ways were also sought. Controlling the entry of new members to an association, admitting only those deemed to be acceptable, was one method of creating and reproducing a special identity, even if such a device were not strictly permissible. General meetings and elections of officers were, of course, the formal ways in which associations reproduced themselves but many also adopted informal methods, most usually providing occasions for its members to socialise, as at funerals, Masses, banquets, parades and other public occasions. But the social harmony of a society was sometimes broken by personal rivalries and antagonisms: in some cases the disruption thus caused was short-lived, but in others it endured and even endangered the very existence of a society. Potentially more harmful for the wider community within which a society operated was the disruption which ensued from a conflict between a society and the mayor (and sometimes the council) of its commune or between a society and the sub-prefect or even prefect. Political or religious issues often underpinned such tensions. These were more obviously foregrounded when both right-wing and left-wing associations of the same kind were established in a single commune, as direct rivals with each other: in such cases, competition and conflict were ironic by-products of the spirit of fraternalism. But it was in such associations and during such conflicts that many peasants might have had their first encounter with the increasingly secular and democratic world of rural France during the nineteenth century.

Secularisation and democratisation

Monitoring the rise of voluntary associations in Loir-et-Cher during the nine-teenth century is more readily achieved than is measuring the decline of the role of religion within the rural social economy. The State monitored the former much more meticulously than it did the latter. Nonetheless, there can be little doubt that both of these changes took place, although not unin-terruptedly in time nor uniformly in space.

The matter was not as straightforward as might appear to have been sug-gested, for two reasons. First, the Catholic Church was itself involved pater-nalistically with the promotion and creation of voluntary associations. It became involved both in particular circumstances and in particular places throughout the nineteenth century, and the encouragement of voluntary associations was to be part of the general programme of social Catholicism during the closing decades of the century. But such activity can itself be inter-preted as representing an attempt on the part of the Church to check its decline, to maintain its hold over the moral life of rural populations by sup-porting social organisations – such as mutual aid societies and agricultural syndicates – which might be perceived by the peasantry to be of more direct value to them in their daily lives, in their struggle for survival, than the institu-tions and practices of the Church itself. While this was indeed the case, in rural Loir-et-Cher the role and influence of the Catholic Church in the formation and organisation of instrumental voluntary associations was overshadowed by those of secular, republican inclinations. The second qualification which needs to be made relates to the role of the Protestant Church in Loir-et-Cher. There had been a long history of Protestantism in one particular locality within Loir-et-Cher, stretching from Mer in the Val de Loire on to the plateau of the Petite Beauce towards the Forest of Marchenoir, and there were still Protestants there in significant numbers during the nineteenth century. This was also a locality in which voluntary associations had both a precocious and a strong development. This hearth of Protestantism appears to have favoured the growth of voluntary associations, although the exact nature of the link cannot readily be established. The Protestant ethic might be seen itself as favouring all forms of self-help and mutual aid; and, given their existence within a generally Catholic society, the Protestants in the locality of Mer had developed a strong sense of their separate identity, a strong sense of community which likewise might have been fertile ground for the creation of voluntary associations. It has been claimed more generally that Protestants were better educated, more ambitious and less routinised than Catholics and the eminent French historian, Jules Michelet (1798-1874) saw them as a progressive force in the struggle between liberty and fatalism which, for him, constituted the meaning of history.[14]

Even bearing in mind those two specific qualifications, it nonetheless remains plausible to argue more generally that in rural Loir-et-Cher during the nineteenth century the gradual decline of magic and superstition, and of religion, was accompanied by a gradual acceptance of scientific reasoning, of ideas emanating from the natural and social sciences. There was a turning away from ecclesiastical assurance and towards secular insurance as a more effective form of risk management – and both processes involved men more substantially than they did women. Slowly, people's ideas about their relations to nature and to other people were transformed, as was also confidence in their own ability to effect a greater degree of control over those relations. It is in this broad intellectual context that the growth of instrumental voluntary associations in rural Loir-et-Cher during the nineteenth century needs to be interpreted.

Voluntary associations not only provided material benefits to their members; they also encouraged a greater degree of belief in the ability of peasants to control their own lives, to create their own histories. Voluntary associations cultivated peasant empowerment. They enabled those with initiative, ability and the trust of their fellow members to participate in organising associations and in running them on a day-to-day basis. With their general meetings, their committee structures and their elections, indeed with their emphasis upon the responsibility of each member of an association and of the accountability of its officers, associations provided immediate and practical demonstrations of the workings and benefits (as well as problems) of a democratic system. Knowledge and experience gained through membership of, and more especially active participation in, a voluntary association could be transferred into the wider arena of the rural social economy. Thus associations provided lessons in individual empowerment and in local democracy which could be seen as encouraging many of their members to engage with wider, regional and even national, issues affecting them and also as enabling some of them to seek election to local municipal councils. In varied ways, voluntary associations contributed to the politicisation of the peasantry of Loir-et-Cher: they helped to make many more people aware of the social, economic and political structures within which they lived their daily lives and to make some people more aware of their own potential roles as agents able to work towards the control and transformation of those structures.

Change and continuity

The provision of insurance and protection by voluntary associations could be seen as a new, 'modern', way of managing risk in rural Loir-et-Cher during the nineteenth century. Associations could be interpreted as representations of a fundamental transformation of peasant *mentalités*, as products of the adoption of a 'rational' way of thinking and acting. That scientific principles underpinned insurance associations is made clear, for example, in G. Hubbard's book on the organisation of mutual aid societies, published in 1852. Hubbard argued that such societies should be both administered and financed on a sound, scientific basis and his book included an appendix of thirty-eight pages setting out the statistical calculations (of, for example, age-related mortality and sickness rates) which, in his view, were necessary for the proper functioning of a mutual aid society.[15] In practice, few societies in Loir-et-Cher in the mid-nineteenth century had the level of statistical sophistication advised by Hubbard. Nonetheless, its associations were institutionalised as well as 'rationalised' forms of risk management and as such they were a new feature within the rural social economy. Moreover, their methods became more rigorous though time. They did, therefore, constitute a fundamental, in that sense revolutionary, change in the ways in which peasants confronted hazards.

But they also demonstrated some significant continuities, certainly in terms of their early functioning and perhaps in terms of their historical antecedents. There had been in rural Loir-et-Cher, as in many peasant societies, a tradition of informal mutual aid and co-operation. Individuals would come to the assistance of neighbours in need, offering their time and labour on the understanding that they could in turn call upon their neighbours for similar help when they themselves were in need. It was an informal system based upon the reciprocity of labour assistance. Some of the early voluntary associations seem to have built upon that tradition, in the sense that they obliged members to contribute their labour to the association without requiring them to make any cash payments. The cost to a member was thus the opportunity cost of his time. Clearly, the volunteer fire-fighting corps were generally based on that principle, but so too were many mutual aid societies, especially those of *vignerons*. Similarly, many of the early livestock insurance societies imposed only limited cash obligations upon their members, doing so principally when a claim for indemnity was made by a member and even on such occasions when a member's obligation to make a cash payment was balanced by the benefit of his receiving in exchange a joint of meat from the animal's carcass. Gradually during the nineteenth century, as associations generally grew in number and in size and as the cash economy expanded, the payment of subscriptions or premiums on a regular annual or monthly basis became more common and eventually the normal form of a member's obligations. A concomitant of the growth in size of associations and of their reliance on cash subscriptions must have been that they became to some extent less personal and more institutional.

The question of continuity also deserves consideration on a longer time-scale. Voluntary associations in rural Loir-et-Cher in the nineteenth century might have had earlier antecedents. Specifically, some of the mutual aid societies included in this present study seem to have evolved from earlier *confréries de métiers* (confraternities of craftsmen), from *confréries de Saint-Vincent* (confraternities of St Vincent) which were associations of *vignerons*, or possibly from *confréries pieuses* (devotional confraternities). More generally, many such *confréries* existed in the department in the sixteenth, seventeenth and eighteenth centuries and even earlier. Until recently there only existed isolated studies of these associations in Loir-et-Cher, but a systematic examination of them in the archdeaconries of Blois and Vendôme during the sixteenth to eighteenth centuries has now been completed by Marc Bouyssou. Together those two archdeaconries encompassed the central districts of what was later to become Loir-et-Cher, straddling the valleys of the Loire and the Loir and incorporating parts of their adjacent *pays*. Bouyssou has demonstrated that *confréries* were both absolutely and relatively more numerous in the archdeaconry of Blois than in that of Vendôme; that they were more common in parishes with the largest populations; and that they were especially numerous and active in the Val de Loire, created there by communities of *vignerons* and of merchants using *confréries* to make statements about their own separate and distinctive identities.[16]

L'esprit d'association?

Many of Loir-et-Cher's *confréries* had chequered histories and it would not be sensible to argue in terms of any direct and uninterrupted continuity between them and the voluntary associations which developed in Loir-et-Cher during the nineteenth century. It may, however, be conjectured that there might have persisted in the Val de Loire and especially in its viticultural districts *un esprit d'association* (a spirit of association, a fraternalist sentiment) which favoured the development of voluntary associations in the valley during the nineteenth century. Indeed, one could go further. Roger Dion, in his historical geography of the epic embanking of the Loire, has shown how the construction, maintenance and repair of the *levées* of the Loire required co-operative efforts on the part of its riverine communities over centuries.[17] An *esprit d'association* might likewise have been constructed in the Val de Loire and especially among its *vignerons* throughout that period and adapted to new circumstances in the nineteenth century.

The Loire valley was certainly through history a significant corridor of communication along which new information and ideas, as well as people and commodities, flowed. It encouraged contact and exchange. It continued to do so in the nineteenth century, when improved roads and new railways almost obliterated the river itself as a major means of communication. The Loire valley, it could be argued, had historically been open to new ideas and practices, connected into the wider world. To a lesser extent that was also true of the valley of the Cher, which merged with that of the Loire a little downstream from the department, at Tours, and also of the valley of the Loir. 'Open' rather than 'closed' rural social economies were likely to develop in such situations and an *esprit d'association* may be seen as a component of 'openness'.

Support for republicanism was at times manifested strongly in these three valleys, as well as in the Petite Sologne and in the Petite Beauce between Mer and Marchenoir. Those localities of the department which had large numbers of small farmers (and especially of *vignerons*) and/or large numbers of agricultural labourers and wood-cutters were those in which a rural socialism grew during the nineteenth century, often nurtured by local men from the liberal professions and the *petite bourgeoisie* as well by ideologically committed outsiders, all of whom were united in their anti-clericalism and some of whom were Freemasons. These were the rural localities of Loir-et-Cher in which the French Communist Party was to see a remarkable development in the early twentieth century.[18] There are perhaps signs here of a 'long march' from *confréries* to *communisme*. While the spirit of fraternalism need not be linked exclusively to the transformation of rural Loir-et-Cher after the Revolution of 1789, it was explicitly debated and given practical expression there during the nineteenth century. Voluntary associations themselves were sometimes explicitly based on the fraternalist notion of a harmonious *monde rural* (rural society) or at least of a single *classe paysanne* (peasant class). But behind such unifying concepts there always lurked divisive practices and it has to be conceded that the rhetoric of fraternity was often ahead of its practice.

Although the actuality of fraternalism might have lagged behind its theoretical exposition, it was nonetheless real. The act of joining a voluntary association was a self-confessed statement of belonging to a specific group and of accepting its objectives. An association produced and reproduced a sense of identity among its members. Such an identity could be based upon a shared occupation or experience or upon a shared place or space. During the first half, perhaps three-quarters, of the nineteenth century most voluntary associations in Loir-et-Cher were grounded within local communities, and were based substantially upon personal, face-to-face contacts. During the second half of the nineteenth century some voluntary associations saw considerable expansions in their memberships, both numerically and geographically, so that social relations within them became less personal, more abstract. In particular, livestock insurance societies developed as federations of geographically separate groups of members and the Syndicat des

Agriculteurs de Loir-et-Cher combined farmers from across the whole of the department. Such practices must have contributed to the development of a sense of class: the spirit of fraternity was not bounded locally and found truer expression when it associated people with a shared interest but not known to each other personally. Voluntary associations thus contributed to the making of the 'peasant class' in Loir-et-Cher but they did not go so far as to result in a single class. There were too many differences among 'peasants', not only in terms of the sizes, tenures and orientations of their farms but also in terms of their politico-religious beliefs. But voluntary associations did help to foster a new sense of place as well as a new awareness of class: they encouraged their members to look beyond their immediate localities, communities and families. In effect, they created a new social institution intermediate between the individual and the State. But for the most part, they excluded women from their activities: the practices of association extended the boundaries of fraternity in some respects but they did not in general go so far as to include women.

These conclusions and conjectures, derived from this study of voluntary associations in Loir-et-Cher, can now be placed comparatively within broader contexts, first of the Val de Loire and then of France.

Fraternity among the French peasantry 1815–1914

The revelation of similarities, as well as differences, among associations within Loir-et-Cher itself inevitably leads one to expect there to be both similarities and differences between those of that department and those elsewhere in France. Each can be made to illuminate our understanding of the other. Most studies of voluntary associations in rural France during the nineteenth century have approached their problem thematically, examining a single association or a single set of associations analytically. There have been virtually no other studies which have adopted the approach employed in this present study, investigating an entire category of associations within a given place and period synthetically. The only one to do so at all thoroughly has been that by Pierre Goujon of the associations of Mâconnais and Chalonnais between 1848 and 1914. While being explicitly concerned with the geography of associations within an entire region, Goujon's study was however selective in terms of the associations considered: it focused upon some expressive associations (devotional associations, musical societies and sports clubs) and two types of instrumental associations (mutual aid societies and agricultural syndicates), rather than providing a comprehensive study of either category of associations.[19] Nonetheless, both the approach and the issues raised by Goujon's excellent study are pertinent to this present work and considerable comparative use will be made of it in due course, when the discussion moves away from the Loire valley to consider France in general.

La vie associative in the Val de Loire

As has been noted earlier, Roger Dion was particularly struck by what he saw as the precocious awakening from the late 1870s of the spirit of agricultural association in the Val de Loire within Loir-et-Cher.[20] While this present study has shown agricultural associations to have been much more numerous, of much earlier origin, and part of a much wider associative movement than was recognised by Dion, his identification of that section of the valley around Blois as being of particular significance in the development of agricultural associations in the Val de Loire would seem to have been validated, both positively by this present study and negatively by others which have found much less evidence of such associations elsewhere in the Val de Loire.

A study by John Shaffer of agrarian change and household organisation in the Nivernais, on the northern margin of the Central Massif, between 1500 and 1900 made an important contribution to our understanding of the differing economic and demographic strategies pursued by landowning peasants, share-croppers and tenant farmers. It focused upon family and household structures but had nothing to say about the role of any kind of voluntary association in the department of the Nièvre's 'agricultural revolution'.[21] Gregor Dallas, in his comparative study of the rural economies of the regions around Nantes and Orléans between 1800 and 1914, portrayed peasants as responding to crises and as developing strategies to cope with the pressures of population on resources. One such response, what Dallas termed a 'survival tactic', was the formation after the association law of 1884 of agricultural syndicates, less so in the Orléanais than in the Nantais where they were more actively promoted by the Catholic authorities. Even so, Dallas found evidence of only thirteen syndicates in the Nantais and of even fewer, only six, in the Orléanais, far fewer than the number identified in Loir-et-Cher. Although acknowledging that 'historians have been sparing in their comments about the early days of rural associations', Dallas accepted the general view that 'the agricultural syndicates of this period were really only glorified boutiques' selling farming supplies. He also accepted, however, that the agricultural syndicate 'was a popular, locally based organisation'. But Dallas' account was somewhat confused: in his view, a syndicate 'won favour by confining its activities to the fulfilment of some very basic economic needs of the small cultivator. Naturally, once that need was satisfied, the syndicate no longer served a useful purpose, which is why so few attained anything like a permanent foothold in the country.' It is difficult to see how a farmer's 'needs' – for supplies – could ever be fully and finally satisfied as Dallas seemed to envisage. If by 'need' he was meaning that of meeting the economic crisis of the late nineteenth century, it could be argued that syndicates could then have played a key role in the survival of small farms and could (and did) continue to do so during much of the twentieth century. In addition to the syndicates, Dallas noted also that 'there were the odd insurance groups' relating to protection from river flooding but such organisations did not have a wide, popular base.[22] A more thorough survey of agricultural syndicates in part of the upper Val de Loire since the late nineteenth century has been provided by Jean Vercherand, who ascribes to them both material and ideological roles.[23]

A different perspective has been provided by J-L. Marais' study of sociability in Anjou, Maine and Touraine from the eighteenth century to the present day. 'Different', because this is in practice a study of expressive associations established for men to play *boule de fort* (bowls, using biased bowls). They first appeared, it seems, in the eighteenth century, perhaps as a result of English influence, but most of their growth came after the 1830s. Marais argued that these societies spread from towns into the countryside, and that most were established in the 'open' regions where religion was little practised and where republican sentiment was strong, notably in viticultural localities. There would seem to have been some affinities here with the voluntary associations of nineteenth-century Loir-et-Cher, even though those studied by Marais were expressive rather instrumental societies. This suggests that in due course the expressive associations of Loir-et-Cher – such as musical societies, library associations, sports clubs and gymnastic societies – might merit the close attention which has been given here to the department's instrumental associations.[24]

General perspectives upon la vie associative

Moving away from the Val de Loire to consider *la vie associative* in rural France as a whole during the nineteenth century must involve, to be manageable, the selection of some general perspectives based upon a critical review of the literature. A number of qualifications need to be made, however, in relation to this consideration. First, no claim can be made to have covered the literature on instrumental voluntary associations in its entirety, because it has not been possible to take into account many studies no doubt published in the plethora of local and regional journals which exist in France. Second, the more readily available literature indicates that some aspects of *la vie associative* and in particular some kinds of association have received more attention than others. Third, any attempt at generalisation about instrumental voluntary associations in rural France during the nineteenth century must of course be qualified at the outset by recognition of their historical and geographical specificities: they were varied and variable in character, they were not standard and stable but diverse and dynamic. Nonetheless, sufficient literature has been reviewed, sufficient studies have been undertaken, and sufficient similarities as well as differences identified to permit a few general, comparative points to be made while acknowledging that some will be more conclusive and some more conjectural than others.

Studies of instrumental voluntary associations in rural France during the nineteenth century have for the most part focused upon the development and activities of agricultural syndicates since the mid-1880s. They have been viewed essentially as local and regional economic associations of small farmers created in order to give them protection in an increasingly competitive, capitalist and global environment. They have thus been interpreted as a functional response to the agricultural crisis which developed in France from the mid-1870s, a response promoted as part of their agrarian programmes by political parties of both the right and the left as well as both by the Church and by the State.[25] Studies of agricultural syndicates themselves have a long history but other forms of association have received much less attention. It is only relatively recently that mutual aid societies have been examined in rural areas, where they have often been interpreted not as an indigenous development but as a spin-off from a much more important urban movement.[26] Livestock insurance societies, anti-phylloxera syndicates and *corps de sapeurs-pompiers* have received least attention.[27] What follows, therefore, will be based principally upon studies of agricultural syndicates and of mutual aid societies. What follows must also be provisional, for there remains much work to be done on the history and geography of voluntary associations in rural France during the nineteenth century.

The timing of voluntary associations
In the history of most kinds of association it is possible to identify an initial period of slow development, then a period of rapid growth, and finally a period of consolidation. These correspond, of course, in broad terms to the 'origin' phase, the 'adoption' phase and the 'saturation' phase which characterise the temporal pattern of the adoption of many innovations. The main challenge presented by such a picture is to endeavour to explain why the 'birth-rate' of associations was so high in particular periods.

The growth of agricultural syndicates in the late nineteenth and early twentieth centuries has often been directly attributed to the economic conditions of that period of agricultural depression. While it would be foolish to deny the link between economic pressures upon small farms and the formation of agricultural syndicates, that connection was not as direct as it has often been argued or assumed to have been. For example, Raphael Faucon's study of agricultural syndicalism in the Nord region stressed its origins in the economic crisis of the late-nineteenth century and Gabriel Désert and Robert Specklin retained an exclusively economic interpretation of agricultural syndicates in their overview of French agriculture between 1880 and 1914.[28] But to an economic interpretation of such syndicates must also be admitted the role of cultural (and especially ideological) factors in their creation.[29] For example, Hubscher's excellent monograph on agriculture and rural society in the Pas-de-Calais between about 1850 and 1914 saw the region's syndicates primarily as direct responses to economic difficulties but recognised that both personal and political factors were also at work.[30] Syndicates had social as well as economic origins and agendas.[31]

It is similarly possible to exaggerate the influence of the legislative structures within which associations developed. Because the theory of association was such an important part of political discourse in France during the nineteenth century and because an elaborate legislative structure was built to contain them in practice, it might be expected that periods which saw high 'birth-rates' for associations were those in which the discourse was especially active, legislation about associations particularly prominent, and official advocacy for them markedly energetic. Such connections have often been assumed, and some-times – but by no means always – demonstrated. The evidence from this study of Loir-et-Cher, like that by Pascal Bousseyroux of mutual aid societies in the Puy-de-Dôme,[32] suggests that such discourse, legislation and official advocacy were neither necessarily nor sufficiently connected to the creation of fraternal associations to account for their historical geography. Studies which have hith-erto assumed but not proved such a connection must be viewed with some scepticism. So too should similar claims by contemporaries. For example when on 5 July 1908 Joseph Ruau, the Minister of Agriculture, delivered a speech at Blois to the second national conference on mutual credit and agri-cultural co-operation he understandably emphasised but nonetheless exagger-ated the role of recent legislation in the formation of agricultural syndicates, insurance associations and mutual aid societies. Ruau argued, inevitably and rightly, that the Republic was doing much to promote a rural, property owning democracy. But in stating that his government's actions had meant the death of individualism and the detachment of the peasantry from their supersti-tions, Ruau was both overstating the extent to which solidarity had permeated the French countryside by the beginning of the twentieth century and under-stating the extent to which fraternalism had been developing there since the early nineteenth century.[33]

Most studies of associations have focused upon just one category and have thus had only one temporal pattern to interpret. This present study of all of the instrumental voluntary associations in rural Loir-et-Cher has demon-strated a sequence for their formation and emphasised that different kinds of association had different periods when their 'birth-rates' were especially high. The three 'innovation waves' identified in Loir-et-Cher between 1815 and 1914 were, in fact, to be followed by others, especially between 1918 and 1939 when many agricultural processing and marketing co-operatives were founded. The development of fraternal associations in rural France could be expected to have been a cumulative, self-reinforcing process, with the perceived success of one association or of one set of associations stimulating the formation of others. As the Comte de Rocquigny observed in 1903, the growth of mutual aid groups cultivated within the countryside a spirit of solidarity which was likely in turn to promote a wide range of fraternal associations.[34] Most studies of rural associations have focused principally upon agricultural societies and syndicates. Consequently they have tended to argue that agricultural associations only developed from the mid-1880s onwards, neglecting or ignoring anti-phylloxera syndicates and, more importantly, livestock insurance societies and mutual aid societies whose origins predated agricultural syndicates by at least forty years.[35] The formation sequence for fraternal associations seems to have been poorly understood, even by some contemporaries, for it has often been stated that agricultural syndicates constituted the first associations, sub-sequently giving rise to the creation of other agricultural associations, including those providing personal and livestock insurance.[36] In his study of some expressive associations in the Mâconnais and Chalonnais, Goujon identified a sequence in their formation, with musical societies being very actively created between 1860 and 1885, sports clubs in the 1880s, and associations linked to schools (such as old pupils' associations) in the late 1890s. The peak periods of formation of societies were identified by Goujon as being 1863–5, 1876–86, and 1896–1901. Those peaks, he argued, did not correspond to any economic or social trends; instead, he identified them as being linked 'aux temps forts de la lutte républicaine' ('to periods of strong republican pres-sures').[37] National political debates resonated within rural localities. There clearly remains considerable scope for further, integrated studies of waves of voluntary associations, while also recognising that different interpretations of peak periods in the formation of associations are likely to be both necessary and appropriate. The *conjontures* with which each such period corresponded could themselves be very different while still being connected to an earlier sequence in the formation of associations.

The spacing of voluntary associations
The geography of voluntary associations in rural France during the nine-teenth century has been examined at a variety of scales, from the local to the national. The distribution of associations was clearly very uneven, at every scale, and some of the reasons for individual patterns were specific to partic-ular places. In the Breton department of Ille-et-Vilaine, for example, all of the mutual aid societies formed in the northern section of the *arrondissement* of Fougères were connected to the granite quarrying activity of the region.[38] In the heavily wooded Vosges, in eastern France, there was developed at the end of the nineteenth century a distinctive form of mutual aid society, the *mutu-alité scolaire forestière*, which created through purchases and plantings of woodlands assets whose management as forestry schools produced substantial incomes that could be invested and allocated to members as long-term pension funds.[39] More generally, in their pioneering survey of mutual aid societies in the four departments of Lorraine, Françoise Birck and Michel Dreyfus emphasised the specificities and varieties of societies within that region.[40] In a

study of mutual aid societies in the Bourbonnais, in the department of Allier in the northern Massif Central, Pascal Demoulin argued that the mutualist movement did not get under way until the late nineteenth century, and that it was predominantly an urban phenomenon. He stressed that its diffusion was not related to political, religious or agrarian structures or indeed to other characteristics of the social economy; he explained it instead in terms of the activities of specific militant individuals, of particularly influential agents.[41]

But general interpretations have also been offered. The geographical distribution of associations has often been, or claimed to have been, structured by settlement patterns. It has often been argued that associations were more prominent in towns than in the countryside, in large rather than in small centres of population, reflecting in part the need for a 'critical mass' of people from among whom an association could draw for its members. By extension, the idea and practice of association has been seen as essentially an urban phenomenon, with the extent to which associations developed in rural areas depending upon the proximity of those localities to towns and upon the extent to which rural settlements had artisans and professionals engaged in essentially 'urban' functions.

While this interpretation undoubtedly illuminates the geography of associations, its relatively uncritical acceptance has had two consequences. First, the degree to which voluntary associations were developed in rural areas and their contribution to rural culture have been played down dramatically. This seems to have been the case especially in relation to mutual aid societies. Second, an 'urban' interpretation tends to ignore the fact that within a given region not all of the towns and not all of the areas characterised by nucleated settlements developed associations to the same extent, and similarly not all of the rural areas and not all of the areas characterised by dispersed settlements were equally disadvantaged in terms of the development of associations.

In relation to *corps de sapeurs-pompiers*, for example, Hubert Lussier has argued that the distribution of brigades in Seine-et-Marne, and indeed in France generally, was only in part related to the distribution of populous and wealthy communes and localities, to the nucleation of population in towns and large rural centres with the financial and human resources to support brigades. Such places, according to Lussier, provided the necessary but not the sufficient conditions for the formation of fire-fighting brigades.[42] Additional interpretations need to be sought, and not only in relation to the *corps de sapeurs-pompiers*. One interpretation of geographically specific patterns of voluntary associations argues more abstractly in terms of the stronger existence in some localities than in others of an *esprit d'association*, as identified for example by Pierre Barral in the department of Isère during the Third Republic.[43] Lussier's study of fire-fighting brigades also invoked this spirit of association in the particular guise of an *esprit pompier*, which he claimed was strongest in those regions and localities which had 'un goût et un respect de la chose militaire' ('a taste and respect for military matters'), such as carrying out orders, marching in step, wearing uniforms and insignia, and bearing arms and flags.[44] Conversely, the predominance of an *esprit de l'individualisme* in a locality can be called upon to explain the weak development of associations there, as argued for example by Jean-Yves Chalvignac in relation to part of Ille-et-Vilaine.[45] Such invocations can, of course, beg as many questions as they answer.

Another interpretation is that associations were most developed in the most connected and most commercially advanced districts, in effect in those localities which were most closely linked into regional, national and even global cultures: such localities were more 'open' to new ideas than were others.[46] The diffusion of the new idea of socialism in the Provençal department of the Var during the nineteenth century has been shown by Tony Judt to have been related in part to settlement structures and to a tradition of sociability, but its spread was also shown to coincide with those areas with improved communications and expanded markets (notably for wine).[47] That a key role was played by *vignerons* and hence of viticultural localities is often emphasised in historical geographies of associations, as it was for example by Louis Devance in his study of mutual aid societies in Burgundy.[48] Such an emphasis is often accompanied by an insistence upon the underlying role of settlement structures. For example, Emmanuel Le Roy Ladurie in his recent survey of the European peasantry argued that almost everywhere *vignerons* were on the left politically and that they were great believers in justice and democracy (and, by implication, fraternity). Underpinning such radical views, he claimed, there was often a form of urban acculturation to democratic and petits-bourgeois values within the large viticultural villages which were, in effect, small towns.[49]

That the special role of *vignerons* in the formation of fraternal associations will need to be considered further is made clear by Goujon's examination of the very uneven geographical distribution of voluntary associations in Mâconnais and Chalonnais. Goujon set out explanations for the geography of associations in that region in terms of their increasing probability, and emphasised their cumulative interpretative power. The development of associations was most marked in the larger and nucleated settlements; in communes closest to towns; and in viticultural communes where alongside the large population of *vignerons* was also to be found a large, non-agricultural population. Goujon thus favoured a model of hierarchical spatial and social diffusion, with a key role played by the large viticultural villages which both imitated *la vie associative* of the bourgeoisie in nearby towns and relayed it out to peasants in neighbouring rural communes.[50]

The significance of voluntary associations
The roles of voluntary associations in rural France during the nineteenth century were clearly numerous and complex. I want here, however, to consider

them in relation to just two broad concepts. It has been the argument of this book that the emergence of instrumental voluntary associations in rural France during the nineteenth century can be best interpreted as the development of a novel form of risk management. This offers a new perspective upon these associations, one which is not to be found within the existing literature on their history and geography. That literature has largely ignored the responses of peasants to risks in nineteenth-century France and it has in particular made scant reference to the resort to insurance as risk management – when such reference has been made, it has been argued or assumed that insurance came to be offered by agricultural syndicates as an extension of their services, rather than predating them (in the case of Loir-et-Cher, by at least half a century).[51] That literature therefore needs to be reinterpreted from this new perspective. This will be done using as an organising framework the two general *structures* of secularisation and survival.

The growth of instrumental voluntary associations can be seen as the obverse side to the decline of the Catholic Church. The former can be more readily measured than the latter, but there is nonetheless general agreement that the nineteenth century saw a considerable de-Christianisation of rural France.[52] Popular religion had relied upon the Church for protection against the many hazards of daily life in the countryside. As faith in religion declined and greater reliance came to be placed upon reason and science, so rural populations turned increasingly to secular forms of risk management. Religious assurance came to be replaced by secular insurance. Of course, the Catholic Church itself promoted livestock insurance societies, mutual aid societies and agricultural syndicates as one means by which it could prolong its influence in rural France, an influence which was increasingly challenged by the State. In addition, Protestants were involved in promoting voluntary associations, especially in eastern France.[53] In the department of the Tarn, on the south-western margin of the Massif Central, most of the 156 mutual aid societies established by 1900 were secular foundations, with only a few having a religious basis – twenty-six of them Catholic and ten Protestant. During the nineteenth century clerical influence on societies in the Tarn declined, with newly founded societies no longer including references to religion in their statutes and no longer incorporating the name of a saint into their titles, while some of the older-established societies changed their titles, deleting references to saints.[54] A similar secularising trend has been noted in the Bordeaux region.[55]

The totalising role of the Catholic Church during the ancien régime had been (with some important local and regional exceptions) virtually unchallenged in the countryside: it was the dominant ideology and one which taught a conservative acceptance of the social and economic position into which an individual had been born and a belief that material suffering should be relieved by charitable acts by the Church and by the better endowed members of a community. In effect, the Church offered just one, dogmatic view of utopia. But in post-revolutionary France a vibrant political discourse developed multiple and often competing visions of utopia. The general concept of fraternity implied social cohesion within communities but the specific debates about it meant that in practice it produced considerable social conflict. While there was, in effect, one Church there were many possible States and debates conducted on the national stage were echoed in many local arenas, not only in contentious regions such as the Vendée.[56]

The ideology of the Church was intended to be all-encompassing, with priests seeing to the care of the individuals in their parishes from the cradle to the grave. But the decline of religion and with it the decline of the parish as an institution was paralleled by the growth of the municipality, of the commune, and so of the State. Mayors and schoolteachers came to play key roles within their communities, as did men of science such as doctors, vets and pharmacists, often challenging and even displacing the traditional source of knowledge and so of power, the priest.[57] Until the law of 1901, commune councils and especially their mayors were closely involved in monitoring the formation and activities of voluntary associations, whether they were being promoted and organised by the Church or by a secular group. Exercising that control was just one of the many and increasing number of ways in which the municipality during the nineteenth century became more visible within communities, in the process gradually eclipsing the power of the Church. Both the Church and the municipality – as well as individuals – could promote voluntary associations, but in all cases it was the commune council and the mayor which acted as the control mechanism on behalf of the regional and national authority. In the Médoc region, the Catholic Church ardently promoted mutual aid societies and it did so with marked success, developing societies in most communes during the 1850s and 1860s, but where communes were slow to establish such societies it was often because of local conflicts between the *curés* and the municipal councils (and especially their mayors).[58] Moreover, only the commune council – as an arm of the State – was able to establish a *corps de sapeurs-pompiers*. The monitoring of associations by commune councils contributed to the secularisation of rural communities.

Voluntary associations came to be promoted by a variety of interests. Peasant agricultural syndicalism from the 1880s is often seen as having been animated by a political traditionalism of the right, as providing not much more than fertiliser boutiques promoted by *notables*, conservatives and the Catholic Church. Emmanuel Le Roy Ladurie has recently reiterated this view and argued that although left-wing agricultural syndicalism did exist, especially among *vignerons*, it did so as a minority movement. He saw socialist syndicalism as being associated much more with the radicalisation of the rural proletariat, of agricultural labourers and wood-cutters.[59] There is, no doubt, much truth in that orthodox interpretation but not the whole truth. Joining

unions and strikes were, of course, options open more to landless rural labourers than to owner-occupiers or tenant farmers,[60] but a republican agricultural
syndicalism and rural fraternalism is also readily identifiable in some parts of
France,[61] as in Loir-et-Cher. As long ago as 1907, Augé-Laribé drew attention
to the very clear and, as he believed, very original socialist sentiments which
had by then animated the creation of seven voluntary associations among the
peasantry of just one commune in southern France.[62] He has long argued that
the first agricultural syndicates in France, including the Syndicat des
Agriculteurs de Loir-et-Cher, were the creations of working farmers; others
were established by landowners, *notables* and *grands bourgeois* who wanted to
check the movement, whose aim was to lead peasants into supporting a democratic Republic.[63] It is this latter group of syndicates – the syndicalism of
dukes and squires – which seems to have attracted more attention from scholars, leading to the wider acceptance of an essentially paternalistic rather than
of a democratic interpretation of agricultural associations. Traditional power
brokers, including the clergy, did indeed lie behind the creation of many agricultural syndicates,[64] but so too did new republicans, men from the liberal professions, progressive and market-oriented farmers and departments'
professors of agriculture, as well as many small farmers.[65] In many regions,
such as the Pas-de-Calais, conservative, Christian and republican forces were
all at work simultaneously promoting syndicates.[66]

While in broad terms the power of the Church declined during the nineteenth century and the power of the State grew, the latter did not simply
replace the former. The contest continued into the twentieth century, even
after the formal separation of Church and State in 1905. But in this context
voluntary associations played an important role, providing a social organisation which was intermediate between the individual and the Church and especially between an individual and the State, with its large and growing body of
officials. Associations were significant in shaping new ways of thinking and
living, important in refashioning social relationships.

The growth of the municipality in rural France during the nineteenth
century was itself paralleled by the growth of democracy and the empowerment of individuals. Voluntary associations provided for many peasants their
first real encounter with democracy; they brought democratic practices and an
awareness of the relation between local and national issues to the most remote
hamlets and to the most isolated individuals. Voluntary associations provided
opportunities for democracy among equal citizens and the empowerment of
individuals to be experienced by their many members, if not exactly daily then
at least often. Many more men came to be actively involved in the running of
their secularising communities than had previously been the case. The same
point could be made in relation to women, but to a much lesser degree than
was the case with men. The memberships of instrumental associations were
dominated by men and such associations were organised almost exclusively by

men. In the Médoc region, for example, mutual aid societies were numerous, totalling almost 700; but only one society was established specifically for working mothers of families, and only a dozen or so recognised women as having the same rights to mutuality as men. In the vast majority of societies women were sometimes admitted or 'tolerated' as members, but they were not required to pay entry fees, only monthly subscriptions; and while they could attend their society's general meetings, they were not permitted to speak or to vote. In effect, women were admitted to such societies as spouses of male members.[67]

Many voluntary associations were initially created with, or soon developed, firm roots in their localities, providing opportunities for active participation in them by local people for local people, although the picture was not a static one. Those mutual aid societies established in the Mâconnais and Chalonnais before the middle of the nineteenth century were products of initiatives by local *vignerons* but increasingly thereafter the mutualist movement was, as Goujon has described it, 'captured' by both old and new *notables* as well as by municipal councils.[68] Agricultural syndicates in the Pas-de-Calais initially emerged at the scale of the cantons and *arrondissements* under the patronage of pre-existing agricultural societies, but they only 'took-off' when the idea of association was put into practice at the scale of the commune, which corresponded more closely to the daily, local experience of peasants. Local syndicates were later federated into unions covering larger areas.[69] A similar process operated in relation to mutual aid societies.[70] To the extent that associations were indeed economic organisations, providing goods and services for their members, then the underlying principle of economies of scale favoured their evolution from small to large memberships and from local to regional spheres of operation (whether discretely over a region or discontinuously as a federation). As agricultural syndicates grew in the size and in the geographical and functional scales of their operations, their character changed. They developed increasingly into pressure groups, acting as intermediaries between the private peasantry and the public authorities, seeking for their members policies and practices which would benefit them and enable them to survive and perhaps to prosper in an increasingly competitive economy.[71] At the same time, as a syndicate grew in size and became more professional, there was also the possibility that it would become less personal. In practice, however, a syndicate relied upon a dense network of local agents, individuals who themselves acted as intermediaries between the syndicate as an organisation and its members as individuals.

By 1914 agricultural syndicates in France had more than a million members and there was a syndicate in at least one in six of the country's communes.[72] Taken together, livestock insurance societies and agricultural syndicates which developed during the seventy or so years prior to 1914 (and especially since the early 1880s) constituted an important set of institutions within rural society, producing and reproducing a sense of peasant identity, perhaps even of class consciousness. In theory, most syndicates – whatever their ideological basis – were founded on the social concept of a single *classe paysanne*, in which the interests of capital and of labour, of landowner and labourer, were united; in practice, most syndicates developed economic functions which meant that few included significant numbers of farm labourers within their memberships, so they could be interpreted as hardening class divisions within rural societies. In addition, as the geographical basis of syndicates moved away from the locality to the region and beyond, so syndicates were forging economic and ultimately social relations among members of the same class but of different *pays*. The geographical basis of fraternalism in rural societies was itself being transformed. Societies and syndicates ultimately contributed towards both the decline of a sense of community among the peasantry and the rise of a sense of class (or at least of solidarity – the peasant world remained too diverse and too fragmented for class consciousness to develop and to be institutionalised in the rural world to the same extent that it was in urban and industrial contexts).[73] As Tombs has pointed out, by 1914 'the agricultural interest' in France had come to be more comprehensively organised and more powerful than any other sectional group, a fact which was to have a significant impact not only upon the survival of the peasantry of France but also upon the development of the economy and society of Europe as a whole.[74]

The growth of voluntary associations in rural France during the nineteenth century was, of course, but one thread in a complex tapestry of economic, social and political changes, as the impacts of the Revolution of 1789 and of the Industrial Revolution merged and interacted. Material and ideological changes occurred on a scale never before encountered. Communications were vastly improved: better-surfaced and more roads and then new canals and especially railways and tramways proliferated, connecting hitherto remote and isolated rural regions and their peoples to each other and especially to Paris; postal and telegraphic services increased the volume and accelerated the flow of information; with increasing literacy and better schooling, provincial and national newspapers flourished as also did magazines, books and libraries; medical, veterinary and pharmaceutical services multiplied; political parties and trade unions were organised; the local municipality had extended powers locally as well as better contacts with regional and national authorities – the list of cultural changes witnessed in rural France during the nineteenth century was clearly extensive and the one just provided is by no means exhaustive. Some historians have found it convenient to describe that web of changes as the 'modernisation' of rural France, while some have seen voluntary associations as part of that process.[75] Curiously, the principal advocate of such an interpretation of the changing character of rural France during the nineteenth century, Eugen Weber, had virtually nothing to say about the role of voluntary associations in that process. Weber has argued that the peasantry were slow to politicise; and, in part because others have claimed that voluntary associations contributed to the politicisation of the peasantry,[76] Weber has preferred to see voluntary associations not as part of the process of 'modernisation', claiming instead that they were simply contributing to traditional sociability and interdependence.[77] Whether associations represented continuity or constituted discontinuity, the scale and scope of their development, spread and social significance in rural France during the nineteenth century – and well before Weber's watershed of 'modernisation' in 1870/80 – can no longer be either ignored or lightly dismissed. Associations were instruments of acculturation, contributing to the adoption of the French language in the countryside, to an enhanced awareness of time and space as cultural constructs rather than natural determinants, to the acceptance of new social disciplines and structures, and to the revelation of the many threads which tied the local and the national to each other. They also provided opportunities for individual emancipation.[78] But rather than regarding associations as part of some controversial 'modernising' process, or even as a sign of 'modernity',[79] it might be more judicious to view them more cautiously as a component of secularisation. They were part of what Bousseyroux has termed 'the silent cultural revolution' in attitudes towards social security,[80] a revolution in attitudes which produced a growing institutionalisation of fraternity.

To view voluntary associations in rural France during the nineteenth century as a component of secularisation is to accept them as part of the panoply of social, economic, political and intellectual changes related to the Revolution of 1789 and to the Industrial Revolution. The changes yoked to those two major sets of events and processes were tumultuous and far-reaching. For the peasantry, as no doubt for others, they provided both opportunities and threats. While it would be possible to produce a long and detailed inventory of those changes and of peasant responses to them, that is not necessary for present purposes. My concern here is principally with the impact of secularisation upon peasant *mentalités* and in particular upon their attitude towards risks.

During the nineteenth century peasants had to deal with an enlarged range of risks. For example, the growing use of matches added to the risk of fires; the extension of a market economy and of better communications involved much greater movement of people, of livestock and of plants, multiplying the risks from contagious diseases; and the introduction of new tools, equipment and machines in agriculture increased the risk of personal accident and injury to those working farms. But while the range of risks was growing, one of the major traditional means of managing them – reliance upon the Church – was declining. For as long as the Church provided virtually the only, certainly the dominant, ideology in rural France peasants turned to it for security, and in turn created their own versions of a popular religion which they could harness in their own struggle for survival.[81] In her fascinating account of the perception of time and history in a Burgundian village, Françoise Zonabend argued that having for decades had to face the problems of the moment, such small communities found in themselves latent powers of adaptation and survival. Because the future was uncertain and belonged to nobody, it was 'dangerous to talk about it and risky to trust in it. So in the village the future has no fixed shape or real definition, yet people think of it all the time, and they work and prepare for it.'[82]

Seeing peasants as survivors has been forcefully advocated by the humanist John Berger, who has made a multifaceted attempt to understand peasants, doing so as a painter, as an art critic, as a novelist, as a writer of film and television scripts, and as an anthropologist and geographer living in a French Alpine commune.[83] Berger sought to articulate peasant experience without romanticising it – and certainly without denigrating it. He has done so principally in *Pig Earth* (1979), a collection of stories about peasant experience – using that literary form because storytelling was itself such a vital part of the peasant experience. In the 'historical afterword' to his collection of stories, Berger presented his view of peasants as a class of survivors and peasant culture as a culture of survival. Moreover, he saw peasants as actively adopting survival strategies, not passively adopting a fatalistic conservatism. He argued that peasants lived in a world of scarcity not of surplus: they themselves would not have seen their relationship to the wider economy as that of 'surplus extraction', because the concept of 'surplus' assumes that the peasant's and his family's needs have been met and that the extraction takes place at the end of his labours. What economic theory might term a surplus was to the peasant, Berger pointed out, a 'preliminary obstacle', so that the peasant and his family 'had to survive the permanent handicap of having a "surplus" taken from them' as rents and taxes before and not after they had secured their own livelihood. In addition, they had to survive 'all the hazards of agriculture . . . and social, political and natural catastrophes'. Peasants lived in a world of scarcity and in a world of change: there was no 'constant given to their lives except the constant necessity of work. Around this work and its seasons they themselves create[d] rituals, routines, and habits in order to wrest some meaning and continuity from a cycle of remorseless change'; a cycle which was in part 'natural and in part the result of the ceaseless turning of the millstone of the economy within which they live'. Thus Berger argued that peasant conservatism has been much misunderstood: for him it was neither resistance to change nor a defence of privilege; instead, it represented 'a depository (a granary) of meaning preserved from lives and generations threatened by continual and inexorable change'. Furthermore, in Berger's view, the peasant's conception of a just and egalitarian society was not in terms of leisure and plenty, it was in terms of community: 'The peasant ideal of equality recognises a world of scarcity, and its promise is for mutual fraternal aid in struggling against this scarcity and a just sharing of what the work produces'.[84]

318   *Fraternity among the French peasantry*

Thus the peasant way of life had for centuries both constituted an unending struggle for survival and acknowledged a role of mutual aid in that contest. In the dramatically and radically altering circumstances of the nineteenth century the character of the peasants' struggle itself changed: forming and joining instrumental voluntary associations became a new way of continuing the struggle for survival, of defending and protecting peasants' interests. Secular insurance, in essence a mutualist mechanism, came to be increasingly appreciated as a way of affording such defence and protection, of managing the risks to property and to person which were the core of peasant experience. Such associations promised their members protection against hazards; they offered an institutionalised security for the individual; they provided financial insurance and social assurance. Peasants were accustomed to dealing with risks, and insurance societies came to be seen as a new way of handling them. Insurance is a rational method of managing risk. Theoretically, its purpose is to substitute certainty for uncertainty about the economic costs of particular and potentially disastrous events. Insurance is a system in which the insurer (in this case, an association), for a consideration, undertakes to render services or to reimburse an insured person if defined and accidental occurrences result in losses during a defined time period. The practice of insurance relies considerably on the 'law of large numbers', on the assumption that in large and homogenous populations it is possible to calculate the normal frequency of common events such as deaths, accidents and fires. The losses can therefore be predicted with acceptable accuracy and with an accuracy which increases as the size of the insured group expands. Insurance associations spread and reduce risk, and larger ones are able to do so more effectively than small ones. Gradually during the nineteenth century peasants in many parts of France came to appreciate the benefits of joining such associations, some of which (such as the livestock insurance societies and the mutual aid societies) provided a very direct form of insurance, while others (such as the anti-phylloxera and agricultural syndicates) offered a broader kind of protection.

No doubt many instrumental voluntary associations were created and acquired members as pragmatic responses to threats and risks, realising as it were by accident the aims of some of the utopians who had elaborated the theory of fraternalism and formulated the principles of socialism and solidarism. But the fraternalist spirit underpinned all such associations and in many cases their formation can be directly attributed to it. Fraternal associations represented the socialisation of risks. To a pragmatic mutual aid born of peasant experience was added a theoretical justification for the practice, and as a national political discourse filtered more and more into the day-to-day lives of those in rural localities, so increasingly the theoretical underpinnings and the political implications of voluntary associations came to be better appreciated.

Photograph 9.1 A *vigneron* and his horse at work on the valley side of the Val de Loire, in the commune of Saint-Gervais-la-Forêt, to the south of Blois
Source: 3 Fi 8031 photo, Archives Départementales de Loir-et-Cher

The most vulnerable among the peasantry were, other things being equal, those with small farms and it was certainly among the small farmers that the practice of creating and of joining voluntary associations was most marked during the nineteenth century. In addition, this study of Loir-et-Cher has echoed some others in noting the special role played by *vignerons* in the formation and activities of associations (photograph 9.1). Mutual aid societies of *vignerons* were especially numerous by the end of the nineteenth century in Burgundy and in the Val de Loire (in Blésois and Touraine).[85] Just why that group of small farmers so actively promoted *une vie associative* and so energetically expressed *un esprit d'association* in many (but not all) viticultural localities has not always been explained by those who have described it. The most commonly used argument is that employed by Goujon for the Mâconnais and Chalonnais and by Le Roy Ladurie for Europe in general, as already noted: that the large *bourgs* of viticultural districts were in effect small towns connected into the commercial and intellectual currents of the larger towns and their regions, and also enriched by the non-agricultural artisans and professionals living in those *bourgs*. Under this interpretation *vignerons* are almost seen as being 'honorary' townsmen – indeed, Goujon's admirable book is titled *Le vigneron citoyen*. But such an interpretation might itself be urban biased, representing the French peasantry yet again from an urban perspective as has so often been the case during the last 200 years. In any event, some mutual aid societies were established by *vignerons* in sparsely populated communes and themselves had only a dozen or so members.[86]

A number of other suggestions can be made. Eugène Fournière emphasised that association is a spatial expression of solidarity and Eric Hobsbawm noted that 'the greatest peasant movements all appear to be regional, or coalitions of regional movements'.[87] It might be, therefore, that development of a spirit of fraternalism among the peasantry in France in the nineteenth century was most favoured where there already existed a strong mutualist tradition. This seems to have been the case, for example, in the southern Jura and northern French Alps, where there were cheese-making co-operatives in the Middle Ages and a strong mutualist movement in the nineteenth century.[88] There was also a strong tradition of mutuality elsewhere in France in wine-producing regions, among whose *vignerons* it might also have been reinforced by a strong sense of place, of belonging to a particular *pays*, locality or commune. The type and quality of wines produced by *vignerons* varied over very short distances and promoted a fierce pride in the local product and thus, it could be argued, in the locality. The product and the locality became synonymous, encouraging a social bond, a collective sense of identity, among producers of the same wine within a given locality. Furthermore, viticulture was both labour intensive and environmentally sensitive, and it was often practised on very small, economically specialised, holdings: the *vignerons'* position was especially vulnerable and their struggle for survival particularly acute. This in turn might have encouraged a heightened awareness of the parallel need for co-operation among groups of them to enable individuals to continue to exist. *Vignerons* had a strong sense both of locality and of community. In addition, of course, their product – wine – itself encouraged sociability and conviviality. Whether for these reasons or for others, many *vignerons* embraced fraternalism in its varied forms, both before the Revolution in *confréries* and afterwards in voluntary associations. Some viticultural districts of France took fraternity a step or two further in the late nineteenth and early twentieth centuries, supporting radicalism and socialism in their various guises.[89] By the beginning of the twentieth century, when the phylloxera crisis had devastated French vineyards and in conjunction with foreign competition had forced many *vignerons* to turn to other products or to leave the land altogether, the idea and the practice of association had become deeply engrained in the countryside and as a movement was to continue to flourish until well after 1914.[90]

Notes

1 Peasants and peasantry in nineteenth-century France

1 Hubscher (1983).
2 Goldberg (1954).
3 Wright (1964) 3–6.
4 Cobban (1949) 429.
5 Zeldin (1973) 131–71.
6 Weber (1977).
7 Le Roy Ladurie (1994) 20.
8 Soboul (1956) 78.
9 Jones (1988) 1–29.
10 Agulhon (1976) 20.
11 Bloch (1956) 173.
12 Soboul (1956) 82; Jones (1988) 17–21.
13 Soboul (1956) 83.
14 Hoffman (1996) 21–34.
15 Soboul (1956) 82; Houssel (1976) 39–67; Jones (1988) 21–9.
16 Soboul (1956) 88–9; Agulhon (1976) 46, 87–8.
17 Jones (1988) 124. See also Dupeux (1976) 105–9.
18 Jones (1988) 266.
19 *Ibid.* 167–205; Bages, Druhle and Nevers (1976).
20 Agulhon (1976) 22.
21 Jones (1988) 212–15.
22 Agulhon (1976) 47–8.
23 Corbin (1987) 413.
24 Wright (1964) v.
25 McPhee (1992a) 15–55.
26 For example, Cholvy (1974); Hubscher (1983); Lehning (1995) 11–34.
27 Barral (1966); Rogers (1987).
28 Hubscher (1994) 183.
29 Young (1950). Quotation from 107. See also Gazley (1973); Sexauer (1976).
30 Agulhon (1976). Young is cited on 203, 264, 265, 271, 273; Marx is cited on 19.
31 Marx and Engels (1950) 159.
32 Duggett (1975) 170.
33 Marx (1926) 132–3.
34 *Ibid.* 133
35 Mitrany (1951); Duggett (1975).
36 Vernois (1962); Barral (1968) 128–36; Ponton (1977); Barral (1988).
37 Guiral *et al.* (1969) 71.
38 Ponton (1977) 63.
39 Lehning (1995) 16.
40 Ponton (1977) 65.
41 Barral (1988) 203.
42 In addition to the novels themselves, see the following critiques: Blanchard (1931); Guy (1952); Bastier (1978); Marcilhacy (1957); Goldin (1984); Prévost (1994).
43 Lagrave (1976).

44 Roger (1966).
45 Weber (1983) xv–xvi.
46 Guillamin (1912) 181.
47 Hélias (1975).
48 Lynch (1901) 177, 190.
49 Duclaux (1905) 120.
50 Prothero (1915) 94, 97.
51 Wolff (1900) 252 and 258.
52 Clark (1973a; 1973b).
53 Herbert (1970); Juneja (1987–8); Brettell and Brettell (1983); Grew (1988); Lacambre (1994).
54 Brettell and Brettell (1983) 39.
55 Chamboredon (1977).
56 Brettell and Brettell (1983) 59–61.
57 Lacambre (1994) 195–201.
58 Juneja (1987–8) 450–1; Brettell and Brettell (1983) 62–72; Marbot (1994); Zeyons (1992).
59 Juneja (1987–8) 446–8.
60 Brettell and Brettell (1983) 34.
61 *Ibid.* 36–7; Sturges, Weisberg, Bourrut-Cacouture and Fidell-Beaufort (1982).
62 Van Tilborgh (1989) 28.
63 Berger (1979a) 189–91; Berger (1980) 69–78; Arts Council of Great Britain (1976) 87–90; Pollock (1977).
64 Walker (1981); Van Tilborgh (1989); Clark (1973b); Arts Council of Great Britain (1978); Berger (1979a) 196–8; Berger (1980) 134–41.
65 Brettell and Brettell (1983) 75–117, quotation from 75–6.
66 *Ibid.* 119–35.
67 Arts Council of Great Britain (1976) 11.
68 Grew (1988) 205, 221.
69 Arts Council of Great Britain (1976) 14.
70 Arts Council of Great Britain (1978) 15–16.
71 Berger (1979a) 198.
72 Arts Council of Great Britain (1978) 17 and 119.
73 Juneja (1987–8) 454, 459.
74 *Ibid.* 463–4.
75 *Ibid.* 464–5.
76 Brettell and Brettell (1983) 84.
77 *Ibid.* 140.
78 Clout (1977); Heywood (1992) 44–50.
79 Cameron (1965); Clapham (1936); Kemp (1971).
80 Toutain (1961); Pautard (1965); Morineau (1971); Newell (1973); Grantham (1978).
81 Toutain (1992–3).
82 Overton (1996).
83 Cameron (1958); Marczewski (1963); Clough (1946); Hohenberg (1972); Kemp (1971).
84 Clout (1983); Price (1983); Grantham (1989b); Hubscher (1994).
85 Hoffman (1996); Bairoch (1988).
86 Pautard (1965).
87 Pautard (1965) 43–8; Clout (1980); Demonet (1990).
88 Bairoch (1988) 21; Van Zanden (1991).
89 Pinkney (1953); Goreux (1956); Hohenberg (1974); Price (1975); Ogden (1980); Heywood (1981).
90 McPhee (1981–2; 1989); Warner (1975).
91 Clout (1983); Magraw (1983) 106–19 and 318–26; Price (1983).
92 Ruttan (1978); Hubscher (1994) 192.
93 O'Brien and Keyder (1978).
94 Moulin (1988).
95 Hubscher (1994) 193.
96 Weber (1977).
97 Pinkney (1958).
98 Ogden (1980).
99 Agulhon (1978a); Weber (1983).
100 Margadant (1979a; 1984); Tilly (1979); McPhee (1981).
101 Tilly (1979) 39.
102 Margadant (1979a) 646.
103 Tilly (1979) 20–2 and 38–9.

104 McPhee (1981) 19–21.
105 Magraw (1983) 106–19 and 318–53.
106 Lehning (1995) 204–10.
107 Berenson (1987).
108 Agulhon (1970; 1973); Margadant (1979b); Huard (1982); Berenson (1984); Weber (1977; 1980; 1982); Jones (1985).
109 Judt (1979).
110 Burns (1984) 116.
111 See, for example, Rogers (1991).
112 See, for example, Baker (1992a).
113 See, for example, Singer (1983).
114 Soboul (1956) 91.
115 *Ibid.* 92–3.
116 *Ibid.* Quotations from 78, 89 and 93.
117 Rosenberg (1988).
118 Zeldin (1973) 138–43, 191–2.
119 Bercé (1974).
120 Agulhon (1976) 336.
121 *Ibid.* 32.
122 Berger (1972) 7–8.
123 Wright (1964) 18–19.
124 Weber (1977).
125 *Ibid.* 276–7.
126 Moulin (1988, translation 1991) 17–18, 75, 103–6.
127 Agulhon (1976) 157.
128 Baker (1992a).
129 Planhol (1994) 293–326; McPhee (1992b) 221–43.
130 Edelstein (1992).
131 Merriman (1979).
132 Gratton (1971); Frader (1991).
133 Tombs (1996) 215–17; 226–32; 234–7; 285–8; 296–8. Quotations from 285–6, 234 and 298.
134 Le Roy Ladurie (1979); Blum (1982).
135 Berger (1977); Berger (1979b) 200.
136 Agulhon, Désert and Specklin (1976) 78–9, 158; Devlin (1987); Houssel (1976) 200–2, 225, 389–90; Weber (1977) 340, 345, 354–5; Magraw (1983) 335; Price (1987)

50, 299.
137 Planhol (1994) 291–3; Ozouf and Ozouf (1997).
138 Warner (1975) 5.
139 For example, Lévi-Strauss and Mendras (1973–4); Barral (1969) 3–4; Margadant (1979a) 647–8; Soulet (1988).
140 Dupeux (1962); Dion (1934a).
141 Pinkney (1991).
142 Chartier (1996).

2 The theory and practice of fraternal association in nineteenth-century France

1 Sabine (1952) 469–70.
2 Agulhon (1986) 20.
3 Zeldin (1973) 433.
4 A. de Tocqueville *Democracy in America*. Cited in Smith and Freedman (1972) 34.
5 Tombs (1996) 61–87.
6 *Ibid.* 64.
7 *Ibid.* 66.
8 Zeldin (1973) 433; Charlton (1963) 65–95; Lovell (1992); Tombs (1996) 75–80.
9 Zeldin (1973) 433–8; Charlton (1963) 65–79.
10 Zeldin (1973) 438–49; Charlton (1963) 81–2.
11 Zeldin (1973) 449–52; Charlton (1963) 80.
12 M. Dommanget *et al.*, *Babeuf et les problèmes du babouvisme* (1963) 55 ff., cited in Zeldin (1973) 450–1.
13 Zeldin (1973) 452–4; Charlton (1963) 82–7.
14 Zeldin (1973) 466.
15 *Ibid.* 725–87.
16 *Ibid.* 640–82.
17 E. Durkheim, *De la division du travail social* (Paris 1893), cited in Rose (1954) 115.
18 Fournière (1907).
19 This section draws upon Debbasch and Bourdon (1990).

20 Agulhon (1981) 21.
21 Bruneau (1988) 32; Roudet (1988) 13–14.
22 Agulhon (1978a; 1981; 1988).
23 Agulhon (1976) 48 and 518–19; Agulhon (1977); Chaline (1986; 1995).
24 Agulhon (1976) 154.
25 Jardin and Tudesq (1983) 50–2.
26 *Ibid.* 181–2.
27 Agulhon (1983) 18–19.
28 *Ibid.* 88–9 and 91–2.
29 Agulhon (1981) 25–6; Barral (1968).
30 Augé-Laribé (1955) 259.
31 Désert and Specklin (1976) 417–18.
32 Zeldin (1973) 155.
33 *Ibid.* 169.
34 *Ibid.* 192.
35 Augé-Laribé (1955) 258.
36 See, for example, Barral (1968); Cleary (1989) 21–47.
37 Gratton (1971; 1972).
38 Zeldin (1973) 654.
39 *Ibid.* 660–4.
40 Agulhon (1978a).
41 Agulhon (1966; 1968; 1977; 1979).
42 Agulhon (1978a; 1981; 1986; 1988; 1990).
43 Agulhon (1986; 1990).
44 François (1986); Levasseur (1990).
45 Chamboredon (1984).
46 François and Reichardt (1987).
47 Poujol (1978).
48 Ion (1990).
49 Anderson (1971) 215–19.
50 Rose (1954) 50–71.
51 *Ibid.* 72–115.
52 Gallagher (1957).
53 Babchuk and Warriner (1965).
54 Sills (1968).
55 Rose (1954); Sills (1968); Smith and Freedman (1972).
56 Berelson and Steiner (1964) 364.
57 Smith and Freedman (1972) viii.
58 Levasseur (1990) 9–15.
59 Smith and Freedman (1972) 33–56.

60 *Ibid.* 57–85.
61 Debbasch and Bourdon (1990) 11–19.
62 For example: Barral (1969) 8; Margadant (1979a) 647–8; François and Reichardt (1987) 466; Soulet (1988).
63 François and Reichardt (1987) 470–2.
64 This present project was conceived in the early 1980s, before publication of the paper by François and Reichardt (1987): see, for example, Baker (1980; 1983; 1984a; 1986a and b).
65 Goujon (1981; 1993).

3 Loir-et-Cher during the nineteenth century: period, place and people

1 Young (1792).
2 Dion (1934b) 7–10, especially figure 1.
3 Angeville (1836).
4 Demangeon (1946).
5 Chartier (1996).
6 Ozouf-Marignier (1989).
7 Joanne (1869); Asfaux (1955); Dupeux (1962); Gobillon (1976); Brette (1980); Boucher (1984); Fauchon (1990); Baker (1995).
8 This section draws upon Faupin (1909); Paumier (1941); Dion (1934a); Dupeux (1962) 69–73. For contemporary descriptions of the *pays* of Loir-et-Cher, see, for example, Faupin (1909) and Pétigny (1863).
9 Dion (1934a).
10 Fénelon (1963).
11 Even a small river valley like that of the Cisse has been internally differentiated as possessing 'three faces': Robinet (1968).
12 Vidal de la Blache (1903) 8.
13 This section draws upon Gallon (1949); Dupeux (1962) 83–97 and

551–7; Vassort (1985) 289–91 and 309.

14 Dupeux (1962) 83–97.

15 *Ibid.* 557.

16 Bomer (1958 and 1959); Fénelon (1963); Baker (1968); Chaumier and Gillardot (1974); Dupeux (1962) 85–6.

17 Dupeux (1962) 88–92.

18 Gallon (1949) 44–5.

19 Vassort (1985) 309.

20 The proportion increased from 12.8 per cent of total marriages 1870–7 to 21.9 per cent 1919–24.

21 Dupeux (1962) 552–4.

22 Vassort (1985) 309.

23 Dupeux (1962) 555–7.

24 Clapham (1936) 53.

25 Dupeux (1962) 557.

26 *Ibid.* 557; data from the 1911 census from the *Annuaire de Loir-et-Cher 1912.*

27 Weber (1977) 493–4.

28 Dupeux (1962) 217–20; Sutton (1969; 1971); Campbell and Barr (1988); Fee and Prendergast (1990); Halstead and Hooper (1990); Pickup and Walters (1990); Sowerby and Stacey (1990); Anon. (1994).

29 Martin-Demézil (1980).

30 Sutton (1973); Anon. (1994) 79–80.

31 Dupeux (1962) 216; Dion (1934b).

32 Dupeux (1962) 213–17; *La Vie du Rail* no. 849, 3 June 1962; Crozet (1939); Rigollet (1966; 1990); Riffaud (1973 and 1978); Poitou (1985) 62–4; Steinmetz and Guellier (1987); Nickson and Martin (1988); Paul (n.d.); Baker (1992b); Anon. (1994) 55–69.

33 Rigollet (1990). The role of the *tramways à vapeurs* in rural Loir-et-Cher from their inception through to their closure in 1934 is being researched by Richard Turner under my supervision for a higher degree at the University of Cambridge. See Turner (forthcoming). Beck (1987; 1988) provides a useful study of the impact of a new railway line upon a rural area in the adjacent department of Indre-et-Loire.

34 Campbell and Barr (1988); Fee and Prendergast (1990); Halstead and Hooper (1990); Pickup and Walters (1990); Sowerby and Stacey (1990).

35 Vassort (1977; 1978).

36 For general surveys of schooling in nineteenth-century France, see: Weber (1977) 303–38; Price (1987) 308–20.

37 This section draws upon Dupeux (1962) 160–6 and 424–6; Martin-Demézil (1980); Olivier (1962); Chadwick (1988); Clark (1988).

38 Bazin (1988).

39 Olivier (1962) 18.

40 *Ibid.* 31 and 94.

41 Gobillon (1983; 1991); Gaillot, Gaillot and Pisani (1981a and 1981b).

42 Olivier (1962) 28 and 79; Delecluse (1986).

43 Bourdin (1980); Baker (1992a); Gaillot, Galliot and Pisani (1981a and 1981b).

44 In addition to works already cited in relation to schooling, see: Bourdin (1980); Béraud (1981); Chollet (1981); Gaillot (1981); Laure (1981); Paumier (1969).

45 Vassort (1977; 1978) Furet and Ozouf (1977).

46 Dupuy (1974); Baker (1990b and 1992b).

47 For a table of conscripts' literacy by cantons between 1827 and 1890, see Baker (1992b) 211.

48 Weber (1977) 292–302; Bozon (1981 and 1987).

49 Weber (1977) 302.

50 Tooth and Shepherd (1987); Barnet and Edwards (1992); Farmer (1988). The smaller stature of conscripts in the Sologne in the 1820s was noted a few decades later: Pétigny (1863); Poitou (1985) 197–200 and Beauchamp (1990) provide more recent commentaries upon the physical infirmities and general physical condition of the inhabitants of the Sologne in the nineteenth century.

51 Baker (1998).

52 Vassort (1985) 318; Doucet (1994); Diot (1976); Martin (1984) 13; Lobier (1970).

53 As indicated in the text, this section draws substantially upon Dupeux (1962) and Vassort (1985), but see also Poitou (1985) on the Sologne and Vassort (1995) on Vendômois.

54 Dupeux (1956).

55 Sutton (1969; 1971); Chaumier and Gillardot (1974); Poitou (1985) 7–168.

56 See chapter 7 below.

57 Touvet (1955).

58 Muller (1974 and 1976); Gobillon (1975) 339–69 and (1985); Poitou (1985) 241–9; Bizeau and Notter (1996).

59 Le Meur (1976).

60 Dupeux (1962) 170–4.

61 Muller (1974 and 1976).

62 Marcilhacy (1962); Gobillon (1996).

63 Notter (1981).

64 Boulard (1982) 219; Martin (1984) 22–3.

65 AD 201 M 110.

66 Anon. (n.d); Belton (1888); Lelièvre (1913).

67 This section is based upon Dupeux (1962) 319–407, 449–89 and 502–69; Dupuy (1961 and 1980); Martin (1984) 85–226; Vassort

(1985) 297–8 and 314–19.

68 Ponteil (1966) 30–3, 156–63, 282–5 and 378–81; Gobillon (1975) 370–92 and (1976); Brette (1980); Singer (1983) 37–107.

69 Baker (1992a).

70 Silver (1973; 1980–1). The quotation is from Silver (1980–1) 290.

71 Jones (1987); Packham (1987).

72 Vassort (1985) 294; Muller (1974).

73 Vivier (1980).

74 Vassort (1985) 271.

75 *Ibid.* 298.

76 Gobillion (1989).

77 Motheron (1985).

78 Chaline (1986).

79 On the Society of Agriculture, see Vassort (1985) 284, 287, 292 and 296; on the *comice agricole* of Vendôme, see Silver (1973) 33–6.

80 Vassort (1985) 304.

81 Dion (1934a) 705–11.

82 Touvet (1955).

83 Fénéant (1986; 1990).

84 Prudhomme (1982; 1984).

85 Gobillon (1985) 18. See also Bouyssou (1992).

86 Martin (1984) 29–49.

87 Muller (1974); Notter (1981).

88 Prudhomme (1985).

4 Insurance societies

1 See section on 'Religion' in chapter 3 above.

2 AD M 267 cote 12.

3 See chapter 8 below.

4 AD X Sociétés d'assurances diverses.

5 AD X Sociétés d'assurances diverses: statistiques, correspondance 1902–1914.

6 AD X Compagnies d'assurances 1839–1859.

7 This section draws upon the following records in the Archives Départementales de Loir-et-Cher:

7M 360 Agriculture: Assurances
mutuelles agricoles. Etats de situation, subventions, correspondance
1899–1925; a collection of files
each with the general title
Agriculture: Assurances mutuelles
agricoles. Dossiers des Sociétés and
then listing a specific set of
communes 7M 361 Courmemin –
Mareuil 1900–1914; 7M 362
Chailles – Cour-sur-Loire
1900–1914; 7M 363 Authon –
Chambon 1900–1914; 7M 364
Naveil – Saint Sulpice 1900–1914;
7M 365 Marolles – Muides
1900–1914; 7M 366 Sargé – Vineuil
1900–1914; also 7M 367
Agriculture: Assurances mutuelles
agricoles. Circulaires, instructions,
statuts 1901–1923; 7M 368
Agriculture: Assurance mutuelles
agricoles. Subventions et
statistiques 1923–1940. Only in
cases which particularly require it
is the specific *liasse* noted
hereafter: if no special note is
made, then the statement is based
upon sources contained within the
liasses listed here.

8 Dupeux (1962) 81.
9 *Ibid.* 81.
10 AN F[11] 2705 Enquête agricole de
1862.
11 *Ibid.*
12 *Ibid.* (There is no response to the
question about the normal method
of ploughing recorded in relation
to the two cantons of Ouzouer-le-
Marché and Romorantin: it is
highly unlikely that cattle were
being used for ploughing at that
time in the former, but quite
possible in the the latter.)
13 Livestock costs have been derived
from valuations in the dossiers of
livestock insurance associations;
for a family of five's living expenses

1846–1914, see Dupeux (1962) 290.
14 Augé-Laribé (1950); Barral (1968)
67–103.
15 *L'Agriculture Pratique* 2 (19
February 1895) 114.
16 *L'Agriculture Pratique* 2 (19
February 1895) 117 and (26
February 1895) 124–5.
17 *L'Agriculture Pratique* 2 (19 March
1895) 146.
18 *L'Agriculture Pratique* 2 (2 April
1895) 160.
19 *L'Agriculture Pratique* 2 (9 April
1895) 170.
20 AD 7M 367.
21 AD 7M 360.
22 Vezin (1910b) 563.
23 *L'Agriculture Pratique* (21
December 1913, 4 January and 18
January 1914).
24 De Rocquigny (1903).

5 Mutual aid societies

1 Mitchell (1991) 172.
2 Gueslin (1987) 193–5.
3 Gibaud (1986) 14.
4 Birck and Dreyfus (1988) 10.
5 Dreyfus (1988) 11–13.
6 Goldman (1983) 69.
7 Gueslin (1987) 116–212.
8 For example, Bennet (1962) on the
Isère; Birck and Dreyfus (1988) on
the four departments of Meurthe-
et-Moselle, Meuse, Moselle and
Vosges; Bousseyroux (1990) on
Puy-de-Dôme; Chabrol-Chardon
(1981) on the Médoc region;
Chalvignac (1985) on the *pays* of
Fougères in Ille-et-Vilaine;
Demoulin (1993) on Allier;
Devance (1984) on the Côte-d'Or;
Fortin (1950) on Pas-de-Calais;
Guimbretière (1985) on the
Vendée; Marchal (1979) on
Meurthe-et-Moselle; Navarro
(1983) on the Tarn; Valette (1979)

on Gironde.

9 Sewell (1984); Sheridan (1984); Gouda (1996) 255.

10 Hayward (1959).

11 This summary of the orthodox view of mutual aid societies is based upon the following: Saint-Jours (1982) 104–5, 115–16; Sheridan (1984); Valette (1984); Gueslin (1987) 115–212; Price (1987) 50, 65, 68, 205–6, 218, 214–47, 271, 300; Mitchell (1991); McPhee (1992b) 134, 140, 182, 186, 204, 212.

12 Mitchell (1991) 173.

13 Laurent (1865) *tome* I, 426 and Leroy-Beaulieu (1914) 381, cited in Gueslin (1987) 196; Goldman (1983) 73–4.

14 Gueslin (1987) 125 and 196.

15 Goldman (1983) 69, 74–7.

16 Augulhon, Désert and Specklin (1976); Weber (1977); Magraw (1983) 318–53; Moulin (1988); Lehning (1995).

17 Goldman (1983) 70–1; in Hubbard (1852) an appendix of thirty-eight pages sets out the basic principles for undertaking the statistical calculations needed for the proper functioning of a mutual aid society.

18 This discussion of the legal context draws particularly upon Saint-Jours (1982), Goldman (1983), Weiss (1983–4) and Gueslin (1987).

19 I am, of course, especially grateful to the staff of the Archives Départementales de Loir-et-Cher for allowing me access to the uncatalogued materials on mutual aid societies. A list of those *liasses* consulted is included in the bibliography. In the notes for this chapter, I have only provided a documentary reference for an individual society when using material other than that to be found in the relevant 'alphabetical' *liasse* for the commune concerned.

20 Gueslin (1987) 195.

21 *Ibid.* 195.

22 AD 11J 31.

23 AD 4M 202.

24 Contres had a population of 2,575 in 1851. The 49 members included 6 stone-cutters, 4 carpenters, 4 masons, 4 clog-makers, 3 bakers, 3 metal-workers, 2 inn-keepers, 2 rope-makers, 2 roofers, 2 long-sawmen, 2 blacksmiths, 2 coopers, and 2 weavers, and a miscellany of other individuals, including a doctor, a teacher, a harness-maker and a wig-maker.

25 AN F^{12} 5372. The director of the farm school at that time was Edouard Malingié (Baker 1996).

26 Muller (1974) 'La Seconde République' 39–41.

27 AD X Sociétés de secours mutuels: affaires diverses de 1853 à 1870 and Sociétés de secours mutuels: états récapitulatifs 1905–1907–1908.

28 AN F^{12} 5372.

29 Chissay (1853); Mennetou-sur-Cher (1858); Saint-Romain (1862); Mesland (1867); Onzain (1867); Saint-Julien-de-Chedon (1869); Chaumont-sur-Loire (1872); Onzain (1878); Cheverny (1880); Les Montils (1886); Rilly-sur-Loire (1886); Ouchamps (1899).

30 Marchenoir (1863); Saint-Georges-sur-Cher (1866); Herbault (1872); Chaumont-sur-Tharonne (1873); Cormeray (1873); Couture (1900).

31 AD 4M 203 Associations dissoutes 1904–1946.

32 For the society at Romorantin, see AN F^{12} 5372.

33 Data on population are for 1866 and 1901, the censal years nearest to 1868 and 1902.

34 AD X Sociétés de secours mutuels: états récapitulatifs 1905–1907–1908.
35 These percentage calcuations have had to ignore the small number of boys and girls who are listed collectively, not separately.
36 In 1902 there were 85 men, 98 women, and 69 children; in 1909, 68 men, 80 women, and 82 children.
37 AD 4M 202.
38 Morée; Pezou; Saint-Martin; Souday; Villedieu.
39 Dauzé; Selommes.
40 Morée; Sougé; Pezou.
41 Bourray.
42 Morée.
43 Selommes; Villedieu.
44 See above, p. 162.
45 AD 4M 202.
46 AD X Sociétés de secours mutuels: dossier sociétés dissoutes 1852–1924.
47 *Ibid.*
48 The two were the Société Vendômoise dite des ouvriers and the Société Vendômoise d'assistance pour les femmes.
49 The four were the societies at Chissay, Fougères and Feings, Beauchêne, and Saint-Aignan.
50 The three societies were the Société Générale de secours mutuels for men of all occupations at Blois, the Vendômoise for men of all occupations at Vendôme, and the Société d'assistance mutuelle for women at Vendôme.
51 The three societies were those at Beauchêne (with only 20 members), at Onzain (with 96 members) and the society of St Joseph in the Blois suburb of Vienne (with 428 members).
52 This was explicitly the case with societies at Mesland, Mennetou-

sur-Cher and Saint-Romain and was implicitly the case with seven other societies for whom the 1868 survey does not list a daily rate of benefit in cash terms.
53 The starting age at which pensions would be paid varied from society to society, between fifty and sixty-five years; eligibility for a pension also varied, requiring between ten and twenty years of membership.
54 One in four (27%) of the sickness societies and two in five (45%) of the sickness and pension societies actually paid out funeral costs in 1907.
55 The societies were those at Monteaux (des Vignerons, with 21 members), Santenay (with 38 members), Villemardy (38), Chaumont-sur-Loire (44), La Chapelle-Vendômoise (La Fraternelle, 49), Cour-Cheverny (des Vignerons, 57), Ouzouer-le-Marché (94), Saint-Dyé (104), Contres (Les Amis Réunis, 135) and Theillay (235).
56 AD X Sociétés de secours mutuels: dossier sociétés dissoutes 1852–1924.
57 *Ibid.* Theillay (1888).
58 *Ibid.*
59 In 1868 there were societies with fifty or fewer ordinary members in the following communes (with the number of ordinary members in 1868 and the communes' populations in 1866 shown in brackets): Chissay (17; 1,055); Beauchêne (20; 424); Chitenay (20; 883); Droué (21; 1,037); Saint-Romain-sur-Cher (25; 1,414); Mesland (31; 688); Maslives (32; 508); Montlivault (35; 976); Marchenoir (37; 720); Montoire (38; 3,193); Chouzy-sur-Cisse (46; 1,465); Mondoubleau (46; 1,585); Suèvres (46; 1,991);

Pontlevoy (48; 2,436); Pouillé (50; 797); and Saint-Georges-sur-Cher (50; 2,345). In 1907 there were societies with fifty or fewer ordinary members in the following communes (with the number of ordinary members in 1907 and the communes' populations in 1906 shown in brackets): Huisseau-en-Beauce (16; 385); Cour-sur-Loire (25; 319); Courbouzon (28; 479); Couffi (14; 762); Pouillé (14; 762); Mesland (14; 800); Saint-Romain-sur-Cher (17; 1,608); Saint-Lubin (18; 518); Cellettes (20; 1,066); Choussy (21; 355); Chissay (22; 1,036); Faverolles (23; 728); La Chaussée-Saint-Victor (25; 750); Seur (27; 270); Muides (28; 566); Avaray (31; 606); Cormeray (31; 726); Neung-sur-Beuvron (31; 1,450); Cormeray (32; 726); Yvoy-le-Charron (32; 810); Thoré (32; 863); Chissay (32; 1,036); Villetrun (33; 283); Pezou (33; 1,030); Saint-Viâtre (33; 1,906); Onzain (35; 2,321); Saint-Secondin (36; 798); Fontaines-en-Sologne (36; 939); Saint-Loup (38; 335); Villemardy (38; 382); Santenay (38; 620); Saint-Amand (38; 817); Mont-Près-Chambord (39; 1,478); Lamotte-Beuvron (39; 2,333); Séris (41; 560); Châtillon-sur-Cher (41; 1,765); Choussy (42; 355); Saint-Gervais (42; 578); Vouzon (42; 1,564); Cellettes (43; 1,066); Pierrefitte (43; 1,601); Nouan-le-Fuzelier (45; 2,005); Lamotte-Beuvron (47; 2,333); Faverolles (48; 728); Choue (48; 981); La Chapelle-Vendômoise (49; 467); and Thézée (50; 1,397). Not included here in the 1907 list are the six societies in Blois and the three societies in Vendôme each with fifty or fewer members:

admission to them was generally restricted by occupation.

60 Such societies were established at Chailles (1887), Chaumont-sur-Loire (1872), Cheverny (1880), Chissay (1853), Mennetou-sur-Cher (1857), Mesland (1867), Les Montils (1886), Montlivault (1862), Onzain (1864), Onzain (1876), Ouchamps (1899), Saint-Georges-sur-Cher (1875), Saint-Julien-de-Chedon (189) and possibly in some other communes.

61 AD X Sociétés de secours mutuels: dossiers sociétés dissoutes 1852–1924.

62 *Ibid.*

63 AN F^{12} 5372.

64 *Ibid.*; AD X Sociétés de secours mutuels: dossiers sociétés dissoutes 1852–1924.

65 AD X Sociétés de secours mutuels: dossiers sociétés dissoutes 1852–1924.

66 AN F^{12} 5372.

67 *Ibid.*

68 AD X Sociétés de secours mutuels: dossiers sociétés dissoutes 1852–1924.

69 Mitchell (1991) 180 and 185.

70 *Ibid.* 186.

6 Fire-fighting corps

1 Weber (1977) 16–17; Thomas (1973) 17–20.

2 Prudhomme (1985) 16–17.

3 AD 6M 1093, 1087, 1097 and 1100.

4 The best and most recent account of the history of fire brigades in France is that of Lussier (1987), a detailed study of *corps de sapeurs-pompiers* of the department of Seine-et-Marne but set within the national context.

5 This chapter is based on

uncatalogued collections of correspondence, circulars, reports and copies of the rules of *corps de sapeurs-pompiers* in AD Série R. The *liasses* consulted all carry the general title *Sapeurs-Pompiers*, with individual *liasses* then being titled as follows: Organisation des services d'incendie 1807–1842; Achat de pompes 1841–1851; Pompes à incendie 1857–1864; Armements 1854–1866; Corps dont la réorganisation n'a pu avoir lieu 1860–1910; Caisses communales de secours 1864–1869; Instructions, enquêtes, effectifs, matériel, accidents, subventions 1876–1897; Armements 1852–1872. *Liasses* relating to the fire brigades of particular communes are titled as follows: Averdon-Chitenay 1875–1902; Autainville-Brévainville 1876–1918; Cellettes-Chapelle Vendômoise 1862–1913; Chaumont-sur-Loire-Conan 1876–1914; Cormeray-Fréteval 1860–1914; Cormeray-Menars 1852–1901; Marcilly-en-Gault-Moisy 1876–1914; Mulsans-Oucques 1876–1919; Mer-Orchaise 1864–1900; Molineuf-Muides 1865–1920; Ouzouer-le-Doyen-Saint Amand 1870–1919; Sambin-Suèvres 1858–1914; Saint Léonard-Vineuil 1867–1900; Ouchamps-St Laurent-des-Eaux 1852–1900; Saint Bohaire-Salbris 1877–1920; Villechauve-Yvoy-le-Marron 1860–1920; Talcy-Villebarou 1876–1920. Unless otherwise noted, statements in the text referring to particular brigades are based on documents located in the relevant *liasse* for the commune concerned.

6   Dion (1934a); Dupeux (1962);

Poitou (1985). For a brief newspaper article on the origins of the *corps de sapeurs-pompiers* in Loir-et-Cher, see *Le Chercheur* (1932).

7   Prudhomme (1985). For a comparison with Loir-et-Cher, see the study by Lussier (1987) of the fire brigades of Seine-et-Marne in the nineteenth century.

8   In this analysis, a brigade is regarded as having been 'established' by a given date if it then had in existence a recognised group of officers and men: in the absence of such information, a brigade is regarded as having been 'established' by a given date date if it had by then acquired a pump or had been approved by the prefect or had been awarded a subsidy by the department. In determining the date of 'establishment' of a brigade, no use has been made of the date of the decision by a commune council either to create a brigade or to purchase a pump, because there was often a considerable time lag between such a decision and a brigade's becoming operational.

9   AD 6M 1087, 1093, 1097 and 1100.

10   AD R Sapeurs-pompiers: organisation de services d'incendie 1807–1842 and Sapeurs-pompiers: achats de pompes, correspondance diverses 1841–1851; Prudhomme (1985) 27–8.

11   AD R Sapeurs-pompiers: organisation de services d'incendie 1807–1842 and Sapeurs-pompiers: achats de pompes, correspondance diverses 1841–1851; Prudhomme (1985) 28.

12   AD R Sapeurs-pompiers: achats de pompes, correspondance diverses 1841–1851; Prudhomme (1985) 58.

13 AD R Sapeurs-pompiers: caisses communales de secours, 1864–1869; Prudhomme (1965) 39.

14 AD R Sapeurs-pompiers: achats de pompes, correspondance diverses, 1841–1851; Prudhomme (1985) 47.

15 AD R Sapeurs-pompiers: achats de pompes, correspondance diverses 1841–1851 and Sapeurs-pompiers: armement 1852–1872; AD Dépôts des Communes no. 543, Registres des délibérations du conseil municipal de Marchenoir; Prudhomme (1985) 44.

16 AD R Sapeurs-pompiers: achats de pompes, correspondance 1841–1851; Prudhomme (1985) 47.

17 AD R Sapeurs-pompiers: armement 1852–1872.

18 AD R Sapeurs-pompiers: organisation de services d'incendie 1807–1842.

19 This was the case at Chambord (1852); La Chapelle Vendômoise (1895); Coulanges (1865); Courbouzon (1842); Dhuizon (1860); Les Montils (1847); Monthou-sur-Bièvre (1893); Prénouvellon (1846); Seur (1805); and Talcy (1851).

20 For those at Avaray (1852 and 1860); La Chapelle-Saint-Martin (1867); Courbouzon (1852); Cour-sur-Loire (1869); Lestiou (1864); Maves (1854); Mer (1852); Montoire-sur-le-Loir (1866); Mulsans (1862); Romorantin (1805); and Vendôme (1855).

21 Lussier (1987) 77–9.

22 AD R Sapeurs-pompiers: organisation de services d'incendie 1807–1842.

23 AD R Sapeurs-pompiers: 1852–1872 F-M.

24 Copies of the rules exist for the brigades at Romorantin (1806);

Saint-Dyé-sur-Loire (*c.* 1860); Montoire (1863); and Mulsans (1862).

25 'Strict dress uniform is compulsory for all men in the Company; it comprises for Firemen: a uniform jacket with epaulettes, a helmet with crest and plume, a black collar, a belt with cartridge pouch, sabre bayonet and bayonet sheath, blue trousers, boots or shoes, white gloves. For the Rescuers: a busby with plume, bib, apron, and axe. Collars, belts, sabre bayonets, jackets, trousers and footwear the same as for Firemen. Everything very clean and neat.'

26 AD R Sapeurs-pompiers: caisses communales de secours, 1864–1869.

27 AD R Sapeurs-pompiers 1852–1872 F-M.

28 AD R Sapeurs-pompiers: armements 1854–1866.

29 Commune councils frequently gave *corps de sapeurs-pompiers* grants towards the costs of such festivities: see, for example, the 1882 and 1894 entries in the Registres des délibérations du conseil municipal of Mesland (AD Dépôts des communes no. 510). It must also be recognised that on such such occasions latent social tensions and conflicts also surfaced, as at Chitenay in 1884: Prudhomme (1985) 42.

30 AD R Sapeurs-pompiers: caisses communales de secours, 1864–1869.

31 Prudhomme (1985) 42–3.

32 The Franco-Prussian War disrupted the functioning of the fire brigades of at least the following communes: La Chapelle Saint-Martin, Josnes, Landes-le-Gaulois, Marchenoir, Noyers, Oucques,

Saint-Bohaire, Seur, and Villerbon.
33 AD R Sapeurs-pompiers: organisation de services d'incendie
   1807–1842.
34 *Ibid.*
35 Such *caisses* were established, for
   example, in 1863 for the brigade at
   Vendôme and 1867 for the brigade
   at Blois: AD R Sapeurs-pompiers:
   caisses communales de secours,
   1864–1869.
36 Lussier (1987) 147–50.
37 AD R Sapeurs-pompiers: instructions, enquêtes, effectifs, matériel,
   accidents, subventions 1876–1897.
38 See below, chapter 8; Baker
   (1986c).
39 Baker (1992a) 282.

7 Anti-phylloxera syndicates

1 'A voluntary association has been
   established by the undersigned for
   the purpose of protecting against
   the spread of phylloxera their vines
   located in the commune of Mer
   and its immediate neighbourhood.
   It will monitor and treat vines in a
   circle, centred on Mer, with a
   radius of 5 kilometres.'
2 Ordish (1972); Stevenson (1980);
   Dion (1959); Unwin (1991).
3 Smart (1994) 725.
4 Barral (1968) 75–7.
5 Augé-Laribé (1907): cited in Barral
   (1968) 75.
6 Barral (1968) 76 and 121; Garrier
   (1973) vol. I, 417–27; Garrier
   (1989); Smart (1994) 725–7.
7 AD 7M Agriculture: Comités
   d'études et de vigilance contre le
   phylloxéra 1874–1894.
8 Dupeux (1962) 205.
9 This chapter is based in part upon
   published annual reports of the
   prefect – AD N Conseil Général:
   rapports du préfet (1877–1906) –

but principally upon unpublished
documents in AD 7M Agriculture
as follows: 175 Comités d'études et
de vigilance contre le phylloxéra
1874–1894; 176 Phylloxéra 1874–
1889; 177 Phylloxéra 1877–1891;
178 Phylloxéra 1880–1890; 179
Phylloxéra 1880–1890; 180
Syndicats anti-phylloxériques:
délibérations, comptes de gestion
1883–1896; 181 Syndicats anti-
phylloxériques: délibérations,
comptes de gestion A–J
1884–1898; 182 Syndicats anti-
phylloxériques: délibérations,
comptes de gestion N–S
1884–1892; 183 Syndicats anti-
phylloxériques: délibérations,
comptes de gestion S 1884–1889;
184 Syndicats anti-phylloxériques:
délibérations, comptes de gestion
T–V 1884–1889; 185 Syndicats
anti-phylloxériques: délibérations,
comptes de gestion M 1884–1892;
186 Syndicats anti-phylloxériques
1884–1892; 187 Syndicats anti-
phylloxériques 1884–1892; 188
Phylloxéra: syndicats anti-
phylloxériques, états phylloxériques
1885–1901; 191 Syndicats anti-
phylloxériques 1890–1902; 192
Syndicats anti-phylloxériques,
comptes de gestion 1890–1897;
194 Enseignement agricole,
phylloxéra, rapports sur les
services, personnel, 1893–1900.
Hereafter, individual files are not
cited unless it seems to be
particularly important to do so.
10 Dupeux (1962) 78–9 and 205; Dion
   (1934a) 655–61; AD 7M 176
   Agriculture: Phylloxéra 1874–1889
   – manuscript notes by M. Tanviray,
   the department's professor of
   agriculture, on the *Service du
   Phylloxéra*; AD N Conseil
   Général: rapport du préfet et

procès-verbaux des délibérations 1906.

11 AD N Conseil Général: rapport du préfet et procès-verbaux des délibérations 1877: the prefect's report of 24 December 1877.

12 AD N Conseil Général: rapport du préfet et procès-verbaux des délibérations 1881–1884.

13 AD 7M 239.

14 Letter of 28 September 1888 from Trouard-Riolle to the prefect: AD 7M 176.

15 Syndicates have been defined as being 'in operation' in those calendar years during which membership subscriptions were collected and/or vines were treated.

16 Dupeux (1962) 205; Ordish (1972) 90–3.

17 Clapham (1936) 184.

18 Dupeux (1962) 472–541.

8 Agricultural associations

1 Augé-Laribé (1955) 258; Augé-Laribé (1926).

2 Cleary (1989) 35.

3 Berthonneau (1928) 59. Berthonneau was director of the principal agricultural syndicate in Loir-et-Cher.

4 Augé-Laribé (1955) 260.

5 For general surveys, see: Augé-Laribé (1955) 258–82; Barral (1968) 105–40; Barral (1969) 8–12; Cleary (1989) 33–47. For exemplary studies of agricultural syndicates, see: Anon. (1921); Cleary (1979; 1982; 1987); Faucon (1966); Garrier (1969); Hubscher (1979) 614–35; Lesueur (1905); Leveau (1969); Mesliand (1969); Rebiffe (1948); Rinaudo (1980).

6 Augé-Laribé (1955) 258–9.

7 *Ibid.* 259; Hubscher (1979)

191–251; Chaline (1986); Roux (1986); Mesliand (1989) 207–11.

8 Flaubert (1950) 144–67; Goldin (1984).

9 Trouillard (1908); Gobillon (1975) 101–2.

10 The remainder of this section on the Society of Agriculture and the *comices agricoles* is based principally upon the following dossiers: AD 7M 246, 249, 264, 266 and 267. Hereafter individual documents are only referenced when their significance requires it. See also Trobert (1949) and Cuevas-Delétang (1991).

11 *Annuaire de Loir-et-Cher* (1819) 122–5.

12 Baker (1996).

13 For example, in Pas-de-Calais and in Vaucluse: Hubscher (1979) 614 and Mesliand (1989) 211.

14 For example, in Gironde: Roux (1986).

15 Augé-Laribé (1926) 107.

16 Dion (1934a) 705–11.

17 Nivault (1950).

18 Braux (1954) 62.

19 Dupeux (1962) 145, 211, 213.

20 Vassort (1985) 284, 287, 292, 296, 304.

21 This section on agricultural syndicates and the following section on the Syndicat des Agriculteurs de Loir-et-Cher draw principally upon the following unpublished sources: AD 7M 56, 227, 228, 239, 244, 245, 247, 248 and 265; 10M 55, 56, 57, 58 and 59. Hereafter, individual sources are only noted if it is considered especially necessary to do so.

22 Dupeux (1962) 181–289.

23 Vezin (1901); Anon. (1909) 11–12.

24 Tanviray and his successors as professors of agriculture submitted to

the prefect annual reports on their work and on the condition of agriculture in the department. Published reports have been consulted in this connection for the period 1880–1914: AD Série N Conseil Général: rapport du préfet et procès-verbaux des délibérations.

25 Berthonneau (1905) 3–7.
26 Wolff (1900). In preparing his paper, Wolff visited Blois and discussed with Jules Tanviray and his colleagues their ideas about agricultural syndicalism: 253.
27 AD 201 M 145.
28 Cleary (1982; 1987).
29 Prugnaud (1963) 82–99; Barral (1968) 105–13.
30 Their peak memberships ranged from 25 to 340. The membership of SALC has been excluded from the calculation of the average size of syndicates.
31 Anon. (1950); Lévigne (1928).
32 This section draws upon Baker (1986a).
33 Hitier (1930).
34 Anon. (1920) 13; AD 7M 244.
35 Berthonneau (1905) 5.
36 Dupeux (1962) 183–8.
37 Baker (1986a) 53–5.
38 Baker (1980).
39 Dupeux (1962) 78–81 and 97–119; Baker (1983) 206.
40 Dion (1934a) 706.
41 For details, see Baker (1986a) 54–8.
42 Negrin (1983) 62–3.
43 Berthonneau (1903) and (1905) 29–33.
44 *Le Progrès de Loir-et-Cher* (16 July 1909).
45 Vezin (1910).
46 AD 7M 56 and 367; AD X Sociétés d'assurances diverses.
47 AD 7S 12 and Z 448; AN F[10] Agriculture 4651–5.

9 Synthesis: conclusions, comparisons and conjectures

1   Dupeux (1962) 209.
2   Cited in Trobert (1949) 49.
3   AD 7M 267 cote 9.
4   Dupeux (1962) 209.
5   Vassort (1995) 450–64.
6   See chapter 3 above; Dupeux (1962).
7   Berger (1979b) 195–213; quotation from 203.
8   Dupeux (1962) 213.
9   Baker (1986b).
10 Data on area under vines from AD Série P Evaluation des propriétés foncières: contributions directes, tableaux des communes de l'arrondissement de Blois-Est 1879; de Blois-Ouest 1879; de Romorantin 1879; de Vendôme 1879. Data on farm size in 1862 from AN F[11] 2705 Enquête agricole de 1862. Data on population in 1881 from AD 201M 128. Data on distances from Blois from the *Annuaire de Loir-et-Cher* for 1885.
11 AD 3M 480, 481 and 482.
12 Baker (1986c).
13 McPhee (1992b) 245–63.
14 Léonard (1955) 59; Joutard (1977) 270 cited in Kofman (1980) 8.
15 Hubbard (1852).
16 Nouel (1888); Edeine (1974) vol. II 783–95; Gobillon (1985) 18; Notter (1981); Bouyssou (1991 and 1992).
17 Dion (1934a) 317–413.
18 Lemaire (1980). See also Baker (1984a).
19 Goujon (1981; 1993).
20 Dion (1934a) 705–11.
21 Shaffer (1982).
22 Dallas (1982) 270–9 and 325–6.
23 Vercherand (1989).
24 Marais (1986); Baker (1986c);

Cabaret and Cabaret (1995)
176–211.

25 Augé-Laribé (1926); Barral (1968);
Nicolas (1988); Cleary (1989).

26 Birck and Dreyfus (1988);
Bousseyroux (1990); Chabrol-
Chardon (1981); Chalvignac
(1985); Demoulin (1993); Devance
(1984); Fortin (1950); Goldman
(1983); Guimbretière (1985);
Marchal (1979); Navarro (1983);
Valette (1979).

27 I am not aware of a book-length
treatment of livestock insurance
societies or of anti-phylloxera syndicates in the nineteenth century.
For such a treatment of *corps de
sapeurs-pompiers*, see Lussier
(1987).

28 Faucon (1966); Désert and
Specklin (1976) 417–18.

29 Barral (1968) 104–40.

30 Hubscher (1979) 614–23.

31 Cleary (1989) 33–6.

32 Bousseyroux (1990) 17–47 and
57–9.

33 Ruau (1908) 17–30.

34 Rocquigny (1903).

35 Traimond (1979) identified twenty-
three organisations providing livestock insurance in the three
departments of Landes, Pyrénées-
Atlantiques and Hautes-Pyrénées
between 1590 and 1833.

36 Wolff (1900) 254–5.

37 Goujon (1993) 70–3.

38 Chalvignac (1985) 37.

39 Birck and Dreyfus (1988) 115–18.

40 *Ibid.* 134–5.

41 Demoulin (1993) 28–30.

42 Lussier (1987) 24–37.

43 Barral (1962) 53–60.

44 Lussier (1987) 37–8.

45 Chalvignac (1985) 24, 32 and 37.

46 As argued, for example, by
Hubscher (1979) 619–23.

47 Judt (1975; 1979).

48 Devance (1984) 49, 51, 55–6.

49 Le Roy Ladurie (1994) 20.

50 Goujon (1993) 73–6.

51 Hubscher (1979) 627–9.

52 For general surveys, see Cholvy
(1974) 159–61; Cholvy and Hilaire
(1985); Price (1987) 261–306;
Gibson (1989) 227–67. For specific
regional examples, see Cholvy
(1973); Corbin (1975) 619–93;
Marcilhacy (1962); and Goujon
(1993) 18–51.

53 Birck and Dreyfus (1988) 31.

54 Navarro (1983) 51.

55 Valette (1979) 295.

56 Guimbretière (1985) 33 and 121.

57 Singer (1985).

58 Chabrol-Chardon (1981) 20–5.

59 Le Roy Ladurie (1994) 22–3.

60 Braque (1963); Gratton (1969;
1970; 1971).

61 For example, in the Puy-de-Dôme:
Bousseyroux (1990) 63–72.

62 Augé-Laribé (1907).

63 Augé-Laribé (1926) 8–13 and
101–24; (1955) 260.

64 See, for example: Faucon (1966);
Garrier (1969); Garrier (1973)
518–22; Mesliand (1969); Mesliand
(1989) 211–14; Cleary (1982; 1987).

65 See, for example: Faucon (1966);
Leveau (1969).

66 Hubscher (1979) 616.

67 Chabrol-Chardon (1981) 27–8.

68 Goujon (1993) 51–61.

69 Hubscher (1979) 614–19. See also
Rebiffe (1948).

70 Bousseyroux (1990) 61–3.

71 Hubscher (1979) 599–614;
Mesliand (1989) 216–25.

72 Cleary (1989) 35.

73 Baker (1984b) 23–7; Cleary (1989)
40–5.

74 Tombs (1996) 298.

75 Tilly (1979) 22; Burke (1992) 133.

76 Margadant (1979b) 138, 152 and ch. 7 *passim.*
77 Weber (1977) 272 and (1980).
78 Goujon (1993) 90–100.
79 Goujon (1981) 143; Lussier (1987) 39 and 161–2.
80 Bousseyroux (1990) 71.
81 Weber (1977) 339–74; Devlin (1987) 1–42; Gibson (1989) 134–57.
82 Zonabend (1984) 202–3.
83 Berger (1977; 1979a; 1979b; 1980). Kaye (1982) provides a critical review of Berger's way of seeing peasants.
84 Berger (1979b) 195–213.
85 Goldman (1983) 85.
86 Devance (1984) 51.
87 Fournière (1907) 54; Hobsbawm (1973–4) 9.
88 Augé-Laribé (1955) 258; Lebeau (1955); Bennet (1962).
89 See, for example: Girault (1969); Loubère (1974); Judt (1975; 1979); Vallin (1985); Frader (1991).
90 Cleary (1989) 48–149.

Bibliography

Unpublished primary sources

Archives Départementales de Loir-et-Cher (AD)

Dépôts des communes
Registres des délibérations du conseil municipal
510 Mesland 1816–1913
543 Marchenoir 1788–1879
546.2 Oucques 1827–1874

Série J Dons, legs, achats, dépôts
11J 31 Travaux de la Société des Sciences et Lettres de Loir-et-Cher: statistique
générale du département, statistique des sociétés de secours mutuels

Série M Administration générale et économie depuis 1800

3M Elections
480 Elections législatives, arrondissement de Blois. Listes d'enscription des votants 8
février 1871
481 Elections législatives, arrondissement de Romorantin. Listes d'enscription des
votants 8 février 1871
482 Elections législatives, arrondissement de Vendôme. Listes d'enscription des
votants 8 février 1871

4M Police
196 Enquête sur les sociétés et les associations 1831–1929
197 Déclarations des sociétés 1901–1939
198 Récépisses des constitutions des sociétés 1902–1930
201 Cercles diverses: recensements 1844–1895
202 Cercles diverses (par commune) 1837–1895
203 Associations dissoutes 1904–1946
204 Sociétés secrètes 1831–1862
228 Coalitions d'ouvriers, grèves 1830–1898

338

(uncatalogued file)
Police administrative: Grèves, coalitions d'ouvriers, troubles, réunions publiques, sociétés ouvrières (mutuelles), syndicats et coopératives 1809–1880

6M Statistiques
1074 Instruction publique: statistiques des cours d'adulte 1866–1867
1075 Instruction publique: statistiques des cours d'adulte 1865–1866
1087 Statistiques générales: sinistrés 1907, 1909
1100 Statistiques générales: sinistrés 1874–1876
1093 Statistiques des incendies 1866–1867
1097 Statistiques des incendies 1861–1864
1126 Statistiques agricoles: enquête parlementaire de 1866

7M Agriculture, eaux et forêts
56 Sociétés de crédit agricole: enquêtes, instructions, correspondance 1898–1922
57 Office National du Crédit Agricole. Avances à long terme 1904
143 Sociétés d'assurances mutuelles contre la mortalité du bétail 1891–1904
175 Comités d'études et de vigilance contre le phylloxéra 1874–1894; Sociétés agricoles 1852–1872
176 Phylloxéra 1874–1889
177 Phylloxéra 1877–1891
178 Phylloxéra 1880–1890
179 Phylloxéra 1880–1890
180 Syndicats anti-phylloxériques: délibérations, comptes de gestion 1883–1896
181 Syndicats anti-phylloxériques: délibérations, comptes de gestion A–J 1884–1898
182 Syndicats anti-phylloxériques: délibérations, comptes de gestion N–S 1884–1892
183 Syndicats anti-phylloxériques: délibérations, comptes de gestion S 1884–1889
184 Syndicats anti-phylloxériques: délibérations, comptes de gestion T–V 1884–1889
185 Syndicats anti-phylloxériques: délibérations, comptes de gestion M 1884–1892
186 Syndicats anti-phylloxériques 1884–1892
187 Syndicats anti-phylloxériques 1884–1892
188 Phylloxéra: syndicats anti-phylloxériques, états phylloxériques 1885–1901
191 Syndicats anti-phylloxériques 1890–1902
192 Syndicats anti-phylloxériques, comptes de gestion 1890–1897
194 Enseignement agricole, phylloxéra, rapports sur les services, personnel, 1893–1900
227 Agriculture: project de création d'une ferme école. Primes d'honneur. Prix pour les écoles 1868–1931
228 Fermes écoles de la Charmoise (Pontlevoy) et de Maray 1847–1865
239 Conférences agricoles 1885–1892
244 Syndicats professionels et agricoles: états des syndicats existant dans le départment 1887–1902
245 Société d'Agriculture de Loir-et-Cher, comices, expositions, concours 1808–1901
246 Comices agricoles 1828–1852
247 Comices agricoles: correspondances diverses, encouragements à l'agriculture 1845–1848
248 Comice agricole du département 1847–1876
249 Comices agricoles 1853–1864
264 Sociétés d'agriculture et comices agricoles 1842–1882

265 Projets de créations des sociétés d'agriculture, comices, comités consultatifs et conseils d'agriculture an XIII-1840
266 Société d'Agriculture de Loir-et-Cher 1805–1852: cotes 1 à 6
267 Société d'Agriculture de Loir-et-Cher 1805–1852: cotes 7 à 13
360 Assurances mutuelles agricoles. Etats de situation, subventions, correspondance 1899–1925
361 Assurances mutuelles agricoles: dossiers des sociétés Courmemin-Mareuil 1900–1914
362 Assurances mutuelles agricoles: dossiers des sociétés Chailles-Cour-sur-Loire 1900–1914
363 Assurances mutuelles agricoles: dossiers des sociétés Authon-Chambon 1900–1914
364 Assurances mutuelles agricoles: dossiers des sociétés Naveil-Saint-Sulpice
365 Assurances mutuelles agricoles: dossiers des sociétés Marolles-Muides 1900–1914
366 Assurances mutuelles agricoles: dossiers des sociétés Sargé-Vineuil 1900–1914
367 Assurances mutuelles agricoles: circulaires, instructions, statuts 1901–1923
368 Assurances mutuelles agricoles: subventions et statistiques 1923–1940

10M Travail
23 Syndicats professionnels 1880–1886
55 Syndicats industriels, commerciaux, agricoles 1908–1912
56 Syndicats professionnels: enquêtes, statistiques 1910–1930
57 Syndicats professionnels: dossiers de syndicats dissouts 1889–1930
58 Syndicats professionnels: enquêtes 1919–1926
59 Syndicats professionnels: enquêtes et statistiques 1929–41

201M Dénombrement de la population
35–8 Dénombrement de 1851
49–51 Dénombrement de 1856
55–7 Dénombrement de 1861
81–3 Dénombrement de 1866
108–9 Dénombrement de 1872
121–4 Dénombrement de 1876
128–31 Dénombrement de 1881
136–7 Dénombrement de 1886
143–5 Dénombrement de 1891
158–60 Dénombrement de 1896
166 Dénombrement de 1901
171–2 Dénombrement de 1911

Série P Finances, cadastre, postes depuis 1800
(uncatalogued files)
Evaluation des propriétés foncières, contributions directes 1879

Série R Affaires militaires, organismes de temps de guerre depuis 1800
(uncatalogued files)
Listes départementales du contingent: 1827, 1831, 1839, 1840, 1841, 1847, 1848, 1849, 1860, 1870, 1880, 1890
Sapeurs-pompiers: organisation des services d'incendie 1807–1842

Sapeurs-pompiers: achat de pompes, correspondances diverses 1841–1851
Sapeurs-pompiers: 1852–1872 A–F
Sapeurs-pompiers: 1852–1872 F–M
Sapeurs-pompiers: armement 1852–1872
Sapeurs-pompiers: armement 1854–1866
Sapeurs-pompiers: pompes à incendie 1857–1864
Sapeurs-pompiers: corps dont la réorganisation n'a pu avoir lieu 1860–1910
Sapeurs-pompiers: caisses communales de secours 1864–1869
Sapeurs-pompiers: Averdon–Chitenay 1875–1902
Sapeurs-pompiers: Instructions, enquêtes, effectifs, matériel, accidents, subventions
    1876–1897
Sapeurs-pompiers: Autainville–Brévainville 1876–1918
Sapeurs-pompiers: Cellettes–Chapelle Vendômoise 1862–1913
Sapeurs-pompiers: Chaumont–sur–Loire-Conan 1876–1914
Sapeurs-pompiers: Cormeray–Fréteval 1860–1914
Sapeurs-pompiers: Cormeray–Ménars 1852–1901
Sapeurs-pompiers: Marcilly–en-Gault-Moisy 1876–1914
Sapeurs-pompiers: Mulsans–Oucques 1876–1919
Sapeurs-pompiers: Mer–Orchaise 1864–1900
Sapeurs-pompiers: Molineuf–Muides 1865–1920
Sapeurs-pompiers: Ouzouer-le-Doyen–Saint-Amand 1870–1919
Sapeurs-pompiers: Sambin–Suèvres 1858–1914
Sapeurs-pompiers: Saint-Léonard–Vineuil 1867–1900
Sapeurs-pompiers: Ouchamps–Saint-Laurent-des-Eaux 1852–1900
Sapeurs-pompiers: Saint-Bohaire–Salbris 1877–1920
Sapeurs-pompiers: Villechauve–Yvoy-le-Marron 1860–1920
Sapeurs-pompiers: Talcy–Villebarou 1876–1920

Série S Travaux publics et transports depuis 1800
7S 12 Associations syndicales, curage, dessèchement des marais: instructions, tableaux
    des associations 1856–1873

Série X Assistance et prévoyance sociale depuis 1800
(uncatalogued files)
Sociétés d'assurances
Sociétés d'assurances diverses
Sociétés d'assurances mutuelles de Loir-et-Cher 1817–1838
Compagnies d'assurances 1839–1859
Sociétés de secours mutuels: affaires diverses O à Saint 1847–1900
Sociétés de secours mutuels: affaires diverses B 1852–1898
Sociétés de secours mutuels: affaires diverses C–F 1852–1898
Sociétés de secours mutuels: affaires diverses H à M 1852–1899
Sociétés de secours mutuels: affaires diverses S à V 1852–1898
Sociétés de secours mutuels: dossiers sociétés dissoutes 1852–1924
Sociétés de secours mutuels: affaires diverses de 1853 à 1870
Sociétés de secours mutuels, sportives antérieure à 1902: affaires diverses, Blois à Mont
Sociétés de secours mutuels, sportives antérieure à 1902: affaires diverses, Montoire à
    Villebarou

Sociétés d'assurance diverses: statistiques, correspondance, 1902–1914
Sociétés de secours mutuels. Etats récapitulatifs 1905, 1907, 1908
Sociétés d'assurances: enquête de 1907

Série Z Sous-préfectures depuis 1800
448 Syndicats

Archives Nationales(AN)

F^{10} 4651–5 Agriculture: police des eaux: curages, cours d'eau et syndicats, Loir-et-Cher, 1819–1948
F^{11} 2705 Enquête agricole de 1862
F^{11} 2717 Enquête agricole de 1882
F^{12} 5372 Sociétés de secours mutuels: demandes d'autorisation, statuts, contentieux 1850–1912

Published primary sources

L'Agriculture Pratique du Centre 1894–1914 (AD Pér. 7).
Annuaire de Loir-et-Cher 1815–1914 (AD Pér. 15).
L'Avenir 1861–1914 (AD Pér. 20).
Bulletin de la Société d'Agriculture de Loir-et-Cher 1843–1851 (AD Pér. 281).
Bulletin Annuel: Société d'Agriculture de Loir-et-Cher – Comice de l'arrondissement de Blois 1883, 1900, 1901, 1902 (AD Pér. 281).
Bulletin Mensuel du Syndicat des Agriculteurs de Loir-et-Cher March 1887–October 1894.
Conseil Général: rapports du Préfet et procès-verbaux des délibérations 1876–1914 (AD N).
L'Indépendant de Loir-et-Cher 1869–1914 (AD Pér. 126).
Recueil des Actes Administratifs de Loir-et-Cher 1821–1914 (AD Pér. 197).
Société d'Agriculture du Départment de Loir-et-Cher. Procès-verbaux des séances 1808–1841 (AD Pér. 348).

Angeville, A. d'. 1836 *Essai sur la statistique de la population française, considérée sous quelques-uns de ses rapports physiques et moraux* (Bourg-en-Bresse; reprinted The Hague 1969).
Anon. 1909 *Syndicat des Agriculteurs de Loir-et-Cher. Fête du 23e Anniversaire* (Blois).
Augé-Laribé, M. 1907 *Le problème agraire du socialisme. La viticulture industrielle du Midi de la France* (Paris).
Belton, L. 1888 'Notes sur l'histoire des Protestants dans le Blésois' *Mémoires de la Société des Sciences et Lettres de Loir-et-Cher* 11, 61–218.
Berthonneau, J. 1903 'Syndicats de battage' *L'Agriculture Pratique du Centre* (16 August 1903) 259–60.
    1905 *Monographie du Syndicat des Agriculteurs et des Caisses de Crédit Mutuel Agricole de Loir-et-Cher* (Blois).
Bourgeois, A. 1892 'Les métiers de Blois' *Mémoires de la Société des Sciences et Lettres de Loir-et-Cher* 13, 2 vols.
Decharme, P. 1908 *La commune et l'agriculture* (Paris).

Duclaux, M. 1905 *The fields of France* (London 1903; new edn London).

Faupin, E. 1909 *Essai sur la géologie de Loir-et-Cher (Beauce, Sologne, Perche)* (Blois).

Flaubert, G. 1950 *Madame Bovary* (Paris 1857), translated by A. Rusell (Harmondsworth).

Fournière, E. 1907 *L'individu, l'association et l'Etat* (Paris).

Guillamin, E. 1904 *La vie d'un simple* (Paris).

Hubbard, G. 1852 *De l'organisation des sociétés de prévoyance ou de secours mutuels et des bases scientifiques sur lesquelles elles doivent être établies* (Paris).

Joanne, A. 1869 *Géographie, histoire, statistique et archéologique des 89 départements de la France: Loir-et-Cher* (Paris).

Laurent, E. 1865 *Le pauperisme et les associations de prévoyance* (Paris).

Lelièvre, M. 1913 *Armand de Kerpezdron (1772–1854)* (Paris).

Leroy-Beaulieu, P. 1914 *Traité théorique et pratique d'économie politique* (6th edn, Paris).

Lesueur, E. 1905 *L'agriculture et les syndicats agricoles dans le département du Pas-de-Calais* (Paris).

Lynch, H. 1901 *French life in town and countryside* (2nd edn, London).

Nouel, E. 1888 'Extraits des anciens registres de Naveil (2e partie)' *Bulletin de la Société Archéologique Scientifique et Littéraire du Vendômois* 27, 17–47.

Pétigny, F. de. 1863 *Essai sur la population du département de Loir-et-Cher au XIXe siècle* (Paris). Also published as an article in *Mémoires de la Société des Sciences et Lettres de la Ville de Blois* 1 (1863) (AD G/F 115).

Pinet, A. 1860 *Nouvelle géographie historique du département de Loir-et-Cher* (Paris).

Prothero, R. E. 1915 *The pleasant land of France* (London 1908; 3rd edn London).

Rocquigny, R. de.

1900 *Les syndicats agricoles et leur oeuvre* (Paris).

1903 'Le progrès des assurances mutuelles agricoles en France' *Le Musée Social. Mémoires et Documents* 7, 221–4.

Ruau, J. 1908 *La politique agricole de la République* (Paris).

Trouillard, G. 1908 'La Société d'Agriculture de Loir-et-Cher sous le Directoire et le Consulat' *L'Agriculture Pratique du Centre* (10 May) 294–6.

Vezin, A. 1901 'Jules Tanviray' *L'Agriculture Pratique du Centre* (27 January) 28.

1910a 'Le mouvement mutualiste en Loir-et-Cher' *L'Agriculture Pratique du Centre* (20 February) 115–16.

1910b 'L'assurance du bétail en Loir-et-Cher' *L'Agriculture Pratique du Centre* (4 September) 563.

Vidal de la Blache, P. 1903 *Tableau de la géographie de la France*, being volume I part 1 of E. Lavisse (ed.), *Histoire de France* (Paris).

Wolff, H. W. 1900 'The agricultural syndicates of France' *Journal of the Royal Agricultural Society of England* 3rd series 11, 252–62.

Unpublished secondary sources

Asfaux, D. 1955 *La formation du département de Loir-et-Cher* (Diplôme d'études supérieures, University of Poitiers).

Anon. n.d. *Episodes de l'histoire des Protestants de Mer* (manuscript, n.p.) (AD).

Baker, A. R. H. (ed.) 1987 *The modernisation of rural France? Some geographical changes in Loir-et-Cher during the nineteenth century* (field class essays, University of Cambridge).

1990 *Continuity and change in rural France during the nineteenth century: five case studies in Loir-et-Cher* (field class essays, University of Cambridge).

Baker, A. R. H. and Black, I. (eds.) 1988 *Geographical change in nineteenth-century rural France: some case studies of Loir-et-Cher* (field class essays, University of Cambridge).

Baker, A. R. H. and Howell, P. M. R. (eds.) 1992 *Conscripts and councillors. Case studies in nineteenth-century Loir-et-Cher* (field class essays, University of Cambridge).

Barnet, A. and Edwards, M. 1992 'Changing living standards in five cantons in the department of Loir-et-Cher in the nineteenth century' in Baker and Howell (eds.) (1992) 1–7.

Bazin, M. 1988 *Les conséquences de la Loi Guizot dans les cantons de Blois* (mémoire, n.p.) (AD G/F 776).

Béraud, A. 1981 *Ecoles et écoliers d'autrefois à Selles-sur-Cher* (Selles-sur-Cher).

Bourdin, M. and Bourdin, Mme. 1980 'Une école primaire vers 1860' *Centre Départementale Universitaire pour le 3e Age en Loir-et-Cher* (Spring) 22–31.

Brette, D. 1980 *Les maires de l'arrondissement de Blois (cantons d'Ouer-le-Marché et de Saint-Aignan) de 1800 à 1882* (mémoire de maîtrise d'histoire, Université François-Rabelais, Tours).

Campbell, C. and Barr, D. 1988 'An investigation of the expenditure of Mesland commune council 1826–1911' in Baker and Black (eds.) (1988) 39–43.

Chadwick, L. 1988 'Schooling in nineteenth-century France: a case study of Mesland (Loir-et-Cher)' in Baker and Black (eds.) (1988) 9–12.

Chalvignac, J-Y. 1985 *Les sociétés de secours mutuels en pays de Fougères de la fin du Second Empire et au début de la IIIe République* (mémoire de maîtrise d'histoire, Université de Haute Bretagne, Rennes).

Clark, R. 1988 'Peasants into French peasants: the early years of primary schooling in Loir-et-Cher' in Baker and Black (eds.) (1988) 13–20.

Farmer, A. 1988 'The stature of conscripts in the cantons of Savigny, Marchenoir, Salbris and Montrichard 1856–1886' in Baker and Black (eds.) (1988) 1–4.

Fee, D. and Prendergast, A. 1990 'Change in Champigny: a rural commune in nineteenth-century Loir-et-Cher' in Baker (ed.) (1990).

Fénelon, P. 1963 *La Petite Beauce* (typescript; AD CR 242)

Gobillon, M. 1975 *L'évolution de la propriété agraire et de la vie rurale dans le Blésois: aspects (1ère moitié du XIXe siècle)* (thèse 3e cycle lettres, Paris–10 Nanterre).

Halstead, F. and Hooper, P. 1990 'A study of social and economic change in the commune of La Bosse during the nineteenth century' in Baker (ed.) (1990).

Holden, J. 1996 'Military service in the commune of Mulsans during the nineteenth century' in Howell and Baker (eds.) (1996) 64–73.

Howell, P. M. R. and Baker, A. R. H. (eds.) 1996 *Modernity in Mulsans* (field class essays, University of Cambridge).

Jones, S. 1987 'A study of the membership of Mesland municipal council 1836–1881' in Baker (ed.) (1987).

Lobier, D. 1970 *Les hommes sur le théatre d'opérations de la Loire 4 septembre 1870–8 février 1871* (mémoire de maîtrise, Faculté des Lettres, Université de Tours).

Muller, E. 1974 *Les fêtes en Loir-et-Cher 1814–1852* (mémoire maîtrise d'histoire, Université François Rabelais, Tours).

Notter, A. 1981 *Le culte des saints en Sologne aux XIXe-XXe siècles* (thèse d'Ecole de Chartres).

Olivier, P. 1962 *Les étapes de la scolarisation dans le Loir-et-Cher de 1833 à 1910* (mémoire de l'Ecole Normale Supérieure, Blois).

Packham, D. 1987 'Report on Mesland council minutes, 1836–1914' in Baker (ed.) (1987).

Paul, C. n.d. *Naissance et évolution des transports ferroviaires dans le Loir-et-Cher de 1848 à 1934* (mémoire de l'Ecole Normale Supérieure, Blois AD Photocopie 280).

Paumier, M. 1969 'Un instituteur républicain sous le Second Empire' (typescript: AD CR 273).

Pickup, J. and Walters, P. 1990 'Aspects of modernisation and changing *mentalité* in the commune of Ménars' in Baker (ed.) (1990).

Silver, J. 1973 *French rural response to modernisation: the Vendômois, 1852–1885* (Ph.D. thesis, University of Michigan).

Sowerby, J. and Stacey, A. 1990 'A study of the commune of Marchenoir' in Baker (ed.) (1990).

Sutton K. 1967 *The changing land use of the Sologne in the nineteenth century* (M.A. thesis, University of London).

Tooth, C. and Shepherd, M. 1987 'Some observations on the education, stature and occupations of military conscripts in Loir-et-Cher in 1853, 1868, and 1888' in Baker (ed.) (1987).

Trobert, A. 1949 *L'agriculture en Loir-et-Cher de 1789 à 1815 et le monde rural de ce département dans ses rapports avec l'administration consulaire et impériale* (typescript, Paris: AD G/F 400).

Published secondary sources

Agulhon, M. 1966 *La sociabilité méridionale* (Aix-en-Provence).

　1968 *Pénitents et francs-maçons de l'ancienne Provence.*

　1970 *La République au village. Les populations du Var de la Révolution à la IIe République* (Paris), translated as *The Republic in the village: the people of the Var from the French Revolution to the Second Republic* (Cambridge 1982).

　1973 *1848 ou l'apprentissage de la République* (Paris), translated as *The Republican experiment* (Cambridge 1983).

　1976 'L'essor de la paysannerie 1789–1852: La Révolution et l'Empire' in M. Agulhon, G. Désert and R. Specklin (eds.) *Apogée et crise de la civilisation paysanne 1789–1914*, being vol. III of G. Duby and A. Wallon (eds.) *Histoire de la France rurale* (Paris) 19–57.

　1977 *Le cercle dans la France bourgeoise, 1810–1848. Etude d'une mutation de sociabilité* (Paris).

　1978a 'Vers une histoire des associations' *Esprit* 6, 13–18.

　1978b review of E. Weber (1976) *Annales: Economies, Sociétés, Civilisations* 33, 843–4.

　1979 *La République au village. Les populations du Var de la Révolution à la IIe République* (new edn, Paris).

1981 'Les associations depuis le début du XIXe siècle' in M. Agulhon and M. Bodiguel *Les associations au village* (Le Paradou) 9–37.

1983 *The Republican experiment 1848–1852* (Cambridge).

1986 'Introduction: la sociabilité, est-elle objet d'histoire?' in E. François (ed.) *Sociabilité et société bourgeoise en France, en Allemagne et en Suisse* (Paris).

1988 'L'histoire sociale et les associations' *Revue de l'Économie Sociale* 14 (April) 35–44.

1990 'Exposé de clôture' in R. Levasseur (ed.) *De la sociabilité: spécificité et mutations* (Cap-Saint-Ignace, Québec) 327–45.

Agulhon, M. and Bodiguel, M. 1981 *Les associations au village* (Le Paradou).

Agulhon, M., Désert, G. and Specklin, R. (eds.) 1976 *Apogée et crise de la civilisation paysanne 1789–1914*, being vol. III of G. Duby and A. Wallon (eds.) *Histoire de la France rurale* (Paris).

Anderson, R. 1971 'Voluntary associations in history' *American Anthropologist* 73, 209–22.

Anon. 1920 *Monographie du Syndicat des Agriculteurs de Loir-et-Cher* (Vendôme).

Anon. 1921 *Syndicat agricole des arrondissements de Chartres, Châteaudun et Nogent-le-Roye* (Chartres).

Anon. 1950 'L'Union des associations et des coopératives agricoles de Loir-et-Cher' *L'Opinion Economique et Financière* 3 no. 5 80–1.

Anon. 1994 'Se déplacer en Sologne' *Journal de la Sologne* (June).

Arts Council of Great Britain 1976 *Jean-François Millet* (London).

1978 *Gustave Courbet 1819–1877* (London).

Augé-Laribé, M. 1926 *Syndicats et coopératives agricoles* (Paris).

1950 *La politique agricole de la France de 1880 à 1940* (Paris).

1955 *La révolution agricole* (Paris).

Babchuk, N. and Warriner, C. K. 1965 'Introduction' to 'Signposts in the study of voluntary groups' *Sociological Inquiry* 35, 135–7.

Bages, R., Druhle, M. and Nevers, J–Y. 1976 'Fonctionnement de l'institution municipale et pouvoir locale en milieu rural' *Etudes Rurales* 63–4, 31–54.

Bairoch, P. 1988 'Dix-huit décennies de développement agricole français dans une perspective internationale (1800–1980)' *Economie Rurale* 184–6, 13–23.

Baker, A. R. H. 1968 'Etablissements ruraux sur la marge sud-ouest du Bassin parisien dans les premières années du XIXe siècle: description analytique des types' *Norois* 60, 481–92.

1973 'Adjustments to distance between farmstead and field: some findings from the south-western Paris Basin in the early nineteenth century' *Canadian Geographer* 17, 259–75.

1980 'Ideological change and settlement continuity in the French countryside: the development of agricultural syndicalism in Loir-et-Cher during the late nineteenth century' *Journal of Historical Geography* 6, 163–77.

1983 'Devastation of a landscape, doctrination of a society: the politics of the phylloxera crisis in Loir-et-Cher (France) 1866–1914' *Würzburger Geographische Arbeiten* 60, 205–17.

1984a 'Fraternity in the forest: the creation, control and collapse of wood-cutters' unions in Loir-et-Cher 1852–1914' *Journal of Historical Geography* 10, 157–73.

1984b 'Reflections on the relations of historical geography and the *Annales* school of history' in Baker, A. R. H. and Gregory, D. (eds.) *Explorations in historical geography* (Cambridge) 1–27.

1986a 'The infancy of France's first agricultural syndicate: the Syndicat des Agriculteurs de Loir-et-Cher 1881–1914' *Agricultural History Review* 34, 45–59.

1986b 'Les syndicats agricoles de la Vallée de la Cisse (1883–1914)' *Vallée de la Cisse* 8, 59–71.

1986c 'Sound and fury: the significance of musical societies in Loir-et-Cher during the nineteenth century' *Journal of Historical Geography* 12, 249–67.

1990a 'Fire-fighting fraternities? The corps de sapeurs-pompiers in Loir-et-Cher during the nineteenth century' *Journal of Historical Geography* 16, 121–39.

1990b 'On geographical literature as popular culture in rural France, c. 1860–1900' *Geographical Journal* 156, 39–43.

1992a 'Collective consciousness and the local landscape: national ideology and the commune council of Mesland (Loir-et-Cher) as landscape architect during the nineteenth century' in A. R. H. Baker and G. Biger (eds.) *Ideology and landscape in historical perspective. Essays on the meanings of some places in the past* (Cambridge) 255–88.

1992b 'Les bibliothèques des gares en Loir-et-Cher avant 1914' *Mémoires de la Société des Sciences et Lettres de Loir-et-Cher* 47, 207–16.

1995 'Locality and nationality: geopieties in rural Loir-et-Cher (France) during the nineteenth century' in S. Courville and N. Séguin (eds.) *Espace et culture* (Laval) 77–88.

1996 'Farm schools in nineteenth century France and the case of La Charmoise 1847–1865' *Agricultural History Review* 44, 47–62.

1998 'Military service and conscription in nineteenth century provincial France: some evidence from Loir-et-Cher' *Transactions of the Institute of British Geographers* 23, 193–206.

Barral, P. 1962 *Le département de l'Isère sous la Troisième République 1870–1940* (Paris).

1966 'Note historique sur l'emploi du terme "paysan"' *Etudes Rurales* 21, 72–80.

1968 *Les agrariens français de Méline à Pisani* (Paris).

1969 'Aspects régionaux de l'agrarisme français' *Le Mouvement Social* 67, 3–16.

1988 'Littérature et monde rurale' *Economie Rurale* 184–6, 199–204.

Bastier, J. 1978 '*Les Paysans* de Balzac et l'histoire du droit rural' *Revue d'Histoire Moderne et Contemporaine* 25, 396–418.

Beauchamp, C. 1990 *Délivrez-nous du mal! Epidémies, endémies, médecine et hygiène au XIXe siècle dans l'Indre, l'Indre-et-Loire et le Loir-et-Cher* (n.p.)

Beck, R. 1987 'Les effets d'une ligne du Plan Freycinet sur une société rurale' *Francia* 5, 561–77.

1988 *Der Plan Freycinet und die Provinzen* (Frankfurt-on-Main).

Bennet, J. 1962 *L'Isère, haut lieu de la mutualité* (Etampes).

Bercé, Y.-M. 1974 *Croquants et nu-pieds: les soulèvements paysans en France du XVIe au XIXe siècle* (Paris).

Berelson, B. and Steiner, G. A. 1964 *Human behavior: an inventory of scientific findings* (New York).

Berenson, E. 1984 *Populist religion and left-wing politics in France, 1830–1852* (Princeton, NJ).

1987 'Politics and the French peasantry: the debate continues' *Social History* 12, 213–29.

Berger, J. 'A class of survivors' 1977 *New Society* 42, 611–13.

1979a *Permanent red: essays in seeing* (London): 'Millet and labour' 189–91 and 'The politics of Courbet' 196–8.

1979b *Pig earth* (London).

1980 *About looking* (New York): 'Millet and the peasant' 69–78 and 'Courbet and the Jura' 134–41.

Berger, S. 1972 *Peasants against politics. Rural organization in Brittany, 1911–1967* (Cambridge, MA).

Berthonneau, J. 1928 'Les organisations professionelles agricoles' *L'Illustration Economique et Financière,* supplement to no. 6, 59–61.

Birck, F. and Dreyfus, M. 1988 *La mutualité en Lorraine. Etude d'un patrimoine historique* (Paris).

Bizeau, Abbé P. and Notter, A. 1996 'Pratiques religieuses et dévotion au XIXe siècle' in C. Deluz, (ed.) *Blois: une diocèse, une histoire* (Blois) 132–40.

Blanchard, M. 1931 *La campagne et ses habitants dans l'œuvre de Honoré de Balzac. Etudes des idées de Balzac sur la grande propriété* (Paris).

Bloch, M. 1956 *Les caractères originaux de l'histoire rurale française,* vol. II (2nd edn, Paris).

Blum, J. 1982 *Our forgotten past. Seven centuries of life on the land* (London).

Bomer, B. 1958 'Paysages ruraux du Bassin Parisien méridional' *L'Information Géographique* 22, 55–67.

1959 'Paysages ruraux entre Val-de-Loire et vallée du Loir' *Annales de l'Est* 21, 68–78.

Boucher, J-J. 1984 *Histoire du Loir-et-Cher à travers son Conseil Général de 1790 à nos jours* (Paris).

Boulard, F. 1982 *Matériaux pour l'histoire religieuse du peuple français* (Paris).

Bourdin, M. et Mme, 1980 'Une école primaire vers 1860' *Centre Départementale Universitaire pour le 3e Age en Loir-et-Cher* (Spring) 22–31.

Bousseyroux, P. 1990 *La mutualité dans le Puy-de-Dôme au XIXe siècle (1848–1914)* (Paris).

Bouyssou, M. 1991 'Les confréries religieuses en Blaisois et Vendômois XVIe–XIXe siècles' *Annales de Bretagne et des Pays de l'Ouest (Anjou, Maine, Touraine)* 98, 27–49.

1992 'Les confréries religieuses en Blaisois et Vendômois (XVIe–XVIIIe siècle)' *Mémoires de la Société des Sciences et Lettres de Loir-et-Cher* 47, 31–69.

Bozon, M. 1981 *Les conscrits* (Paris).

1987 'Apprivoiser le hasard: la conscription au XIXe siècle' *Ethnologie Française* 17, 291–301.

Braque, R. 'Aux origines du syndicalisme dans les milieux ruraux du centre de la France (Allier-Cher-Nièvre-sud-du-Loiret)' *Le Mouvement Social* 42, 79–116.

Braux, D. 1954 'Le mouvement coopératif en Loir-et-Cher' *Revue Géographique et Industrielle de France,* 62–3.

Brettell, R. and Brettell, C. B. 1983 *Painters and peasants in the nineteenth century* (Geneva).

Bruneau, C. 1988 'L'état des publications sur les associations en France (1930–1985)'

Revue de l'Economie Sociale 14 (April) 29–33.

Burke, P. 1992 *History and social theory* (Cambridge).

Burns, M. 1984 *Rural society and French politics. Boulangism and the Dreyfus affair, 1886–1900* (Princeton, NJ).

Cabaret, M. and Cabaret, J.-P. 1995 *Le chant d'une ville. La musique à Blois du XVe au XIXe siècle* (Blois).

Callon, G. 1949 'Le mouvement de la population dans le département de Loir-et-Cher au cours de la période 1821–1920' *Bulletin de la Société d'Histoire Naturelle de Loir-et-Cher* 21, 3–45.

Cameron, R. E. 1958 'Economic growth and stagnation in France, 1815–1914' *Journal of Modern History* 30, 1–13.

1965 *France and the economic development of Europe, 1800–1914* (2nd edn, Chicago).

Chabrol-Chardon, C. 1981 'Les sociétés de secours mutuels du Médoc 1834–1914' *Actes du 105e Congrès National des Sociétés Savantes 1980. Colloque sur l'histoire de la securité sociale* (Paris) 19–30.

Chaline, J-P. 1986 'Sociétés savantes et académies de province en France dans la première moitié du XIXe siècle' in E. François (ed.) *Sociabilité et société bourgeoise en France, en Allemagne et en Suisse* (Paris).

1995 *Sociabilité et érudition: les sociétés savantes aux 19e et 20e siècles* (Paris).

Chamboredon, J-C. 1977 'Peinture des rapports sociaux et invention de l'eternel paysan: les deux manières de Jean-François Millet' *Actes de la Recherche en Sciences Sociales* 17–18, 6–28.

1984 review of M. Agulhon and M. Bodiguel (1981) in *Annales Économies Sociétés Civilisations* 39, 52–8.

Charlton, D. G. 1963 *Secular religions in France, 1815–1870* (London).

Chartier, R. 1996 'The Saint-Malo-Geneva line' in P. Nora (ed.) *Realms of memory. The construction of the French past* (New York).

Chaumier, J-M. and Gillardot, P. 1974 'L'évolution du paysage rural de la Grande Sologne' *Norois* 21, 393–410.

Chollet, J. 1981 'Instituteurs de père en fils ou 60 années d'enseignement à Villefranche-sur-Cher' *Journal de la Sologne et des ses Environs* 32, 13.

Cholvy, G. 1973 *Religion et société au 19e siècle: le diocèse de Montpellier* (Lille).

1974 'Société, genres de vie et mentalités dans les campagnes françaises de 1815 à 1880' *L'Information Historique* 36, 155–66.

Cholvy, G. and Hilaire, Y-M. 1985 *Histoire religieuse de la France contemporaine, 1800–1880* (Toulouse).

Clapham, J. H. 1936 *Economic development of France and Germany, 1815–1914* (4th edn, Cambridge).

Clark, T. J. 1973a *The absolute bourgeois. Artists and politics in France 1848–1851* (London).

1973b *The image of the people. Gustave Courbet and the 1848 Revolution* (London).

Cleary, M. C. 1979 'Le premier syndicalisme agricole vu par un anglais' *Revue du Rouergue* 33, 43–53.

1982 'The plough and the cross: peasant unions in south-western France' *Agricultural History Review* 30, 127–36.

1987 'Priest, squire and peasant: the development of agricultural syndicates in south-west France 1900–1914' *European History Quarterly* 17, 145–163.

1989 *Peasants, politicians and producers. The organisation of agriculture in France since 1918* (Cambridge).

Clough, S. B. 1946 'Retardative factors in French economic development in the nineteenth and twentieth centuries' *Journal of Economic History* 6, supplement.

Clout, H. D. 1977 'Agricultural changes in the eighteenth and nineteenth centuries' in H. D. Clout (ed.) *Themes in the historical geography of France* (London) 407–46.

1980 *Agriculture in France on the eve of the railway age* (London).

1983 *The land of France 1815–1914* (London).

Cobban, A. 1949 'France – a peasants' republic' *The Listener* 41, 429–30.

Corbin, A. 1975 *Archaisme et modernité en Limousin au XIX siecle, 1845–1880* (Paris).

1987 'Coulisses' in M. Perrot (ed.) *De la Révolution à la Grande Guerre*, being vol. IV of P. Ariès and G. Duby (eds.) *Histoire de la vie privée* (Paris) 413–611.

Croubois, C. (ed.) 1985 *Le Loir-et-Cher de la préhistoire à nos jours* (Saint-Jean-d'Angély).

Crozet, R. 1939 'Contribution à l'histoire de la voie ferrée de Paris à Toulouse et du réseau ferré entre Loire moyenne et Cher' *Revue d'Histoire Moderne* 38, 241–60.

Cuevas-Delétang, P. 1991 'Les comices agricoles de Lamotte-Beuvron au Second Empire. La première fête du Comité central agricole de la Sologne à Lamotte-Beuvron, le 21 septembre 1862' *Bulletin du Groupe de Recherches Archéologiques et Historiques de la Sologne* 13, 64–81 and 95–100.

Dallas, G. 1982 *The imperfect peasant economy. The Loire country, 1800–1914* (Cambridge).

Debbasch, C. and Bourdon, J. 1990 *Les associations* (3rd edn. Paris).

Delecluse, J-M. 1986 'L'école communale et ses maîtres à Coulanges de F. Guizot à J. Ferry' *Vallée de la Cisse* 8, 49–58.

Deluz, C. (ed.) 1996 *Blois: une diocèse, une histoire* (Blois).

Demangeon, 1946 A. *La France* vol. I, part 2 (Paris).

Demonet, M. 1990 *Tableau de l'agriculture française au milieu du XIXe siècle: l'enquête de 1852* (Paris).

Demoulin, P. 1993 *La mutualité bourbonnaise 1870–1930* (Paris).

Désert, G. and Specklin, R. 1976 'L'ébranlement: les réactions face à la crise' in M. Agulhon, G. Désert and R. Specklin (eds.) *Apogée et crise de la civilisation paysanne* 409–52, being vol. III of G. Duby and A. Wallon (eds.) *Histoire de la France rurale* (Paris).

Devance, L. 1984 'Les sociétés de secours mutuels dans le département de la Côte-d'Or au XIXe siècle' *Actes du 108e Congrès Nationale des Sociétés Savantes 1983* (Paris) 49–58.

Devlin, J. 1987 *The superstitious mind. French peasants and the supernatural in the nineteenth century* (New Haven).

Dion, R. 1934a *Le Val de Loire. Etude de géographie régionale* (Tours).

1934b *Essai sur la formation du paysage rural français* (Tours).

1959 *Histoire de la vigne et du vin en France des origines au XIXe siècle* (Paris).

Diot, G. 1976 'Les Prussiens à Landes-le-Gaulois en 1870–1871' *Vallée de la Cisse* 3, 41–8.

Doucet, M. 1994 'Les Loir-et-Cheriens pendant l'année terrible. Les gardes nationales et la mise en défense du département de Loir-et-Cher (August–December 1870),

Mémoires de la Société des Sciences et Lettres de Loir-et-Cher 49, 159–74.

Dreyfus, M. 1988 *La mutualité: une histoire maintenant accessible* (Paris).

Duggett, M. 1975 'Marx on peasants' *Journal of Peasant Studies* 2, 159–82.

Dupeux, G. 1956 'Aspects agricoles de la crise: le département de Loir-et-Cher' *Société de la Révolution de 1848. Etudes. Bibliothèque de la Révolution de 1848* 19, 65–92.

1962 *Aspects de l'histoire sociale et politique du Loir-et-Cher 1848–1914* (Paris).

1976 *French society 1789–1870* (London).

Dupuy, J. 1961 'Le socialisme en Loir-et-Cher de 1795 à 1852' *Bulletin de la Société Archéologique du Vendômois*, 39–54.

1974 'Trente ans d'histoire de la presse blésoise (1851–1881)' *Mémoires de la Société des Sciences et Lettres de Loir-et-Cher* 35, 93–102.

1980 'La vie poitique en Loir-et-Cher de 1870 à 1874' *Mémoires de la Société des Sciences et Lettres de Loir-et-Cher* 36, 81–100.

Edeine, B. 1974 *La Sologne. Contribution aux études d'ethnologie métropolitaine* 2 vols. (Paris).

Edelstein, M. 1992 'Integrating the French peasants into the nation-state – the transformation of electoral participation (1789–1870)' *Journal of European Studies* 15, 319–26.

Fauchon, P. 1990 *1790 en Loir-et-Cher: la naissance d'un département* (Chailles).

Faucon, R. 1966 'Les origines du syndicalisme agricole dans la Region du Nord' *Revue du Nord* 48, 67–89.

Fénéant, J. 1986 *Francs-maçons et sociétés secrètes en Val de Loire* (Chambray).

1990 'Les sociétés secrètes en Vendômois' *Bulletin de la Société Archéologique, Scientifique et Littéraire du Vendômois* 48–55.

Fortin, A. 1950 'Les sociétés de secours mutuels dans le Pas-de-Calais sous le Second Empire' *Revue du Nord* 32, 206–18.

Frader, L. L. 1991 *Peasants and protest. Agricultural workers, politics and unions in the Aude, 1850–1914* (Berkeley).

François, E. (ed.) 1986 *Sociabilité et société bourgeoise en France, en Allemagne et en Suisse* (Paris).

François, E. and Reichardt, R. 1987 'Les formes de sociabilité en France du milieu du XVIIIe au milieu du XIXe siècle' *Revue d'Histoire Moderne et Contemporaine* 34, 453–72.

Furet, F. and Ozouf, J. 1977 *Lire et écrire: L'alphabétisation des français de Calvin à Jules Ferry* (Paris), translated as *Reading and writing. Literacy in France from Calvin to Jules Ferry* (Cambridge 1982).

Gaillot, R. 1981 'Quelques aspects du programme scolaire d'antan' *Journal de la Sologne et ses Environs* 34, 15–18.

Gaillot, P. and Gaillot, R. 1982 'En marge d'un centenaire: St Sulpice et son école' *Vallée de la Cisse* 6, 51–5.

Gaillot, P., Gaillot, R. and Pisani, M. T. (eds.) 1981a *Centenaire de l'école publique 1881–1981* (Blois).

1981b *Regards sur l'école primaire en Loir-et-Cher au XIXe siècle* (Blois).

Gallagher, O. R. 1957 'Voluntary associations in France' *Social Forces* 36, 153–60.

Gallon, G. 1949 'Le movement de la population dans le département de Loir-et-Cher au cours de la période 1821–1920' *Bulletin de la Société d'Histoire Naturelle de Loir-et-Cher* 21, 3–45.

Garrier, G. 1969 'L'Union du Sud-Est des syndicats agricoles avant 1914' *Le Mouvement Social* 67, 17–38.

  1973 *Paysans du Beaujolais et du Lyonnais 1800–1970* 2 vols. (Grenoble).

  1989 *Le phylloxéra. Une guerre de trente ans* (Paris).

Gazley, J. G. 1973 *The life of Arthur Young* (Philadelphia).

Gibaud, B. 1986 *De la mutualité à la securité sociale: conflits et convergences* (Paris).

Gibson, R. 1989 *A social history of French Catholicism* (London).

Girault, J. 1969 'Le rôle du socialisme dans la révolte des vignerons de l'Aube' *Le Mouvement Social* 67, 89–109.

Gobillon, M. 1976 'Les maires d'Onzain au XIXème siècle' *Bulletin du Groupe d'Etudes d'Histoire et de Géographie Locales d'Onzain et de ses Communes Voisines* 1, n.p.

  1977 'Les instituteurs d'Onzain (première moitié du XIXème siècle)' *Bulletin du Groupe d'Etudes d'Histoire et de Géographie Locales d'Onzain et de ses Communes Voisines* 3, 2–6.

  1983 'Les instituteurs du Blésois dans la première moitié du XIXe siècle' *Mémoires de la Société des Sciences et Lettres de Loir-et-Cher* 38, 5–30.

  1985 'La foi des paroisses de la Révolution au Second Empire' *Mémoires de la Société des Sciences et Lettres de Loir-et-Cher* 40, 5–45.

  1989 'La Société des Amis de la Constitution' *Mémoires de la Société des Sciences et Lettres de Loir-et-Cher* 44, 83–111.

  1991 'Charles-Victor Prat, le premier inspecteur des écoles du Loir-et-Cher (1835–1850)' *Mémoires de la Société des Sciences et Lettres de Loir-et-Cher* 46, 109–39.

  1996 'Clericaux et anticléricaux' in Deluz, C. (ed.) *Blois: une diocèse, une histoire* (Blois) 146–56.

Goldberg, H. 1954 'The myth of the French peasantry' *American Journal of Economics and Sociology* 13, 363–8.

Goldin, J. 1984 *Les comices agricoles de Gustave Flaubert. Transcription intégrale et genèse dans le manuscrit g 223* (Geneva).

Goldman, S. 1983 'Les sociétés de secours mutuels en agriculture, de la loi de 1850 à la première guerre mondiale' *Actes du 107e Congrès des Sociétés Savantes 1982* (Paris) 69–95.

Goreux, L-M. 1956 'Les migrations agricoles en France depuis un siècle' *Etudes et Conjonctures* 4.

Gouda, F. 1996 *Poverty and political culture. The rhetoric of social welfare in the Netherlands and France, 1815–1854* (Amsterdam).

Goujon, P. 1981 'Association et vie associative dans les campagnes au XIXe: le cas du vignoble de Saône et Loire' *Cahiers d'Histoire* 26, 107–51.

  1993 *Vigneron citoyen: Mâconnais et Chalonnais (1848–1914)* (Paris).

Grantham, G. W. 1975 'Scale and organization in French farming, 1840–1880' in W. N. Parker and E. L. Jones (eds.) *European peasants and their markets. Essays in agrarian economic history* (Princeton).

  1978 'The diffusion of the new husbandry in northern France' *Journal of Economic History* 38, 311–37.

  1989a 'Capitalism and agrarian structure in early nineteenth-century France' *Research in Economic History* 5, 137–59.

1989b 'Agricultural supply during the Industrial Revolution: French evidence and European implications' *Journal of Economic History* 49, 43–72.

Gratton, P. 1969 'Le communisme rural en Corrèze' *Le Mouvement Social* 67, 123–45.

1970 'Mouvement et physionomie des grèves agricoles en France de 1890 à 1935' *Le Mouvement Social* 71, 3–38.

1971 *Les luttes de classes dans les campagnes* (Paris).

1972 *Les paysans français contre l'agrarisme* (Paris).

Grew, R. 1988 'Picturing the people: images of the lower orders in nineteenth-century French art' in R. I. Rotberg and T. K. Rabb (eds.) *Art and history: images and their meaning* (Cambridge) 203–31.

Gueslin, A. 1978 *Les origines du Crédit Agricole, 1840–1914* (Paris).

1987 *L'invention de l'économie sociale: le XIXe siècle français* (Paris).

Guillamin, E. 1983 *The life of a simple man*, edited and translated by E. Weber (London).

Guimbretière, A. M. 1985 *Racines mutualistes. Sociétés de secours mutuels vendéennes, milieu XIXe–début XXe* (n.p.).

Guiral, P. *et al.* (eds.) 1969 *La société française (1815–1914) vue par les romanciers* (Paris): 'Le monde paysan' 70–82

Guy, R. 1952 *La Terre d'Emile Zola. Etude historique et critique* (Paris).

Hayward, J. E. S. 1959 'Solidarity: the social history of an idea in nineteenth century France' *International Review of Social History* 4, 261–84.

Hélias, P-J. 1975 *Le cheval d'orgueil* (Paris), translated as *The horse of pride* (New Haven, 1978).

Herbert, R. H. 1970 'City vs country: the rural image in French painting from Millet to Gauguin' *Artforum* 8, 44–5.

Heywood, C. 1981 'The role of the peasantry in French industrialisation 1815–1880' *Economic History Review* 34, 359–76.

1992 *The development of the French economy, 1750–1914* (Basingstoke).

Hitier, H. 1930 'A. Riverain' *Procès-verbaux de l'Académie d'Agriculture de France* session 26 February.

Hobsbawm, E. J. 1973–4 'Peasants and politics' *Journal of Peasant Studies* 1, 3–22.

Hoffman, P. T. 1996 *Growth in a traditional society. The French countryside 1450–1815* (Princeton).

Hohenberg, P. 1972 'Change in rural France in the period of industrialisation, 1830–1914' *Journal of Economic History* 32, (1972) 219–40.

1974 'Migrations et fluctuations démographiques dans la France rurale 1836–1901' *Annales Economies Sociétés Civilisations* 29, 461–97.

Houssel, J-P. *et al.* 1976 *Histoire des paysans français du XVIIIe siècle à nos jours* (Roanne).

Huard, R. 1982 *La pré-histoire des partis: le mouvement républicain en Bas-Languedoc, 1848–1881* (Paris).

Hubscher, R. H. 1979 *L'agriculture et la société rurale dans le Pas-de-Calais du milieu du XIXe siècle à 1914* (Arras).

1983 'La France paysanne: réalités et mythologies' in R. Hubscher *et al.* (eds.) *La société* (Paris) 9–151, being vol. I of Y. Lequin (ed.) *Histoire des français XIX–XX siècles.*

1994 'Entre tradition et modernisation' in E. Le Roy Ladurie (ed.) *Paysages, paysans. L'art et la terre en Europe du Moyen Age au XXe siècle* (Paris 1994) 183–93.

Ion, J. 1990 'Les trois formes de la sociabilité associative' in R. Levasseur (ed.) *De la sociabilité: spécificité et mutations* (Cap-Saint-Ignace, Québec) 169–82.

Jardin, A. and Tudesq, A-J. 1983 *Restoration and reaction, 1815–1848, The Cambridge History of Modern France* vol. I (Cambridge).

Jones, P. M. 1985 *Politics and rural society. The southern Massif Central, c.1750–1880* (Cambridge).

1988 *The peasantry in the French Revolution* (Cambridge).

Joutard, P. 1977 *La légende des Camisards* (Paris).

Judt, T. 1975 'The development of socialism in France: the example of the Var' *Historical Journal* 18, 55–83.

1979 *Socialism in Provence 1871–1914* (Cambridge).

Juneja, M. 1987–8 'The peasant image and agrarian change: representations of rural society in nineteenth-century French painting from Millet to Van Gogh' *Journal of Peasant Studies* 15, 445–73.

Kaye, H. J. 1982 'Another way of seeing peasants: the work of John Berger' *Journal of Peasant Studies* 9, 85–105.

Kemp, T. 1971 *Economic forces in French history 1760–1914* (London).

Kofman, E. 1980 'Protestantism and modernisation in France: myth and reality' *Middlesex Polytechnic Geography and Planning Papers* 1.

Lacambre, G. 1994 'La terre: réalité et nostalgie' in E. Le Roy Ladurie (ed.) *Paysages, paysans. L'art et la terre en Europe du Moyen Age au XXe siècle* (Paris) 195–239.

Lagrave, R-M. 1976 'Le travail agraire dans le roman français' *Sociologia Ruralis* 16, 85–102.

Lebeau, R. 1955 *La vie rurale dans les montagnes du Jura méridional* (Lyon).

Le Chercheur 1932 'L'origine des sapeurs-pompiers en Loir-et-Cher' *Dépêche du Centre* (25 June).

Lehning, J. R. 1995 *Peasant and French. Cultural contact in rural France during the nineteenth century* (Cambridge).

Lemaire, D. 1980 'Les origines du P.C.F. en Loir-et-Cher' *Cahiers du Communisme* 12, 68–78.

Le Meur, M-V. 1976 'Le culte des saints dans le diocèse de Blois aux environs de 1840' *Cahiers de l'Institut d'Histoire de la Presse et de l'Opinion* 2, 9–23.

Léonard, E. 1955 *Le protestant français* (2nd edn, Paris).

Le Roy Ladurie, E. 1979 'Peasants' in P. Burke (ed.) *The new Cambridge modern history* vol. XIII *Companion volume* (Cambridge) 115–63.

1994 'Paysans français, paysans d'Europe de la protohistoire à l'époque actuelle' in E. Le Roy Ladurie (ed.) *Paysages, paysans. L'art et la terre en Europe du Moyen Age au XXe siècle* (Paris) 9–23.

Le Roy Ladurie, E. (ed.) 1994 *Paysages, paysans. L'art et la terre en Europe du Moyen Age au XXe siècle* (Paris).

Levasseur, R. (ed.) 1990 *De la sociabilité: spécificité et mutations* (Cap-Saint-Ignace, Québec).

Leveau, R. 1969 'Le Syndicat de Chartres (1885–1914)' *Le Mouvement Social* 67, 61–78.

Lévigne, C. 1928 'L'Union des Associations Agricoles' *L'Illustration Economique et Financière*, supplement 6 (29 December) 62.

Lévi-Strauss, L. and Mendras, H. 1973–4 'Rural studies in France' *Journal of Peasant Studies* 1, 363–78.

Loubère, L. A. 1974 *Radicalism in Mediterranean France: its rise and decline, 1848–1914* (Albany).

Lovell, D. W. 'The French Revolution and the origins of socialism: the case of early French socialism' *French History* 6, 185–205.

Lussier, H. 1987 *Les sapeurs-pompier au XIXe siècle: associations volontaires en milieu populaire* (Paris).

Magraw, R. 1983 *France 1815–1914: the bourgeois century* (Oxford).

Marais, J-L. 1986 *Les sociétés d'hommes. Histoire d'une sociabilité du 18e siècle à nos jours, Anjou, Maine, Touraine* (La Botellerie-Vauchrétien).

Marbot, B. 1994 'La photographie rurale' in E. Le Roy Ladurie (ed.) *Paysages, paysans. L'art et la terre en Europe du Moyen Age au XXe siècle* (Paris) 240–6.

Marchal, M. 'Les sociétés de secours mutuel en Meurthe-et-Moselle avant 1945' *Actes du 104e Congrès National des Sociétés Savantes 1979* (Paris) 179–90.

Marcilhacy, C. 1957 'Emile Zola "historien" des paysans beaucerons' *Annales Economies Sociétés Civilisations* 12, 573–86.

1962 *Le diocèse d'Orleans sous l'épiscopat de Mgr Dupanloup 1849–1878: sociologie religieuse et mentalités collectives* (Paris).

Marczewski, J. 1963 'The take-off hypothesis and French experience' in W. W. Rostow (ed.) *The economics of take-off into sustained growth* (New York).

Margadant, T. W. 1979a 'French rural society in the nineteenth century: a review essay' *Agricultural History* 53, 644–51.

1979b *French peasants in revolt. The insurrection of 1851* (Princeton).

1984 'Tradition and modernity in rural France during the nineteenth century' *Journal of Modern History* 56, 667–97.

Martin, O. 1984 *Les Catholiques sociaux dans le Loir-et-Cher* (Saint-Maur-des-Fossés).

Martin-Demézil, J. 1980 'Analyse d'une grande enquête: l'enseignement primaire en Loir-et-Cher, 1833' *Mémoires de la Société des Sciences et Lettres de Loir-et-Cher* 36, 59–76.

Marx, K. 1852 *The Eighteenth Brumaire of Louis Napoleon* (translation by E. and C. Paul, London 1926).

Marx, K. and Engels, F. 1950 *Selected works in two volumes* (London).

McPhee, M. 1981–2 'A reconsideration of the "peasantry" of nineteenth-century France' *Journal of Peasant Studies* 9, 5–25.

1989 'The French Revolution, peasants and capitalism' *American Historical Review* 94, 1265–80.

1992a *The politics of rural life. Political mobilisation in the French countryside 1846–1852* (Oxford).

1992b *A social history of France 1780–1880* (London).

Merriman, J. M. 1979 'Introduction' in J. M. Merriman (ed.) *Consciousness and class experience in nineteenth-century Europe* (New York) 1–16.

Mesliand, C. 1969 'Le Syndicat Agricole Vauclusien' *Le Mouvement Social* 67, 39–60.

1989 *Les paysans du Vaucluse 1860–1939* (Aix-en-Provence).

Mitchell, A. 1991 'The function and malfunction of mutual aid societies in nineteenth-century France' in J. Barry and C. Jones (eds.) *Medicine and charity before the welfare state* (London) 171–89.

Mitrany, D. 1951 *Marx against peasants* (Chapel Hill).

Morineau, M. 1971 *Les faux-semblants d'un marriage économique: agriculture et démographie en France au XVIIIe siècle* (Paris).

Motheron, A. 1985 'La société populaire, républicaine, sabotière, révolutionnaire des sans-culottes de Montoiore' *Histoire et Traditions Populaires du Bas-Vendômois* 6 no. 2 (n.p.)

Moulin, A. 1988 *Les paysans dans la société française* (Paris), translated as *Peasantry and society in France since 1789* (Cambridge, 1991).

Muller, E. 1976 'Les fêtes en Loir-et-Cher de 1814 a 1852' *Bulletin de la Société d'Art et d'Archéologie de la Sologne* 2, 2–6.

Navarro, J-P. 1983 *La naissance des sociétés de secours mutuels dans le Tarn* (Toulouse).

Negrin, G. 1983 'L'entreprise de battages dans la Petite Beauce' *Mémoires de la Société des Sciences et Lettres de Loir-et-Cher* 38, 47–68.

Newell, W. H. 1973 'The agricultural revolution in nineteenth-century France' *Journal of Economic History* 33, 697–731.

Nickson, G. and Martin, E. 1988 *Le chemin de fer du Blanc à Argent* (Breil-sur-Royal).

Nicolas, P. 1988 'Emergence, développement et rôle des co-opératives agricoles en France: aperçus sur une histoire séculaire' *Economie Rurale* 184–6, 116–22.

Nivault, E. 1950 'L'œuvre su Syndicat des Agriculteurs de Loir-et-Cher' *L'Opinion Economique et Financière* 3 no. 5, 78–9.

O'Brien, P. and Keyder, C. 1978 *Economic growth in Britain and France 1780–1914* (London).

Ogden, P. 1980 'Migration, marriage and the collapse of traditional peasant society in France' in P. White and R. I. Woods (eds.) *The geographical impact of migration* (London) 152–79.

Oesinger, J-C. 1962 'Les C. F. secondaires du Loir-et-Cher' *La Vie du Rail* 849, (3 June) 34–5.

Ordish, G. 1972 *The great wine blight* (London).

Overton, M. 1966 *Agricultural revolution in England: the transformation of the agrarian economy 1500–1850* (Cambridge).

Ozouf, J. and Ozouf, M. 1997 'Le Tour de la France par deux enfants: *The Little Red Book of the Republic'* in P. Nora (ed.) *Realms of memory: the construction of the French past,* vol. II, *Traditions* (New York) 125–48.

Ozouf-Marignier, M-Y. 1989 *La formation des départements: la représentation du territoire français à la fin du 18e siècle* (Paris).

Papin, V. 1962 'Miettes d'histoire ferroviaire dans le Blésois' *La Vie du Rail* 849, (3 June) 22–5.

Paumier, M. 1941 *Les grandes régions géologiques de Loir-et-Cher* (Tours).

Pautard, J. 1965 *Les disparités régionales dans la croissance de l'agriculture française* (Paris).

Pinkney, D. H. 1953 'Migrations to Paris during the Second Empire' *Journal of Modern History* 25, 1–12.

1958 'The dilemma of the American historian of modern France' *French Historical Studies* 1, 11–25.

1991 'Time to bury the Pinkney thesis?' *French Historical Studies* 17, 219–23.

Planhol, X. de. 1994 *An historical geography of France* (Cambridge).
Poitou, C. 1985 *Paysans de Sologne dans la France ancienne: la vie des campagnes solognotes* (Le Coteau).
Pollock, G. 1976 *Millet* (London).
Ponteil, F. 1966 *Les institutions de la France de 1814 à 1870* (Paris).
Ponton, R. 1977 'Les images de la paysannerie dans le roman rural à la fin du 19e siècle' *Actes de la Recherche en Sciences Sociales* 17–18, 62–71.
Poujol, G. 1978 *La dynamique des associations (1844–1905)* (Paris).
Prévost, M-L. 1994 'Le paysan, héros de roman' in E. Le Roy Ladurie (ed.) *Paysages, paysans. L'art et la terre en Europe du Moyen Age au XXe siècle* (Paris) 247–53
Price, R. 1975 'The onset of labour shortage in nineteenth-century French agriculture' *Economic History Review* 28, 260–77.
   1983 *The modernization of rural France. Communications networks and agricultural market structures in nineteenth-century France* (London).
   1984 'Recent work on the economic history of nineteenth-century France' *Economic History Review* 37, 417–34.
   A social history of nineteenth-century France (London).
Prudhomme, A. 1982 'L'opposition républicaine dans la région sous le Second Empire (1858–1863)' *Bulletin du Groupe d'Etudes d'Histoire et de Géographie Locales d'Onzain et de ses Communes Voisines* 13 (n.p.)
   'Les sociétés secrètes républicaines sous le Second Empire (1852–1860)' *Mémoires de la Société des Sciences et Lettres de Loir-et-Cher* 39, 73–99.
   1985 *Histoire des pompiers de Loir-et-Cher 1762–1914* (Vendôme).
Prugnaud, 1963 L. *Les étapes du syndicalisme agricole en France* (Paris).
Rebiffe, J. 1948 *Syndicat agricole départemental d'Eure-et-Loir 1886–1948: soixante ans d'action syndicale et coopérative* (Chartres).
Riffaud, J-C. 1973 'Les tramways du Loir-et-Cher' *Revue Bimestrielle de la Fédération des Amis des Chemins de Fer Secondaires* 116, 2–58; see also 118, (1978) 2–31.
Rigollet, G. 1966 'A propos du centenaire de la ligne Paris-Vendôme' *Bulletin de la Société Archéologique, Scientifique et Littéraire du Vendômois* 34–42.
   1990 *Du char à boeufs au TGV* (Vendôme).
Rinaudo, Y. 1980 'Le syndicalisme agricole dans le Var' *Le Mouvement Social* 112, 79–95.
Robinet, A. 1968 'Passé, présent et avenir de la vallée de la Cisse' *Etudes Ligériennes* 2, 24–35.
Roger, M. 1966 *Emile Guillamin, l'homme de la terre et l'homme de lettres* (Paris).
Rogers, S. C. 1987 'Good to think: the "peasant" in contemporary France' *Anthropological Quarterly* 60, 56–63.
   1991 *Shaping modern times in rural France. The transformation and reproduction of an Aveyronnais community* (Princeton).
Rose, A. M. 1954 *Theory and method in the social sciences* (Minneapolis). Especially chapter 3 'A theory of the function of voluntary associations in contemporary social structure' 50–71 and chapter 4 'Voluntary associations in France' 72–115.
Rosenberg, H. A. 1988 *A negotiated world. Three centuries of change in a French Alpine community* (Toronto).
Roudet, B. 1988 'Bilan des recherches sur la vie associative' *Revue de l'Economie Sociale* 14 (April) 11–18.

Roux, P. 1986 'La Société d'Agriculture de la Gironde' *Actes du 111e Congrès National des Sociétés Savantes 1986. Colloque sur l'histoire de la securité sociale* (Paris 1986) 133–9.

Ruttan, V. W. 1978 'Structural retardation and the modernisation of French agriculture: a skeptical view' *Journal of Economic History* 38, 714–28.

Sabine, G. 1952 'The two democratic traditions' *Philosophical Review* 61, 451–74.

Saillet, J. 1969 'Les composantes du mouvement dans la Marne' *Le Mouvement Social* 67, 80–8.

Saint-Jours, Y. 1982 'France' in P. A. Köhler and H. F. Zacher (eds.) *The evolution of social insurance 1881–1981: studies of Germany, France, Great Britain, Austria and Switzerland* (London).

Sewell, W. H. 1984 'Du compagnonnage aux sociétés de secours mutuel: forme d'organisation ouvrière pendant la première moitié du XIXe siècle' *Revue de l'Economie Sociale* 2, 15–20.

Sexauer, B. 1976 'English and French agriculture in the late-eighteenth century' *Agricultural History* 50, 491–505.

Shaffer, J. W. 1982 *Family and farm. Agrarian change and household organisation in the Loire valley 1500–1900* (Albany).

Sheridan, G. J. 1984 'Aux origines de la mutualité en France: le développement et l'influence des sociétés de secours mutuels, 1800–1848' *Revue de l'Economie Sociale* 1, 17–25.

Sills, D. 1968 'Voluntary associations: sociological aspects' in D. Sills (ed.) *International encyclopedia of the social sciences* vol. XVI (New York) 363–79.

Silver, J. 1980–1 'French peasant demands for popular leadership in the Vendômois (Loir-et-Cher), 1852–1890' *Journal of Social History* 14, 277–94.

Singer, B. 1983 *Village notables in nineteenth-century France: priests, mayors, schoolmasters* (Albany).

Smart, R. E. 1994 'Phylloxera' in J. Robinson (ed.) *The Oxford companion to wine* (Oxford) 725–8.

Smith, C. and Freedman, A. 1972 *Voluntary associations: perspectives on the literature* (Cambridge, MA).

Soboul, A. 1956 'The French rural community in the eighteenth and nineteenth centuries' *Past and Present* 10, 78–95.

Souchon, A. 1915 *Agricultural Societies in France* (n.p.).

Soulet, J-F. 1988 'Une nouvelle approche de la France rurale au XIXe siècle? (A propos de thèses récentes sur les Pyrénées)' *Revue Historique* 279, 381–92.

Steinmetz, G. and Guellier, J-Y. 1987 'Le tramway de Blois à Châteaurenault' *Bulletin du Group d'Etudes d'Histoire et du Géographie Locales d'Onzain et des Communes Voisines* 23.

Stevenson, W. I. 1980 'The diffusion of disaster: the phylloxera outbreak in the département of the Hérault 1862–1880' *Journal of Historical Geography* 6, 47–63.

Sturges, H., Weisberg, G., Bourrut-Cacouture, A. and Fidell-Beaufort, M. 1982 *Jules Breton and the French rural tradition* (Omaha).

Sutter, J. 1958 'Evolution de la distance séparant le domicile des futurs époux (Loir-et-Cher 1870–1954; Finistère 1911–1953)' *Population* 13, 227–58.

Sutton, K. 1969 'La Triste Sologne. L'utilisation du sol dans une région française à l'abandon au début du XIXe siècle' *Norois* 61, 7–30.

1971 'The reduction of wasteland in the Sologne: nineteenth-century French regional improvement' *Transactions of the Institute of British Geographers* 52, 129–44.

1973 'A French agricultural canal – the Canal de la Sauldre and the nineteenth-century improvement of the Sologne' *Agricultural History Review* 21, 51–5.

Thomas, K. 1973 *Religion and the decline of magic* (Harmondsworth).

Tilly, C. 1979 'Did the cake of custom break?' in J. Merriman (ed.) *Consciousness and class experience in nineteenth-century Europe* (New York) 17–44.

Tombs, R. 1996 *France 1814–1914* (London).

Toutain, J-C. 1961 *Le produit de l'agriculture française de 1700 à 1958* (Paris).

1992–3 *La production agricole de la France de 1810 à 1990: départments et régions. Croissance, productivité, structures*, 3 vols. (Grenoble).

Touvet, M. 1955 'La vie économique du vignoble blésois' *Norois* 6, 223–34.

Traimond, B. 1979 'Les assurances de bétail en Gascogne avant la mutualité (1590–1833)' *Actes du 104e Congrés Nationale des Sociétés Savantes, 1979. Colloque sur l'Histoire de la Securité Sociale* (Paris) 277–90.

Turner, R. 'Tramways for French peasants: the rise and fall of the rural tramways of Loir-et-Cher 1880 to 1934' *Journal of Transport History* (forthcoming).

Unwin, T. 1991 *Wine and the vine: an historical geography of viticulture* (London).

Valette, J. 1979 'Note sur les origines et la création des sociétés de secours mutuels en Gironde jusqu'en 1881' *Actes de 104e Congrès National des Sociétés Savantes 1979. Colloque sur l'histore de la securité sociale* (Paris) 291–8.

1984 'Les rapports à l'Empereur sur la situation des sociétés de secours mutuels (1852–1869)' *Actes du 108e Congrès des Sociétés Savantes 1983. Colloque sur l'histoire de la securité sociale* (Paris) 267–73.

Vallin, P. 1985 *Paysans rouges du Limousin: mentalités et comportement politique à Compreignac et dans le nord de la Haute-Vienne (1870–1914)* (Paris).

Van Tilborgh, L. 1989 *Van Gogh and Millet* (Zwolle, The Netherlands).

Van Zanden, J. L. 1991 'The first green revolution: the growth of production and productivity in European agriculture, 1870–1914' *Economic History Review* 44, 215–39.

Vassort, 1977 J. 'L'enseignement primaire en Vendômois à l'époque de la Révolution' *Bulletin de la Société Archéologique, Scientifique et Littéraire du Vendômois* 49–88.

1978 'L'enseignement primaire en Vendômois à l'époque révolutionnaire' *Revue d'Histoire Moderne et Contemporaine* 24, 625–55.

1985 'De la Révolution à la veille de la Seconde Guerre mondiale' in C. Croubois (ed.) *Le Loir-et-Cher de la préhistoire à nos jours* (Saint-Jean-d'Angély).

1995 *Une société provinciale face à son devenir: le Vendômois aux XVIIIe et XIXe siècles* (Paris).

Vercherand, J. 1989 'Un siècle de syndicalisme agricole: le cas de la Loire' *Bulletin du Centre de l'Histoire Economique et Sociale de la Région Lyonnaise* 39–53.

Vernois, P. 1962 *Le roman historique de Georges Sand à Ramuz. Ses tendances et son évolution (1860–1925)* (Paris).

Vivier, J. 1990 'Les grands travaux de mise en valeur de la Sologne au XIXème siècle' *Bulletin de la Société d'Art et d'Archéologie de la Sologne* 102, 5–25.

Vivier M. and Mme 1980 'La vie de deux jeunes beaucerons à l'époque 1900' *Centre départementale universitaire pour le 3 age en Loir-et-Cher* (Spring) 2–12.

Walker, J. A. 1981 *Van Gogh studies: five critical essays* (London): 'Van Gogh as a peasant painter' 47–60.

Warner, C. K. 1975 'Soboul and the peasants' *Peasant Studies Newsletter* 4 no. 1, 1–5.

Weber, E. 1977 *Peasants into Frenchmen. The modernisation of rural France 1870–1914* (London; also Stanford 1976).

　　1980 'The Second Republic, politics and the peasant' *French Historical Studies* 11, 521–50.

　　1982 'Comment la politique vint aux paysans: a second look at peasant politicization' *American Historical Review* 87, 357–89.

　　1983 *La fin des terroirs. La modernisation de la France rurale 1870–1914* (Paris).

Weiss, J. H. 1983–4 'Origins of the French welfare state: poor relief in the Third Republic, 1871–1914' *French Historical Studies* 13, 47–78.

Wright, G. 1964 *Rural revolution in France. The peasantry in the twentieth century* (Oxford).

Young, A. 1950 *Travels in France during the years 1787, 1788 and 1789* (Bury St Edmunds 1792). Quotations from edition edited by C. Maxwell (Cambridge; first edn 1929).

Zeldin, T. 1973 *France 1848–1945* vol. I *Ambition, love and politics* (Oxford).

　　1977 *France 1848–1945* vol. II *Intellect, taste and anxiety* (Oxford).

Zeyons, S. 1992 *La France paysanne. Les années 1900 par la carte postale* (Paris).

Zonabend, F. 1984 *The enduring memory. Time and history in a French village* (Manchester).

Index

372   *Index*

  legislation 27, 240
traditional
  dependance on land 3
  mutual aid 36, 105, 143–5, 300, 320
  portrayal of peasantry 12, 14–15
  risk management 27, 316, 320
  rural society and culture 18, 19, 20, 23,
    24–5, 27, 35, 49, 50, 18, 282–4, 296,
    298–9
  sociability 43, 97, 316
  (*see also* Catholic Church, labour
    substitution, *notables*)
traditionalism 1, 30, 312
tramways 17, 68m, 70–1, 70p, 71p, 315
Trôo 69, 206
Trouard-Riolle, M. 229–30, 231, 253

unions 324, 26, 38, 40, 313, 318 (*see also* trade
    unions)
Union Centrale des Syndicats des
    Agriculteurs de France 257, 261
Union des Associations Agricoles 265
Union Fédérale des Associations Cantonales
    et Communales de France 129, 130,
    133–4
unions plébéiennes (popular associations) 93,
    97
Union Vinicole de Loir-et-Cher 258, 259
urban, activists 22, 295
  and rural 294–6
  contacts 295
  movement 306
  organisations 143
  perspective 319
  practice of association 105, 168, 301
  role of centres 60, 65, 97, 98, 99, 144, 155,
    157–9, 294
  system 16, 17, 49, 164
urbanisation 64, 144
utopia 141, 283, 312, 318
utopians 32, **33–4**, 318

Valencay 162
Van Gogh, V. 13
Var 43, 310
Vassort, J. 86, 94, 97, 252
veillées 24, 43, 89, 97
Vendée 312
Vendôme 58m, 65, 79g, 85, 90, 99, 287, 301
  agricultural syndicates in 255, 259, 265
  anti-phylloxera syndicates in 224–5, 229,
    237
  arrondissement 54, 73, 78, 98, 163, 228, 239
  fire-fighting corps 195, 208
  insurance societies in 134
  mutual aid societies in 152, 153, 154, 157,

    161, 166, 168
  politics 239
  professor of agriculture of 130, 134, 139
  railways 69, 70
  Society of Agriculture 244, 247, 248
  subprefect 54, 154, 158, 172, 173–4, 207,
    295
  surrounds 74, 108, 221, 222, 227, 287, 231
    (*see also* Vendômois)
Vendômois 74, 88, 96, 157, 158, 222, 226, 227,
    229, 230, 231, 255
Vernou 134
vets 103, 114, 117, 119, 123, 125–6, 133, 136,
    238, 312
  promotion of association 295
  treatment 117, 118, 119, 122, 132, 131
Vezin, A. 127, 128–30, 136, 235, 236, 253,
    280
Vibraye, Marquis de 249
Vienne 154, 167
Vierzon 67
vignerons 26, 84, 86, 88, 96, 97, 221, 221p,
    222p, 292–4, 295, 319–20
  agricultural association 257–8, 258–9, 264,
    269–70, 295, 320
  anti-phylloxera response 220, 228–9, 230–8,
    252–3, 257–8, 264, 270
  apathy 228, 230, 231
  confréries 161, 174, 175, 176, 190, 241, 301,
    320
  conservatism 226–8, 230, 233
  fire-fighting corps 208
  mutual aid societies 153, 155–6, 161, 164,
    167, 171–2, 182, 184–5, 262, 300, 314,
    319–20
  politics 93–4, 238, 262, 314: socialism 302,
    310, 312
  substitute labour 167, 181, 184–6, 187,
    188–9 (*see also* labour substitution)
Villebarou 91, 97, 211, 222
Villechauve 67
Villedieu 206
Villefranche-sur-Cher 67, 171, 211
Villejoint 121, 132
Villeneuve-Frouville 73
Villermain 121, 201
Villetrun 168, 222, 225
Villierfins 121, 132
Villiers 206, 228
Vineuil 121, 132, 192, 204, 258, 264, 274, 275,
    276–9, 280
Vineuil-Saint-Claude 258, 259
vineyards 87–8, 223m, 293–4
viticulture 220, 251, 257, 293–4, 305, 310, 319,
    320 (*see also vignerons*)
voluntary associations 24, 25, **47–52**, 53, 85,
    96–100, 101, 121, 136, 144, **284–320**

voluntary associations (*cont.*)
  agricultural association 4–1, 98, **240–81**,
    304
  associationism 241
  definition of 49
  in France 31, 32, 36, 303–20
  geography of 287–94, 308–10
  legal framework 36–8, 47, 286, 307
  as mediator between State and individual
    (*see* State)
  membership 37, 49, 295, 297, 299, 302,
    313–14
  politics 49, 296–8
  social significance of 291, **294–303, 310–20**
  timing of 306–8
  (*see also* fraternal associations)
Vosges 308
Vouzon 87, 104, 201

Warner, C. 28,
Warriner, C. 48

water-borne transport 66–7
water management syndicates 281
Weber, E. 9, 18–20, 21, 24–5, 26, 65, 80–1,
  82–4, 315–16
  criticism of 19–20
Wolff, H, 10
women 12, 13, 72, 168–70, 299
  associations for 153, 164, 168, 314
  benefit claims by 177, 178, 277
  gendered membership 168–70, 182, 184,
    280, 303, 313–4
  and religion 88, 90
working-class (*see* class)
Wright, G. 1, 5, 24

Young, A. 7, 53
  tour of France 7

Zeldin, T. 23, 32, 34, 35, 41
Zola, E. 9
Zonabend, F. 317

Cambridge Studies in Historical Geography

Titles marked with an asterisk * are available in paperback.

Printed in the United States
By Bookmasters